CW00709062

Joint Ventures

Joint Ventures

Mindreading, Mirroring, and Embodied Cognition

Alvin I. Goldman

OXFORD
UNIVERSITY PRESS

Oxford University Press is a department of the University of Oxford.
It furthers the University's objective of excellence in research, scholarship,
and education by publishing worldwide.

Oxford New York
Auckland Cape Town Dar es Salaam Hong Kong Karachi
Kuala Lumpur Madrid Melbourne Mexico City Nairobi
New Delhi Shanghai Taipei Toronto

With offices in
Argentina Austria Brazil Chile Czech Republic France Greece
Guatemala Hungary Italy Japan Poland Portugal Singapore
South Korea Switzerland Thailand Turkey Ukraine Vietnam

Oxford is a registered trademark of Oxford University Press
in the UK and certain other countries.

Published in the United States of America by
Oxford University Press
198 Madison Avenue, New York, NY 10016

© Oxford University Press 2013

All rights reserved. No part of this publication may be reproduced, stored in a
retrieval system, or transmitted, in any form or by any means, without the prior
permission in writing of Oxford University Press, or as expressly permitted by law,
by license, or under terms agreed with the appropriate reproduction rights organization.
Inquiries concerning reproduction outside the scope of the above should be sent to the Rights
Department, Oxford University Press, at the address above.

You must not circulate this work in any other form
and you must impose this same condition on any acquirer.

Library of Congress Cataloging-in-Publication Data
Goldman, Alvin I., 1938–
Joint ventures: mindreading, mirroring, and embodied cognition / Alvin I. Goldman.
p. cm.
ISBN 978–0–19–987418–7 (hardcover : alk. paper)
1. Philosophy of mind. 2. Metaphysics. 3. Cognitive neuroscience. I. Title.
BD418.3.G67 2012
128′.2—dc23
2012036715

1 3 5 7 9 8 6 4 2
Printed in the United States of America
on acid-free paper

CONTENTS

SOURCES

1. Theory of Mind. In E. Margolis, R. Samuels, & S. Stich, eds., *Oxford Handbook of the Philosophy of Cognitive Science*, 402–424. New York: Oxford University Press (2012).
2. Mirror Neurons and the Simulation Theory of Mindreading. *Trends in Cognitive Sciences* 2:493–501 (1998).
3. Simulationist Models of Face-Based Emotion Recognition. *Cognition* 94:193–213 (2005).
4. Mirroring, Mindreading, and Simulation. In Jaime Pineda, ed., *Mirror Neuron Systems: The Role of Mirroring Processes in Social Cognition*. New York: Humana Press, 311–330 (2008).
5. Mindreading by Simulation: The Roles of Imagination and Mirroring. In M. Lombardo, H. Tager-Flusberg, & S. Baron-Cohen, eds. *Understanding Other Minds*, 3rd ed. New York: Oxford University Press (forthcoming).
6. The Psychology of Folk Psychology. *Behavioral and Brain Sciences* 16:15–28 (1993).
7. Empathy, Mind, and Morals. *Proceedings and Addresses of the American Philosophical Association* 66(3):17–41 (1992).
8. Two Routes to Empathy: Insights from Cognitive Neuroscience. In A. Coplan & P. Goldie, eds., *Empathy: Philosophical and Psychological Perspectives*, 31–44. New York: Oxford University Press (2011).
9. Is Social Cognition Embodied? *Trends in Cognitive Sciences* 13(4):154–159 (2009).
10. A Moderate Approach to Embodied Cognitive Science. *Review of Philosophy and Psychology* 3(1):71–88 (2012).
11. The Individuation of Action. *Journal of Philosophy* 68(21):761–774.
12. A Program for "Naturalizing" Metaphysics, with Application to the Ontology of Events. *Monist* 90(3):457–479 (2007).
13. Actions, Predictions, and Books of Life. *American Philosophical Quarterly* 5:135–151 (1968).

Joint Ventures

Introduction

Philosophy interacts with many other fields when problems of common interest invite distinct but complementary approaches. It has built many bridges with cognitive science, specifically, bridges that carry a hefty part of its total interdisciplinary traffic. Most of the chapters (previously published essays) in the present volume are examples of this traffic; they comprise what I refer to as joint ventures. This introduction sets out the volume's organization and the rationale for its interdisciplinary focus. It presents previews of several chapters, often with chunks of historical background.

Part I addresses a topic of interest to both philosophy of mind and cognitive science, variously called mindreading, mentalizing, commonsense psychology, or theory of mind (ToM). The question is how ordinary people, with no scientific training, understand and identify mental states. Earlier philosophy addressed this question under two headings: the problem of "other minds" and the problem of "self-knowledge." In this guise, the question is filed under the heading of epistemology; for example, how are we *justified* in ascribing mental states to ourselves and others? But a closely related question has become important for philosophy of mind and cognitive science: How do we actually *arrive* at such beliefs? How does our cognitive system execute the twin mentalizing tasks of attributing mental states to the self and attributing them to others? What cognitive processes or capacities are utilized in performing these tasks?[1] Cognitive science has generated an imposing array of scientific findings that merit close attention in trying to assess competing accounts of mindreading. The chapters in part I appeal extensively to such scientific evidence, drawing from several subfields of cognitive science, including child development, social psychology, and motor, affective, and social neuroscience.

Part II covers two topics: empathy and embodied cognition. Because my preferred approach to mindreading is the *simulation theory*—also called the empathy theory—there is significant continuity between the chapters of part I and the two empathy chapters of part II, which were written at least fifteen years apart. The first draws chiefly on behavioral evidence, whereas the second makes central use of cognitive neuroscience. This difference reflects the general path of scientific development in recent decades; my own sojourn in this territory also reflects this.

The second pair of chapters in part II tackles a theme that is making waves in contemporary cognitive science and philosophy of mind, namely embodied cognition. Starting from somewhat different directions, assorted practitioners of cognitive science are urging greater attention to the body's role in cognition. Many of the proposals aim to produce radical changes in how cognition is studied. Other proposals focus on what has already been found using orthodox methods of investigation. Both seek to assign greater emphasis to the body in trying to understand the operations of the mind. Chapters 9 and 10 offer an entirely novel perspective on this terrain.

Part III lays the groundwork for a different partnership between philosophy and cognitive science, one that would link metaphysics and cognitive science. Chapter 11 was originally published in 1971, before collaborations between philosophy and cognitive science got off the ground; it is a specimen of pure metaphysics. Chapter 12, in contrast, proposes a (partial) "naturalization" of metaphysics, a naturalization that has a role for cognitive sciences in the customary project of metaphysics. The book's final chapter tackles a very different topic in the metaphysics of action. The topic is a variant of the traditional free-will-versus-determinism problem, viz., whether human actions are in principle perfectly predictable. Could someone's actions be successfully predicted even if the subject were to learn of the prediction and could therefore resolve to falsify it? That is surely within one's power, isn't it? Some fascinating scenarios (which many of my readers have found entertaining) are explored. So this final chapter may serve—depending on the reader's taste—as something like a sweet dessert.

The book's title, *Joint Ventures*, primarily alludes to its interdisciplinary character. Felicitously, however, it also strikes a second chord. The theory of mindreading defended here, the simulation theory, depicts mindreading itself as a species of joint venture. When a mindreader seeks to predict a target person's mental state, she often tries to reenact, replicate, or share in her own mind some initial mental states of the target, in the hope of generating a further state of the self that matches a state of the target. When successful, such a process is a species of joint venture. The mindreader travels the same mental road and arrives at the same destination as the person whose mind is read.

Not all philosophers are advocates of interdisciplinarity. They may be unpersuaded of its necessity. Now I do not endorse one and the same rationale for every branch or problem of philosophy for which cognitive-science partnerships might be considered. Different fields of philosophy have different agendas; no single size fits all. Let us stick to topics explored in this volume, primarily philosophy of mind. Folk psychology as studied by philosophers began as an armchair discipline. Despite being an arch-empiricist in theory, W. V. Quine did not hesitate to pen several pages in support of an empathy-driven theory of mindreading, all of them offered from his armchair.[2] Nor did Donald Davidson leave his armchair when offering his account of interpretation, which appealed primarily to rationality. Neither offered empirical evidence for his views; nor did Wilfrid Sellars (1955), who first advanced the idea of a folk theory as the conceptual underpinning for our grasp of propositional attitudes. Of course, psychological research on this topic was rather sparse when these philosophers were writing. Now, however, when the amount of relevant scientific work is massive, how can philosophers ignore it? It would be intellectually irresponsible to ignore the huge swaths of evidence and theory that science has generated. Even if everyone agreed that the simulation (or empathy) theory is correct, no philosopher would be epistemically warranted in holding this view based solely on Quine's casual observations and conjectures. Interpretation of scientific evidence is always challenging; but no respectable epistemology would countenance blindness to relevant evidence.

If scientific evidence is critical, where does philosophy fit in? There is ample place for philosophers to make theoretical contributions, to argue for this or that theoretical interpretation as the best explanation of the data. This kind of joint venture typifies the volume. How much of a contribution can philosophers really make, however, if they restrict themselves to theorizing rather than experimenting? Can philosophers really help in advancing the subject? In fact, philosophers have played a large role in advancing the study of mindreading. When ToM was in its infancy, philosophers pointed to a crucial feature in the concept of belief (a popular mental state to attribute). Belief, they explained, is a representational state that has the potential to be false. A full understanding of belief must involve an appreciation of this feature. This fundamental insight paved the way for important experiments on childhood mindreading.[3] Philosophical work also played a big role in articulating both of the dominant theories in the field: the theory-theory (TT) and the simulation theory (ST). TT is the idea that naïve cognition about the mind is a species of folk-scientific thought (like naïve physics), in which even toddlers engage in science-like thinking. As indicated, the first to propound this idea was Sellars in 1955 (though he made no claims about toddlers). The rival theory, ST, also drew inspiration from the insights of historical philosophers who wrote on empathic understanding, abundantly represented in the writings of the Scottish

Enlightenment and the German tradition of "*Verstehen*."[4] Recent philosophers have been players in clarifying and elucidating the mirroring phenomena that figure prominently in many chapters of this volume.[5]

In sum, both the philosophical and scientific contributions to our subject areas have been significant and indeed indispensable. Had either group's contributions been ignored, the road of inquiry (in Charles Peirce's phrase) would surely have been blocked. Complex problems should be approached by triangulation, it being impossible to predict which angles will yield the most important clues and insights.

PART I: MINDREADING AND SIMULATION

Chapter 1: Theory of Mind. This is an overview chapter that outlines the central lines of argument and evidence for the major approaches in the territory. Because I favor the simulation theory, this approach receives the lion's share of attention, with special emphasis on the role of mirroring.

Chapter 2: Mirror Neurons and the Simulation Theory of Mindreading (with Vittorio Gallese). ST was first advanced as a theory of mindreading in the 1980s.[6] It was commonly introduced by the metaphor of putting oneself in "another's shoes." In the early literature, a mindreader was described as launching a simulation process by *pretending* to occupy another person's perspective.[7] In a later book-length treatment, I called this *enactment-imagination* (Goldman, 2006). The core of ST is the idea that mindreaders try to reenact a target's thoughts by creating states in their own minds that match, or imitate, those of the target. Such surrogate, or proxy, states would be fed into a mental mechanism (e.g., a reasoning or affect-generating mechanism) to generate additional states in the mindreader's own mind.

This is the first variant of ST. In 1998, however, I heard Vittorio Gallese deliver a lecture about the then-recent discovery of mirror neurons in the ventral premotor cortex of macaque monkeys. Gallese is a member of the laboratory in Parma, Italy, that made this discovery. A certain class of neurons are activated both when a recorded monkey initiates a particular type of action and when it observes another monkey (or human) perform the same type of action. (However, observers do not behaviorally imitate what they see; their mirror-elicited motor commands are inhibited.[8]) When I first heard Gallese, the Parma team was unfamiliar with the field of mindreading. I explained to him the ongoing debate between TT and ST and suggested that mirroring might exemplify the simulation story. He agreed, and we soon published what appears here as chapter 2. We did not propose that macaque monkeys engage in mindreading. Our weaker conjecture was that mirror neurons might represent a "primitive version" or a "precursor in phylogeny" of a simulation heuristic that might underlie human mindreading. This was the first paper to give neuroscientific shape to a version of simulation-based mindreading.[9]

However, questions abound about how to fold mirroring phenomena into a mindreading story. After several more years of thinking about mirroring and mindreading, I concluded that two *levels* of mindreading should be distinguished: high-level and low-level. This was a major theme of *Simulating Minds* (Goldman, 2006: chaps. 6 and 7). The hypothesis was that high-level mindreading is driven by imagination (or pretense), whereas low-level mindreading is largely driven by automatic and unconscious mirror processes (sometimes called "resonance" processes). The present volume revisits this two-level conception in several chapters, including chapters 1 and 5.

The type of mirroring first discovered in Parma was a visuomotor phenomenon in monkeys, but analogous phenomena were soon found in humans, and they were not exclusively motoric. Cognitive neuroscientists soon identified mirror systems in humans across a wide range of phenomena, including feelings of pain and touch, and emotions like fear and disgust. For the mindreading theorist, two questions arise: whether mindreading occurs in connection with these phenomena, and whether it takes a simulationist form. Chapter 3 addresses this subject. I regard the evidence for mirror-based mindreading in *emotions* as more decisive than the evidence for mirror-based mindreading of *motor intentions*. This evidence is presented in chapter 3.

Chapter 3: Simulationist Models of Face-Based Emotion Recognition (with Chandra Sekhar Sripada). A well-studied phenomenon in affective science is the reading of basic emotions via facial expressions (Ekman & Friesen, 1976). Although early mindreading research largely ignored emotions, they are a legitimate class of mental states that we constantly read. The question arises whether the recognition (i.e., classification) of emotions is executed by simulation or by theory. For three emotions, fear, disgust, and anger, there was evidence that deficits in face-based recognition produced by brain damage were paired with deficits in the production, or experiencing, of the same emotion. What could explain such a pattern of paired deficits? One compelling possibility is that *normal* face-based recognition of emotion proceeds by a mirroring process in which someone who observes a target's face undergoing emotion E, *experiences* E herself (usually at a sub-threshold level). If a normal observer's own emotional experience is the basis for accurately classifying the target's emotion as a case of E, this would be fully analogous to a simulation process of the sort described earlier. The chief difference is that there would be no use of imagination.

We argued that this kind of story is indeed the best explanation of the data. Four possible processing models were sketched that might account for the available (paired-deficit) evidence, and we concluded that these are the only tenable ones. Each of these models, however, is a simulationist model, so whichever one is correct, ST is sustained for this class of mindreading tasks. The chapter does not firmly choose among the four models, but further reflection suggests that the fourth is the strongest. It clearly comports with a mirroring process.[10] Moreover, there is independent evidence that mirroring of emotion does occur (in humans) at least with respect to disgust. An fMRI study by Wicker et al. (2003) scanned normal individuals while they watched targets inhaling various odorants and making facial expressions. The scanned observers experienced the same region of neural activation (the anterior insula, primarily) when observing a target display a disgust-expressive face and when inhaling a foul odorant. This study establishes mirroring for disgust, just as mirroring had been established for motor intention, pain, and touch. Separate evidence from brain-damaged patients with paired deficits in experiencing and recognizing disgust combines with the evidence of disgust mirroring to constitute weighty evidence in support of a model of self-emotion-based attribution.

Chapter 4: Mirroring, Mindreading, and Simulation. This chapter pursues definitional and interpretive issues about mirroring, including its exact relationship to mindreading. It illustrates the sort of conceptual points that philosophers of science often make when new scientific constructs call for clarification and critical assessment. This is a traditional role for philosophers of science.

Questions are reasonably raised about what, exactly, is or should be packed into the notion of mirroring. Choices that theorists make can potentially affect the appropriateness or inappropriateness of proposed interpretations of experimental findings. One conceptual question, for example, is whether mirroring automatically *constitutes* mindreading. If creature X undergoes a mental state that mirrors a mental state of creature Y, does this entail that X mindreads Y?[11] Presumably not. Mindreading is a matter of attributing mental states, where an "attribution" is understood to be a *belief* state. If a pair of mirrored states are both, say, *intentions* (rather than, say, beliefs), then from the fact that X and Y undergo content-similar motor intentions, it does not follow that either of them forms a *belief* about the other's motor intention. Hence it doesn't follow that either X or Y mindreads the other. This issue is quite relevant to a study by Iacoboni et al. (2005), discussed in chapters 1 and 4. The results of this study are said by the authors to support the claim that a motor mirror system (in humans) attributes a future intention to a target (i.e., makes a *prediction* about the target's action). However, even if intention-mirroring has indeed occurred in the observer, it doesn't follow from this that the observer makes a prediction of a future intention (or action). A prediction is a kind of belief, and intentions are not beliefs. The upshot is that it takes more than an experimental establishment of a mirroring event to provide evidence for a mindreading event, that is, for an event in which an observer attributes a mental event (current or future) to a target. This is among the many issues raised in chapter 4.

Chapter 5: Mindreading by Simulation: The Roles of Imagination and Mirroring (with Lucy Jordan). This is the volume's most recent chapter on mindreading. It reviews and discusses the latest publications (through 2012) relevant to the duplex approach to simulational mindreading. The central themes will be familiar from preceding chapters, but there is recent work that offers fresh evidence congenial to ST. Here I highlight three such pieces of evidence.

With respect to high-level mindreading, the tenability of ST depends heavily on the power of enactment imagination (understood as a general-purpose cognitive mechanism). Can it really do what ST requires if accurate mindreading of others is to be a high-frequency achievement? Can the imagination imitate a sequence of cognitive states in a target when it does not already share those same states with the target? Is imagination powerful enough to be an accurate replicator? A recent study by Morewedge and colleagues (2010) delivers striking confirmation of such power. Subjects were asked to perform certain actual or imaginary eating actions, such as eating a certain number of M&Ms one by one. Afterward they were invited to eat freely of the same food. It was found that subjects who merely *imagined* eating thirty M&Ms subsequently ate significantly fewer M&Ms than subjects in other conditions. Quite surprisingly, the mere imagination of food consumption had an impact on subsequent eating choices that was remarkably congruent with the impact of actual consumption.

A different finding bears on the much-debated question of when and how mindreading competence is acquired. Onishi & Baillargeon (2005) jumpstarted an early-competence approach to theory of mind by showing that infants as young as fifteen months had false-belief sensitivity in non-verbal tasks. (This contrasted with a previously well-entrenched view that full mindreading competence is not attained until about four years of age.) More recent work by Kovacs et al. (2010) advances the timeline of false-belief sensitivity even further, to seven months. In addition, as Jordan and I interpret their findings, they are consistent with an early timeline for imaginatively tracking the contents of what another agent (falsely) believes without necessarily having a "metarepresentation" of the other's belief as such. We view this as an excellent fit with simulation theory.

A third important topic covered in chapter 5 is the neural basis for a bi-level, or duplex, model of simulational mindreading. Waytz & Mitchell (2011) distinguish two different neural regions responsible for mental simulation: (1) the so-called "default" network, involving the medial prefrontal cortex, precuneus and posterior cingulate, and lateral prefrontal cortex (see Raichle et al., 2001); and (2) the neural networks responsible for various forms of mirroring ("shared representations"). The default network serves as substrate for what Buckner & Carroll (2007) call "self-projection." This region is linked to such cognitive activities as recalling one's own experiences from the past, imagining future experiences, and imagining fictitious events. What these activities have in common is the mental projection of the self from its actual current situation to some different situation involving one's (personal) past, future, or counterfactual situation. Their notion of self-projection corresponds nicely to what I refer to as "enactment imagination," the core mechanism of high-level mindreading. The second region identified by Waytz and Mitchell corresponds to mirroring.

Chapter 6: The Psychology of Folk Psychology. The first five chapters are silent on two important topics in folk psychology. First is the question of how people understand, or conceptualize, mental-state *concepts*. How do they represent mental states picked out by such verbal labels as believe, desire, hope, fear, hurt, tickle, love, and hate? What contents are associated with these labels? Second, how do people deploy these representational contents in the process of self-attribution? How does the cognitive system recognize something as a state of believing, desiring, hoping, or fearing?

The first part of the discussion is mainly negative. A detailed critique is mounted of how the TT approach might try to account for first-person attribution. The critique appeals heavily to what might be feasible or unfeasible for our cognitive system. The second phase of the discussion is positive. It proposes an alternative story of how our cognitive system might (1) represent mental state concepts and (2) use those representations to categorize one's own current states. The positive story proposed in chapter 6 is admittedly unpopular in most of cognitive science (or so I conjecture). It more closely resembles traditional philosophical models of self-attribution involving inner perception or introspection. Nonetheless, it is defended in the spirit of cognitive science. Notions like "interoception" definitely have currency in *some* cognitive-science circles (see chapter 10).

Readers might wonder why a chapter on self-attribution is included here. ST seems to have no dog in this fight, because nobody claims that simulation is used in *self*-attribution. There are two reasons, however, for this inclusion. First, a fully general account of mindreading must explain both first- and third-person attribution, and it needs to show how these (possibly disparate) attribution processes comport with a univocal account of mental concept understanding. Second, ST—of the kind I advocate, at any rate—*does* have a dog in the self-attribution fight. This is because my version of ST embeds a step of self-attribution—or at least self-classification—in the story of *third*-person attribution. Specifically, ST says that after completing a mental imitation of the target, a simulation-based mindreader takes the output of this process (a decision, for example) and attributes it to the target. But how does the mindreader determine which state this is? Simply being *in* the state does not suffice for its being classified by the agent as being in that state. And how can accurate attributions be made to a target (under ST) unless an accurate classification of one's own state is achieved? Thus, a complete simulation story really needs an account of self-attribution. But it seems contrary to the spirit of simulationism to assign this part of the task to theoretical inference. Direct access, or inner sense, looks much more attractive for this (first-person classification) step of the simulation heuristic.[12]

PART II: EMPATHY AND EMBODIED COGNITION

Chapter 7: Empathy, Mind, and Morals. As previously noted, this is the volume's earliest article on the simulation theory of mindreading.[13] Its more distinctive contributions, though, are its material on empathy, motor mimicry, and morals. Philosophers and scientists of earlier eras took careful note of motor mimicry, and some of them linked it to empathy or mental resonance. Ever insightful, Adam Smith reported such phenomena in the following passage:

> When we see a stroke aimed and just ready to fall upon the leg or arm of another person, we naturally shrink and draw back on our leg or our own arm....The mob, when they are gazing at a dancer on the slack rope, naturally writhe and twist and balance their own bodies, as they see him do. (Smith, 1759/1976: 10)

In the latter part of the twentieth century, the psychologist Andrew Meltzoff discovered the phenomenon of infant motor mimicry, in which newborn infants (some of them less than an hour old) imitated facial expressions such as lip protrusion, tongue protrusion, and mouth opening (Meltzoff & Moore, 1983). These findings naturally prompted theories about the cognitive mechanisms that might underlie this innate ability. The subsequent discovery of mirroring, which revealed mental mimicry in motoric, somatic, and affective domains, has drawn widespread attention to phenomena of mental resonance or "empathy" (using the latter term in a thin sense).

Empathy's role in the moral domain is also explored. The central role of empathy in Schopenhauer's moral philosophy, for example, receives considerable attention. Hume could have been discussed as well, although this was not included. Hume's notion of sympathy is equivalent to the modern notion of empathy. It can be explained as a psychological mechanism that enables one person to receive by communication the sentiments of another. Sympathy, of course, plays a prominent role in Hume's moral philosophy.

Chapter 8: Two Routes to Empathy: Insights from Cognitive Neuroscience. This chapter probes more deeply into the contemporary psychology of empathy, employing neuroscience in greater depth than any of the other chapters here. It begins, however, with definitional issues. Current cognitive science employs the term "empathy" in the rather weak sense of isomorphism between a receiver's and a sender's mental states, nothing more. In ordinary usage, however, there is usually an implied further relationship between the receiver and sender: The receiver must "care" or have "concern" for the sender, and at least know or believe that he or she is the source of the receiver's experience.

After navigating such terminological waters, the chapter turns to the *types* of empathy—or, rather, types of *routes* to empathy. The chapter's core thesis is that there are two routes to empathy: a mirroring route and a "reconstructive" route, closely paralleling the low-level/high-level varieties of simulational mindreading. Concerning the mirroring route, it is emphasized that not all mirror processes use a single neural network or cytoarchitectural pathway. So it is not strictly correct to say that there is *one* mirroring process; instead, there are many. However, they have similar properties, so mirroring can be considered one type of process.[14]

What I call reconstructive empathy is closely related to high-level simulational mindreading. As discussed in chapter 5, its neuroanatomy seems to be associated with the default area described by Schacter, Addis, & Buckner (2008) as dedicated to "self-projection." In terms of "operational" characteristics, mirror processes are comparatively automatic, whereas self-projection requires online, effortful construction. The chapter concludes with reflections on which type of empathy is more effective in generating states isomorphic to those of a target.

Chapter 9: Is Social Cognition Embodied? (with Frédérique de Vignemont). This is the first of two chapters on embodied cognition (EC), a family of approaches to the study of cognition that advocates departures from classical cognitive science. One feature that motivates many of these departures is the thought that classical cognitivism does not take the body seriously enough. It takes cognition to be a matter of processing mental representations in abstraction from the kind of body the system inhabits, the environment in which it is situated, and the relation between the two. The remedy associated with many versions of EC is a radical shift in the study of cognition to an approach (inspired by Heidegger, among others) that focuses on human beings considered as practical agents who interact with the world. The proposed solution is to foreground the fact that cognition is a highly embodied or situated activity, so that thinking beings should be understood first and foremost as acting beings.[15]

The distinctive approach to EC formulated in chapter 9 and elaborated upon in chapter 10 departs from this platform in at least two ways. First, typical formulations of EC, like the one given above, assign as much importance to the environment as to the subject's body. It equates *embodied* cognition with *situated* cognition. Our approach focuses exclusively on the body. Second, although our approach is intended to illuminate features of cognitive functioning that classical cognitivism has missed, it does not take aim at the basic methodology or theoretical constructs of orthodox cognitive science. Specifically, it does not reject mental representation as a legitimate construct for the field. Indeed, its core proposal makes a central appeal to a certain class of representations, namely "bodily representations."

Chapter 10: A Moderate Approach to Embodied Cognitive Science. Why focus on bodily representations rather than the body per se? Most arguments for EC focus on the ways in which the body is involved in cognition. For example, Sato et al. (2007), using transcranial magnet stimulation, asked subjects to make judgments of "odd" or "even" after presentation of an Arabic numeral from 1 to 9. They found a modulation of the excitability of muscles only for the right hand, not the left. They took their findings to support an "embodied finger counting strategy," in which participants use fingers on their right hand to count integers up to 5. Clearly, they are referring to the involvement of finger movements themselves as warranting the "embodiment" interpretation. Yes, that is one possible understanding of embodiment. But if we seek to study embodied *cognition*, perhaps we should focus on (cognitive) *representations* of bodily activity that turn up when we don't expect them to. That approach will also support a finding—when combined with a swath of other evidence from cognitive neuroscience—that cognition uses bodily representations far more than classical cognitivism has suggested. Now, the account of embodiment offered in chapter 10 is a bit subtler yet. Beyond appealing to bodily representations, it appeals to a notion of bodily *codes* or *formats* as a family of mental codes or formats that have the original function of representing (parts of) the (agent's own) body. In terms of this notion of a B-format or a B-code, our somewhat novel conception of EC is advanced. If a representation belonging to a B-format is used to execute a cognitive task that is not itself bodily (such as a counting task), it still qualifies as an EC. It is further argued, appealing to both evolutionary theory and assorted studies in cognitive neuroscience, that this is a widespread phenomenon.[16]

PART III: THE METAPHYSICS OF ACTION

Part III differs from parts I and II in its reduced attention to cognitive science. Indeed, chapters 11 and 13 are pieces of pure metaphysics, with no admixture of cognitive science. The earliest part of my philosophical research was on action theory, beginning with a doctoral dissertation on that subject (Goldman, 1965). The dissertation was reworked and expanded into a book, *A Theory of Human Action* (Goldman, 1970), that devoted its first two chapters to the ontology of actions. Chapter 11 was an APA symposium paper based on these two chapters. In contrast, chapter 12 is a much later (2007) paper that proposes an integration of metaphysics with cognitive science, a sparsely explored venture even now. Its jumping-off point, however, is the debate over action individuation explored in chapter 11.

Chapter 11: The Individuation of Action. The 1960s and '70s featured a lively debate over the individuation of events and actions. The "coarse-grained" or "unifier" theory held that flipping a switch and turning on a light (in a single motion) are one and the same action, whereas the "fine-grained" or "multiplier" theory held that they are two distinct actions. Both sides, however, shared the assumption that there is one true account of the nature of actions or events. The debate was over which account that was (is). After defending the multiplier theory in the 1970s, the view that attracted me in 2007 (when chapter 12 was published) was that each theory has part of the truth. Chapter 12 defends this idea with suitable help from cognitive science.

Returning to the fine-grained approach of 1971, what was the envisaged relation between flipping a switch and turning on a light if it isn't that of identity? The answer I offered was this: They are related by *level-generation*. Level-generation (arguably a species of *constitution*) was understood to be is an asymmetric relation, unlike identity. Henry turns on the light by flipping the switch; he doesn't flip the switch by turning on the light. Thus, Henry's flipping the switch is not identical with his turning on the light; the former level-generates the latter.

Given the generation relation, it is a short step to the idea of an *action tree*, a format for representing multiple (same-agent, simultaneous) actions that features a branching structure. In *A Theory of Human Action*, my principal use of action trees was to smooth the way for an account of intentional action and the "in-order-to" relation. Recently, three philosophers have turned to action trees as a fruitful device for systematizing the moral domain or, more precisely, our mental representation of this domain (Mikhail, 2007; Knobe, 2010; Ulatowski, 2012). Mikhail undertakes to embed the action-tree idea into a program for a "universal moral grammar," which is modeled on Chomsky's program in linguistics. In other words, each of these writers makes *cognitive* applications of the action-tree idea. Instead of conceptualizing an action tree as a formal device to describe objective ontological structures, they use it as a theoretical tool for studying moral cognition.

Chapter 12: A Program for "Naturalizing" Metaphysics, with Application to the Ontology of Events. Chapter 12 was also a manifesto for involving cognitive science in metaphysics, a branch of philosophy far less intertwined with psychology (at this writing) than philosophy of mind or philosophy of cognitive science.[17] Thus, part III contains one chapter with a fairly radical proposal for interdisciplinary engagement, more radical than anything in parts I or II.

Chapter 13: Actions, Predictions, and Books of Life. This final chapter is pure philosophy, with no intended interdisciplinary linkage. The core problem is in the general ballpark of the free will/determinism problem, but there are differences. Here the question focuses on what (if anything) is distinctive to human action, deliberation, and voluntariness, as opposed to ordinary physical events. Physical events, one might assume, are governed by physical laws, and hence are open to being predicted by anyone who knows the laws and the relevant antecedent conditions. Could this be equally true of human actions? Can they be accurately predicted even if the agent learns of the prediction and resolves to falsify it? Isn't it within a person's power to falsify predictions of their behavior? And doesn't this imply that it is impossible for there to be anything like a book of one's life written on the basis of natural laws? A number of philosophers have drawn this conclusion (e.g., Taylor, 1964), but chapter 13 presents a different view.

NOTES

1. If one is a process reliabilist in epistemology, as I am, the question of what *processes* are used in executing mentalizing tasks has a direct bearing on questions of whether the output beliefs are justified or instances of knowledge. These further epistemological questions, however, are not pursued here.

2. See Quine (1990: 42–43, 62–63).
3. Philosophers Jonathan Bennett, Daniel Dennett, and Gilbert Harman noted that possessing the belief concept requires an appreciation that beliefs can be false. This led psychologists Heinz Wimmer & Josef Perner (1983) to perform a study of young children that (ostensibly) revealed important deficiencies in attributing false belief prior to age four, a finding that jumpstarted the field of theory of mind.
4. See Stueber (2006: 5–19) for an instructive historical overview of the late-nineteenth- and-early-twentieth-century tradition of "*Verstehen*," or empathic understanding. Chapter 7 of the present volume quotes striking and insightful passages on empathy (or sympathy) from major thinkers such as Adam Smith. Smith's examples were a major influence on my own early attraction and commitment to the simulation theory.
5. Two philosophers who have made contributions to the topic of mirroring (in addition to the present author) are Susan Hurley (2008) and Corrado Sinigaglia (Rizzolatti & Sinigaglia, 2008).
6. Three philosophers published versions of the simulation (or "replication") theory in the 1980s: Robert Gordon (1986), Jane Heal (1986), and Alvin Goldman (1989a). My 1989 article was reprinted in a previous collection of essays (Goldman, 1992a), so it isn't included here.
7. The best example of my early formulations of simulation theory is, as it happens, in chapter 7 (Empathy, Mind, and Morals), which appears in part II rather than part I. This placement is perhaps a bit incongruous, but it was dictated by a desire to pair this chapter with my only other chapter on empathy.
8. For some more details about mirroring and the range of potential applications of mirroring to a variety of cognitive phenomena, see Rizzolatti & Sinigaglia (2008) and Iacoboni (2008).
9. The principal interpretation of mirroring presented by the Parma laboratory, especially by its director, Giacomo Rizzolatti, is that mirroring triggers an understanding of the target's action from an internal, motoric perspective. Some criticisms of the mirroring approach to action-understanding and of the putative connection between mirroring and mindreading can be found in Jacob & Jeannerod (2005), Csibra (2007), Hickok (2008), and Jacob (2012).
10. Chapter 3 does not use the term "mirroring"; it speaks instead of "unmediated resonance."
11. Speaking of mental states as being involved in mirroring might raise readers' hackles. Shouldn't mirroring be a relation restricted to pairs of neural states? However, if each of a pair of mirrored states is a token substrate of one and the same mental type and mental content, we might allow ourselves to say that these mental-state tokens also mirror one another.
12. A specific version of an inner sense story is presented in chapter 6. However, I have since abandoned that story and present a rather different one in Goldman (2006, chaps. 9 and 10). The interested reader should consult those chapters.
13. Two others preceded it: Goldman (1989a, 1992a).
14. For more details on the identification of mirror networks in the human brain, see Iacoboni (2012).
15. I borrow this formulation from Anderson (2003), although similar statements of the core platform of EC can be found elsewhere as well.
16. Notice that the EC thesis proposed in chapters 9 and 10 is weaker than the more familiar version of EC called "concept empiricism," which defends the "grounding" of all concepts in modal, or bodily, formats.

17. I made earlier forays to a similar effect in the 1980s and '90s (e.g., Goldman, 1989b, 1992b). I regard chapter 12, however, as a superior effort in this direction. My hope is to expand this line of thinking in future writings.

REFERENCES

Anderson, M. L. (2003). Embodied cognition: A field guide. *Artificial Intelligence* 149: 91–130.

Bennett, J. (1978). Some remarks about concepts. *Behavioral and Brain Sciences* 1:557–560.

Buckner, R. L., & Carroll, D. C. (2007). Self-projection and the brain. *Trends in Cognitive Sciences* 11:49–57.

Csibra, G. (2007). Action mirroring and action understanding. In P. Haggard, Y. Rossetti, & M. Kawato (eds.), *Sensorimotor Foundations of Higher Cognition: Attention and Performance* XXII. Oxford, UK: Oxford University Press.

Dennett, D. (1978). Beliefs about beliefs. *Behavioral and Brain Sciences* 1:568–570.

Ekman, P., & Friesen, W. V. (1976). *Pictures of Facial Affect.* Palo Alto, CA: Consulting Psychologists Press.

Gilbert, D. (2006). *Stumbling on Happiness*. New York: Knopf.

Goldman, A. I. (1965). Action. Ph.D. dissertation, department of philosophy, Princeton University.

Goldman, A. I. (1970). *A Theory of Human Action*. Englewood-Cliffs, NJ: Prentice-Hall.

Goldman, A. I. (1989a). Interpretation psychologized. *Mind and Language* 4:161–185.

Goldman, A. I. (1989b). Metaphysics, mind, and mental science. *Philosophical Topics* 17:131–145.

Goldman, A. I. (1992a). *Liaisons: Philosophy Meets the Cognitive and Social Sciences*, Cambridge, MA: MIT Press.

Goldman, A. I. (1992b). Cognition and modal metaphysics. In *Liaisons: Philosophy Meets the Cognitive and Social Sciences*. Cambridge, MA: MIT Press.

Goldman, A. I. (2006). *Simulating Minds: The Philosophy, Psychology, and Neuroscience of Mindreading*. New York: Oxford University Press.

Gordon, R. (1986). Folk psychology as simulation. *Mind and Language* 1:158–171.

Harman, G. (1978). Studying the chimpanzee's theory of mind. *Behavioral and Brain Sciences* 1:576–577.

Heal, J. (1986). Replication and functionalism. In J. Butterfield (ed.), *Language, Mind, and Logic*. Cambridge, UK: Cambridge University Press.

Hickok, G. (2008). Eight problems for the mirror neuron theory of action understanding in monkeys and humans. *Journal of Cognitive Neuroscience* 21(7):1229–1243.

Hurley, S. (2008). The shared circuits model (SCM): How control, mirroring, and simulation can enable imitation, deliberation, and mindreading. *Behavioral and Brain Sciences* 31(1):1–22.

Iacoboni, M. (2008). *Mirroring People*. New York: Farrar, Straus and Giroux.

Iacoboni, M. (2012). Within each other: Neural mechanisms for empathy in the primate brain. In A. Coplan & P. Goldie (eds.), *Empathy: Philosophical and Psychological Perspectives*. Oxford, UK: Oxford University Press.

Jacob, P. (2012). Sharing and ascribing goals. *Mind and Language* 27(2):200–227.

Jacob, P., & Jeannerod, M. (2005). The motor theory of social cognition: A critique. *Trends in Cognitive Sciences* 9:21–25.

Knobe, J. (2010). Action trees and cognitive science. *Topics in Cognitive Science* 2: 555–578.

Kovacs, A. M., Teglas, E., & Endress, A. D. (2010). The social sense: Susceptibility to others' beliefs in human infants and adults. *Science* 330:1830–1834.

Meltzoff, A. N., & Moore, M. K. (1983). Newborn infants imitate adult facial gestures. *Child Development* 54:702–709.

Mikhail, J. (2007). Universal moral grammar: Theory, evidence and the future. *Trends in Cognitive Sciences* 11:143–152.

Morewedge, C. K., Huh, Y. E., & Vosgerau, J. (2010). Thought for food: Imagined consumption reduces actual consumption. *Science* 330:1530–1533.

Onishi, K. H., & Baillargeon, R. (2005). Do 15-month-old infants understand false belief? *Science* 308:255–258.

Raichle, M. E., et al. (2001). A default mode of brain function. *Proceedings of the National Academy of Sciences, U.S.A.* 98:676–682.

Quine, W. V. (1990). *Pursuit of Truth*. Cambridge, MA: Harvard University Press.

Rizzolatti, G., & Sinigaglia, C. (2008). *Mirrors in the Brain*. Oxford, UK: Oxford University Press.

Sato, M., Cattaneo, L., Rizzolatti, G., & Gallese, V. (2007). Numbers within our hands: Modulation of corticospinal excitability of hand muscles during numerical judgment. *Journal of Cognitive Neuroscience* 19:684–693.

Schacter, D. L., Addis, D. R., & Buckner, R. L. (2008). Episodic simulation of future events. *Annals of the New York Academy of Science* 1124:39–60.

Sellars, W. (1955/1997). *Empiricism and the Philosophy of Mind*. Cambridge, MA: Harvard University Press.

Smith, A. (1759/1976). *A Theory of Moral Sentiments*. D. D. Raphael & A. L. Macfie (eds.). Oxford, UK: Clarendon Press.

Stueber, K. R. (2006). *Rediscovering Empathy*. Cambridge, MA: MIT Press.

Taylor, R. (1964). Deliberation and foreknowledge. *American Philosophical Quarterly* 1:73–80.

Ulatowski, J. (2012). Act individuation: An experimental approach. *Review of Philosophy and Psychology* 3(2):249–262.

Waytz, A., & Mitchell, J. P. (2011). Two mechanisms for simulating other minds: Dissociations between mirroring and self-projection. *Current Directions in Psychological Science* 20(3):197–200.

Wimmer, H., & Perner, J. (1983). Beliefs about beliefs: Representation and constraining function of wrong beliefs in young children's understanding of deception. *Cognition* 13: 103–128.

PART ONE

Mindreading and Simulation

CHAPTER 1

Theory of Mind

1. INTRODUCTION

"Theory of Mind" (ToM) refers to the cognitive capacity to attribute mental states to self and others. Other names for the same capacity include "commonsense psychology," "naïve psychology," "folk psychology," "mindreading," and "mentalizing." Mental attributions are commonly made in both verbal and nonverbal forms. Virtually all language communities, it seems, have words or phrases to describe mental states, including perceptions, bodily feelings, emotional states, and propositional attitudes (beliefs, desires, hopes, and intentions). People engaged in social life have many thoughts and beliefs about others' (and their own) mental states, even when they don't verbalize them.

In cognitive science, the core question is this: How do people execute this cognitive capacity? How do they, or their cognitive systems, go about the task of forming beliefs or judgments about others' mental states, states that aren't directly observable? Less frequently discussed in psychology is the question of how people self-ascribe mental states. Is the same method used for both first-person and third-person ascription, or are they entirely different methods? Other questions in the terrain include: How is the capacity for ToM acquired? What is the evolutionary story behind this capacity? What cognitive or neurocognitive architecture underpins ToM? Does it rely on the same mechanisms for thinking about objects in general, or does it employ dedicated, domain-specific mechanisms? How does it relate to other processes of social cognition, such as imitation or empathy?

This chapter provides an overview of ToM research, guided by two classifications. The first classification articulates four competing approaches to (third-person) mentalizing, viz., the theory-theory, the modularity theory, the rationality theory, and the simulation theory. The second classification is the first-person/third-person contrast. The bulk of the discussion is directed at third-person mindreading, but the final section addresses self-attribution. Finally, our discussion provides representative coverage of the principal fields that investigate ToM: philosophy of mind, developmental psychology, and cognitive neuroscience. Each of these fields has its distinctive research style, central preoccupations, and striking discoveries or insights.

2. THE THEORY-THEORY

Philosophers began work on theory of mind, or folk psychology, well before empirical researchers were seriously involved, and their ideas influenced empirical research. In hindsight one might say that the philosopher Wilfrid Sellars (1956) jumpstarted the field with his seminal essay "Empiricism and the Philosophy of Mind." He speculated that the commonsense concepts and language of mental states, especially the propositional attitudes, are products of a proto-scientific *theory* invented by one of our fictional ancestors. This was the forerunner of what was later called the "theory-theory." This idea has been warmly embraced by many developmental psychologists. However, not everyone agrees with theory-theory as an account of commonsense psychology, so it is preferable to avoid the biased label "theory of mind." In much of my discussion, therefore, I opt for more neutral phraseology, "mindreading" or "mentalizing," to refer to the activity or trait in question.

The popularity of the theory-theory in philosophy of mind is reflected in the diversity of philosophers who advocate it. Jerry Fodor (1987) claims that commonsense psychology is so good at helping us predict behavior that it's practically invisible. It works well because the intentional states it posits genuinely exist and possess the properties generally associated with them. In contrast to Fodor's intentional realism, Paul Churchland (1981) holds that commonsense psychology is a radically false theory, one that ultimately should be eliminated. Despite their sharp differences, these philosophers share the assumption that naïve psychology, at bottom, is driven by a science-like theory, where a theory is understood as a set of law-like generalizations. Naïve psychology would include generalizations that link (1) observable inputs to certain mental states, (2) certain mental states to other mental states, and (3) mental states to observable outputs (behavior). The first type of law might be illustrated by "Persons who have been physically active without drinking fluids tend to feel thirst." An example of the second might be "Persons in pain tend to want to relieve that pain." An example of the third might be "People who are angry tend to frown." The business of attributing mental states to others consists of drawing law-guided inferences from their observed behavior, stimulus conditions, and previously determined antecedent mental states. For example, if one knows that Melissa has been engaged in vigorous exercise without drinking, one may infer that she is thirsty.

Among the developmental psychologists who have championed the theory-theory are Josef Perner, Alison Gopnik, Henry Wellman, and Andrew Meltzoff. They seek to apply it to young children, who are viewed as little scientists who form and revise their thinking about various domains in the same way scientists do (Gopnik & Wellman, 1992; Gopnik & Meltzoff, 1997). They collect evidence, make observations, and change their theories in a highly science-like fashion. They generate theories not only about physical phenomena, but also about unobservable mental states like belief and desire. As in formal science, children make transitions from simple theories of the phenomena to more complex ones.

The most famous empirical discovery in the developmental branch of ToM is the discovery by Wimmer & Perner (1983) of a striking cognitive change in children between roughly three and four years of age. This empirical discovery is that three-year-olds tend to fail a certain *false-belief task*, whereas four-year-olds tend to succeed on the task. Children watch a scenario featuring puppets or dolls in which the protagonist, Sally, leaves a chocolate on the counter and then departs the scene. In her absence, Anne is seen to move the object from the counter to a box. The children are asked to predict where Sally will look for the chocolate when she returns to the room, or alternatively where Sally "thinks" the chocolate is. Prior to age four, children typically answer incorrectly, that is, that Sally thinks it's in the box (where the chocolate really is). Around age four, however, normal children answer as an adult would, by specifying the place where Sally left the chocolate, thereby ascribing to Sally (what they recognize to be) a false belief. What happens between three and four that accounts for this striking difference?

Theory theorists answer by positing a change of theory in the minds of the children. At age three, they typically have conceptions of desire and belief that depict these states as simple relations between the cognizer and the external world, relations that do not admit the possibility of error. This simple theory gradually gives way to a more sophisticated one in which beliefs are related to propositional representations that can be true or false of the world. At age three, the child does not yet grasp the idea that a belief can be false. In lacking a representational theory of belief, the child has—as compared with adults—a "conceptual deficit" (Perner, 1991). This deficit is what makes the three-year-old child incapable of passing the false-belief test. Once the child attains a representational theory of belief, roughly at age four, she passes the location-change false-belief test.

A similar discrepancy between three- and four-year-olds was found in a second type of false-belief task, the deceptive container task. A child is shown a familiar container that usually holds candy and is asked, "What's in here?" She replies, "Candy." The container is then opened, revealing only a pencil. Shortly thereafter the child is asked what she thought was in the container when she was first asked. Three-year-olds incorrectly answer "a pencil," whereas four-year-olds correctly answer "candy." Why the difference between the two age groups, despite the fact that memory tests indicate that three-year-olds have no trouble recalling their own psychological states? Theory-theorists again offered the same conceptual-deficit explanation. Because the three-year-olds' theory doesn't leave room for the possibility of false belief, they can't ascribe to themselves their original (false) belief that the container held candy; so they respond with their current belief, namely, that it held a pencil.

This explanation was extremely popular circa 1990. But several subsequent findings seriously challenge the conceptual-deficit approach. The early challenges were demonstrations that various experimental manipulations enable three-year-olds to pass the tests. When given a memory aid, for example, they can recall and report their original false prediction (Mitchell & Lacohee, 1991). They can also give the correct false-belief answer when the reality is made less salient, for instance, if they are told where the chocolate is but don't see it for themselves (Zaitchik, 1991). Additional evidence suggests that the three-year-old problem lies in the area of inhibitory control (Carlson & Moses, 2001). Inhibitory control is an executive ability that enables someone to override "prepotent" tendencies, i.e., dominant or habitual tendencies, such as the tendency to reference reality as one knows it to be. A false-belief task requires an attributor to override this natural tendency, which may be hard for three-year-olds. An extra year during which the executive powers mature may be the crucial difference for four-year-olds, not a change in their belief concept. A meta-analysis of false-belief task findings encourages Wellman, Cross, & Watson (2001) to retain the conceptual-deficit story, but this is strongly disputed by Scholl & Leslie (2001).

Even stronger evidence against the traditional theory-theory timeline was uncovered in 2005, in a study of fifteen-month-old children using a nonverbal false-belief task. Onishi & Baillargeon (2005) employed a new paradigm with reduced task demands to probe the possible appreciation of false belief in fifteen-month-old children and found signs of exactly such understanding. This supports a *much* earlier picture of belief understanding than the child-scientist form of theory-theory ever contemplated.

A final worry about this approach can now be added. A notable feature of professional science is the diversity of theories that are endorsed by different practitioners. Cutting-edge science is rife with disputes over which theory to accept, disputes that often persist for decades. This pattern of controversy contrasts sharply with what is ascribed to young children in the mentalizing domain. They are said to converge on one and the same theory, all within the same narrow time-course. This bears little resemblance to professional science.

Gopnik takes a somewhat different tack in recent research. She puts more flesh on the general approach by embedding it in the Bayes net formalism. Bayes nets are directed-graph formalisms designed to depict probabilistic causal relationships between variables. Given certain assumptions (the causal Markov and faithfulness assumptions), a system can construct algorithms to arrive at a correct Bayes net causal structure if it is given enough information about the contingencies or correlations among the target events. Thus, these systems can learn about causal structure from observations and behavioral interventions. Gopnik and colleagues (Gopnik et al., 2004; Schulz & Gopnik, 2004) report experimental results suggesting that two- to four-year-old children engage in causal learning in a manner consistent with the Bayes net formalism. They propose that this is the method used to learn causal relationships between mental variables, including relationships relevant to false-belief tasks.

There are several worries about this approach. Can the Bayes net formalism achieve these results without special tweaking by the theorist, and if not, can other formalisms match these results without similar "special handling"? Second, if the Bayes net formalism predicts that normal children make all the same types of causal inferences, does this fit the scientific inference paradigm? We again encounter the problem that scientific inference is characterized by substantial diversity across the community of inquirers, whereas the opposite is found in the acquisition of mentalizing skills.

3. THE MODULARITY-NATIVIST APPROACH TO THEORY OF MIND

In the mid-1980s, other investigators found evidence supporting a very different model of ToM acquisition. This is the *modularity* model, which has two principal components. First, whereas the child-scientist approach claims that mentalizing utilizes domain-general cognitive equipment, the modularity approach posits one or more domain-specific modules that use proprietary representations and computations for the mental domain. Second, the modularity approach holds that these modules are innate cognitive structures that mature or come online at preprogrammed stages and are not acquired through learning (Leslie, 1994; Scholl & Leslie, 1999). This approach comports with nativism for other domains of knowledge, such as those subsumed under Spelke's (1994) idea of "core knowledge." The core-knowledge proposal holds that infants only a few months old have a substantial amount of "initial" knowledge in domains such as physics and arithmetic, knowledge that objects must trace spatiotemporally continuous paths through space or that one plus one yields two. Innate principles are at work that are largely independent of and encapsulated from one another. Modularists about mentalizing endorse the same idea. Mentalizing is part of our genetic endowment that is triggered by appropriate environmental factors, just as puberty is triggered rather than learned (Scholl & Leslie, 2001).

Early evidence in support of a psychology module was reported by Simon Baron-Cohen, Alan Leslie, and Uta Frith in two studies, both concerning autism. The first study (Baron-Cohen et al., 1985) compared the performance of normal preschool children, Down syndrome children, and autistic children on a false-belief task. All children had a mental age of above four years, although the chronological age of the second two groups was higher. Eighty-five percent of the normal children, 86 percent of the Down syndrome children, but only 20 percent of the autistic children passed the test. In the second study (Baron-Cohen et al., 1986), subjects were given scrambled pictures from comic strips with the first picture already in place. They were supposed to put the strips in order to make a coherent story and were also supposed to tell the story in their own words. The stories were of three types: mechanical, behavioral, and mentalistic. The autistic children all ordered the mechanical strips correctly and dealt adequately with the behavioral script. But the vast majority of autistic children could not understand the mentalistic stories. They put the pictures in jumbled order and told stories without attribution of mental states.

The investigators concluded that autism impairs a domain-specific capacity dedicated to mentalizing. Notice that the autistic children in the 1986 study were not deficient on either the mechanical or the behavioral script, only on the mentalistic one. Conversely, the Down syndrome children, despite their general retardation, were not deficient on the false-belief task. Thus, autism seems to involve an impairment specific to mentalizing, whereas mentalizing need not be impaired by general retardation as long as the ToM-dedicated module remains intact.

These conclusions, however, are not entirely secure. Some children with autism pass ToM tasks, including false-belief tests. The number of children who pass varies from one study to the next, but even a small percentage calls for explanation. If autism involves a failure to develop a ToM, how could these participants with autism pass the tests? Others, therefore, argue that failure on tasks that tap mentalizing abilities may be more directly interpreted in terms of domain-general deficits in either executive functions or language (Tager-Flusberg, 2000).

Nativist modularists adduce additional evidence, however, in support of their view, especially evidence for an appreciation of intentional agency in preverbal infants. A variety of cues are cited as evidence for the attribution of intentionality, or goal-directedness, in infancy, including joint attention behaviors (gaze-following, pointing, and other communicative gestures), imitation, language and emotional referencing, and looking-time studies.

In one study of gaze-following, Johnson, Slaughter, & Carey (1998) tested twelve-month-old infants on a novel object, a small, beach-ball-sized object with natural-looking fuzzy brown fur. It was possible to control the object's behavior from a hidden vantage point, so that when the baby babbled, the object babbled back. After a period of familiarization, an infant either experienced the object reacting contingently to the infant's own behavior or merely random beeping or flashing. Infants followed the "gaze" of the object by shifting their own attention in the same direction under three conditions: if the object had a face, or the object beeped and flashed contingent on the infant's own behavior, or both. These results were interpreted as showing that infants use specific information to decide when an object does or does not have the ability to perceive or attend to its surroundings, which seems to support the operation of a dedicated input system (Johnson, 2005). Woodward (1998) used a looking-time measure to show that even five-month-olds appear to interpret human hands as goal-directed relative to comparable inanimate objects. They looked longer if the goal-object of the hand changed, but not if the hand's approach path to the goal-object changed. This evidence also suggests an early, dedicated system to the detection of goal-oriented entities.

All of the above findings post-date Alan Leslie's (1994) postulation of a later-maturing cognitive module: the "theory-of-mind mechanism" (ToMM). Leslie highlighted four features of ToMM: (1) It is domain-specific, (2) it employs a proprietary representational system that describes propositional attitudes, (3) it forms the innate basis for our capacity to acquire ToM, and (4) it is damaged in autism. ToMM uses specialized representations and computations, and is fast, mandatory, domain-specific, and informationally encapsulated, thereby satisfying the principal characteristics of modularity as described by Fodor (1983).

An initial problem with the modularity theory is that ToMM, the most widely discussed module postulated by the theory, doesn't satisfy the principal criteria of modularity associated with Fodorian modularity. Consider domain specificity. Fodor says that a cognitive system is domain-specific just in case "only a restricted class of stimulations can throw the switch that turns [the system] on" (1983: 49). It is doubtful that any suitable class of stimulations would satisfy this condition for ToMM (Goldman, 2006: 102–104). A fundamental obstacle facing this proposal, moreover, is that Fodor's approach to modularity assumes that modules are either input systems or output systems, whereas mindreading has to be a central system. Next, consider informational encapsulation, considered to be the heart of modularity. A system is informationally encapsulated if it has only limited access to information contained in other mental systems. But when Leslie gets around to illustrating the workings of ToMM, it turns out that information from other central systems is readily accessible to ToMM (Nichols & Stich, 2003: 117–121). Leslie & German (1995) discuss an example of ascribing a pretend state to another person, and clearly indicate that a system ascribing such a pretense uses real-world knowledge, for example, whether a cup containing water would disgorge its contents if it were upturned. This knowledge would have to be obtained from (another) central system. Perhaps such problems can be averted if a non-Fodorian conception of modularity is invoked, as proposed by Carruthers (2006). But the tenability of the proposed alternative conception is open to debate.

4. THE RATIONALITY-TELEOLOGY THEORY

A somewhat different approach to folk psychology has been championed by another group of philosophers, chief among them Daniel Dennett (1987). Their leading idea is that one mindreads a target by "rationalizing" her, that is, by assigning to her a set of propositional attitudes that make her emerge—as far as possible—as a rational agent and thinker. Dennett writes:

> [I]t is the myth of our rational agenthood that structures and organizes our attributions of belief and desire to others and that regulates our own deliberations and investigations.... Folk psychology, then, is *idealized* in that it produces its predictions and explanations by calculating in a normative system; it predicts what we will believe, desire, and do, by determining what we ought to believe, desire, and do. (1987: 52)

Dennett contends that commonsense psychology is the product of a special stance we take when trying to predict others' behavior: the *intentional stance*. To adopt the intentional stance is to make the default assumption that the agent whose behavior is to be predicted is rational, that her desires and beliefs, for example, are ones she rationally ought to have given her environment and her other beliefs or desires.

Dennett doesn't support his intentional stance theory with empirical findings; he proceeds largely by thought experiment. So let us use the same procedure in evaluating his theory. One widely endorsed normative principle of reasoning is to believe whatever follows logically from other things you believe. But attributors surely do not predict their targets' belief states in accordance with such a strong principle; they don't impute "deductive closure" to them. They allow for the possibility that people forget or ignore many of their prior beliefs and fail to draw all of the logical consequences that might be warranted (Stich, 1981). What about a normative rule of inconsistency avoidance? Do attributors assume that their targets conform to this requirement of rationality? That too seems unlikely. If an author modestly thinks that he must have made some error in his book packed with factual claims, he is caught in an inconsistency (this is the so-called paradox of the preface). But wouldn't attributors be willing to ascribe belief in all these propositions to this author?

These are examples of implausible consequences of the rationality theory. A different problem is the theory's incompleteness: It covers only the mindreading of propositional attitudes. What about other types of mental states, such as sensations like thirst or pain and emotions like anger or happiness? It is dubious that rationality considerations bear on these kinds of states, yet they are surely among the states that attributers ascribe to others. There must be more to mindreading than imputed rationality.

Although first inspired by armchair reflection, rationality theory has also inspired some experimental work that—at least at first blush—seems to be supportive. Gergely, Nadasdy, Csibra, & Biro (1995) performed an intriguing experiment that they interpreted as showing that toddlers take the intentional stance at twelve months of age. They habituated one-year-old infants to an event in which a small circle approaches a large circle by jumping over an obstacle. When the obstacle is later removed, the infants show longer looking times when they see the circle take the familiar jumping path as compared with a straight path toward the target. Apparently, infants expect an agent to take the most rational or efficient means to its goal, so they are surprised when it takes the jumping path, although that's what they have seen it do in the past.

The title of their paper, "Taking the intentional stance at 12 months of age," conveyed the influence of Dennett's rationality theory. Their first interpretation of the results articulated this theme, viz., that infants attribute a causal intention to the agent that accords with a rationality principle. Toward the end of their paper, however, they concede that an infant can represent the agent's action as intentional without attributing a mental representation of the future goal state to the agent's mind. Thus, the findings might simply indicate that the infant represents actions by relating relevant aspects of reality (action, goal-state, and situational constraints) through a principle of efficient action, which assumes that actions function to realize goal-states by the most efficient means available. Indeed, in subsequent writings, they switch their description of infants from the "intentional" stance to the "teleological" stance, an interpretational system for actions in terms of means-ends efficiency (Gergely & Csibra, 2003). The teleological stance is a qualitatively different but developmentally related interpretational system that is supposed to be the precursor of the young child's intentional stance. The two stances differ in that teleological interpretation is non-mentalistic—it makes reference only to actual and future states of reality. Developmentally, however, teleological interpretation is transformed into causal mentalistic interpretation by "mentalizing" the explanatory constructs of the teleological stance (Gergely & Csibra, 2003: 232).

This approach raises three problems: First, can the teleological stance really be transformed into the full range of mentalistic interpretation in terms of rationality principles? One species of mindreading involves imputing beliefs to a target based on inferential relations to prior belief states. How could this interpretational system be a transformation of an efficiency principle? Inference involves no action or causal efficiency. Second, the teleological stance might equally be explained by a rival approach to mentalizing, namely, the simulation theory. The simulation theory might say that young children project themselves into the shoes of the acting object (even a circle) and consider the most efficient means to its goal. They then expect the object to adopt this means. Third, as already noted above, there are kinds of mental states and mindreading contexts that have nothing to do with rationality or efficiency. People ascribe emotional states to others (fear or delight, disgust or anger) based on facial expressions. How could these ascriptions be driven by a principle of efficiency? We don't have the makings here of a general account of mindreading; at most, we have a narrow segment of it. And even this narrow segment might be handled just as well by a rival theory (viz., the simulation theory).

5. THE SIMULATION THEORY

A fourth approach to commonsense psychology is the *simulation theory*, sometimes called the "empathy theory." Robert Gordon (1986) was the first to develop this theory in the present era, suggesting that we can predict others' behavior by answering the question, "What would *I* do in *that* person's situation?" Chess players playing against a human opponent report that they visualize the board from the other side, taking the opposing pieces for their own and vice versa. They pretend that their reasons for action have shifted accordingly. Thus transported in imagination, they make up their mind what to do and project this decision onto the opponent.

The basic idea of the simulation theory resurrects ideas from a number of earlier European writers, especially in the hermeneutic tradition. Dilthey wrote of understanding others through a process of "feeling with" others (*mitfuehlen*), "reexperiencing" (*nacherleben*) their mental states, or "putting oneself into" (*hineinversetzen*) their shoes. Similarly, Schleiermacher linked our ability to understand other minds with our capacity to imaginatively occupy another person's point of view. In the philosophy of history, the English philosopher R. G. Collingwood (1946) suggested that the inner imitation of thoughts, or what he calls the reenactment of thoughts, is a central epistemic tool for understanding other agents. (For an overview of this tradition, see Stueber, 2006.)

In addition to Gordon, Jane Heal (1986) and Alvin Goldman (1989) endorsed the simulation idea in the 1980s. Their core idea is that mindreaders simulate a target by trying to create similar mental states of their own as proxies or surrogates of those of the target. These initial pretend states are fed into the mindreader's own cognitive mechanisms to generate additional states, some of which are then imputed to the target. In other words, attributors use their own mind to mimic or "model" the target's mind and thereby determine what has or will transpire in the target.

An initial worry about the simulation idea is that it might "collapse" into theory- theory. As Dennett put the problem:

> How can [the idea] work without being a kind of theorizing in the end? For the state I put myself in is not belief but make-believe belief. If I make believe I am a suspension bridge and wonder what I will do when the wind blows, what "comes to me" in my make-believe state depends on how sophisticated my knowledge is of the physics and engineering of suspension bridges. Why should my making believe I have your beliefs be any different? In both cases, knowledge of the imitated object is needed to drive the make-believe "simulation," and the knowledge must be organized into something rather like a theory. (1987: 100–101)

Goldman (1989) responded that there is a difference between *theory-driven* simulation, which must be used for systems different than oneself, and *process-driven* simulation, which can be applied to systems resembling oneself. If the process or mechanism driving the simulation is similar enough to the process or mechanism driving the target, and if the initial states are also sufficiently similar, the simulation might produce an isomorphic final state to that of the target without the help of theorizing.

6. MIRRORING AND SIMULATIONAL MINDREADING

The original form of simulation theory (ST) primarily addressed the attribution of propositional attitudes. In recent years, however, ST has focused heavily on simpler mental states and on processes of attribution rarely dealt with in the early ToM literature. I include here the mindreading of motor plans, sensations, and emotions. This turn in ST dates to a paper by Vittorio Gallese & Alvin Goldman (1998), which posited a link between simulation-style mindreading and activity of mirror neurons (or mirror systems). Investigators in Parma, Italy, led by Giacomo Rizzolatti, first discovered mirror neurons in macaque monkeys, using single cell recordings (Rizzolatti et al., 1996; Gallese et al., 1996). Neurons in the macaque premotor cortex often code for a particular type of goal-oriented action, for example, grasping, tearing, or manipulating an object. A subclass of premotor neurons were found to fire both when the animal plans to perform an instance of their distinctive type of action and when it observes another animal (or human) perform the same action. These neurons were dubbed "mirror neurons," because an action plan in the actor's brain is mirrored by a similar action plan in the observer's brain. Evidence for a mirror system in humans was established around the same time (Fadiga et al., 1995). Because the mirror system of an observer tracks the mental state (or brain state) of an agent, the observer executes a mental simulation of the latter. If this simulation also generates a mental-state attribution, this would qualify as simulation-based mindreading. It would be a case in which an attributor uses his own mind to "model" that of the target. Gallese and Goldman speculated that the mirror system might be part of, or a precursor to, a general mindreading system that works on simulationist principles.

Since the mid-1990s, new discoveries of mirror processes and mirror systems have expanded remarkably. Motor mirroring has been established via sound as well as vision (Kohler et al., 2002), and for effectors other than the hand, specifically, the foot and the mouth (Buccino et al., 2001). Meanwhile, mirroring has been discovered for sensations and emotions.

Under the category of sensations, there is mirroring for touch and mirroring for pain. Touching a subject's legs activates the primary and secondary somatosensory cortex. Keysers et al. (2004) showed subjects movies of other subjects being touched on their legs. Large extents of the observer's somatosensory cortex also responded to the sight of the targets' legs being touched. Several studies established mirroring for pain in the same year (Singer et al., 2004; Jackson et al., 2004; Morrison et al., 2004). In the category of emotions, the clearest case is mirroring for disgust. The anterior insula is well known as the primary brain region associated with disgust. Wicker et al. (2003) undertook an fMRI experiment in which normal subjects were scanned while inhaling odorants through a mask—either foul, pleasant, or neutral—and also while observing video clips of other people's facial expressions while inhaling such odorants. Voxels in the anterior insula that were significantly activated when a person inhaled a foul odorant were also significantly activated when seeing others make facial expressions arising from a foul odorant. Thus, there was mirroring of disgust.

The critical question for theory of mind, however, is whether mindreading, that is, mental attribution, occurs as an upshot of mirroring. In 2005, two similar experiments in the domain of motor intention were performed by members of the Parma group, and are claimed to provide evidence for mirror-based—hence, simulation-based—prediction of motor intentions. One experiment was done with monkeys (Fogassi et al., 2005), and the other with humans (Iacoboni et al., 2005). I shall sketch only the latter study.

Iacoboni et al.'s study was an fMRI study in which subjects observed video clips presenting three kinds of stimulus conditions: (1) grasping hand actions without any context ("Action" condition), (2) scenes specifying a context without actions, that is, a table set for drinking tea or ready to be cleaned up after tea ("Context" condition), and (3) grasping hand actions performed in either the before-tea or the after-tea context ("Intention" condition). The Intention condition yielded a significant signal increase in premotor mirroring areas where hand actions are represented. The investigators interpreted this as evidence that premotor mirror areas are involved in understanding the intentions of others, in particular, intentions to perform subsequent actions (e.g., drinking tea or cleaning up).

This mindreading conclusion, however, is somewhat problematic, because there are alternative "deflationary" interpretations of the findings (Goldman, 2008). One deflationary interpretation would say that the enhanced activity in mirror neuron areas during observation of the Intention condition involved only predictions of *actions*, not attributions of *intentions*. Because actions are not mental states, predicting actions doesn't qualify as mindreading. The second deflationary interpretation is that the activity in the observer's relevant mirror area is a mimicking of the agent's intention, not an intention *attribution* (belief). Re-experiencing an intention should not be confused with attributing an intention. Only the attribution of an intention would constitute a belief or judgment about an intention. Thus, the imaging data do not conclusively show that mindreading took place in the identified premotor area.

However, the Iacoboni et al. study presented evidence of intention attribution above and beyond the fMRI evidence. After being scanned, subjects were debriefed about the grasping actions they had witnessed. They all reported representing the intention of drinking when seeing the grasping action in the during-tea condition and representing the intention of cleaning up when seeing the grasping action in the after-tea condition. Their verbal reports were independent of the instructions the subjects had been given at the outset. Thus, it is quite plausible that their reported intention attributions were caused by activity in the mirror area. So the Iacoboni et al. study does provide positive evidence for its stated conclusion, even if the evidence isn't quite as probative as they contend.

Where else might we look for evidence of mirroring-based mindreading? Better specimens of evidence are found in the emotion and sensation domains. For reasons of space, attention is restricted here to emotion. Although Wicker et al. (2003) established a mirror process for disgust, they did not test for disgust attribution. However, by combining their fMRI study of normal subjects with neuropsychological studies of brain-damaged patients, a persuasive case can be made for mirror-caused disgust attribution (in normals). Calder et al. (2000) studied patient NK, who suffered insula and basal ganglia damage. In questionnaire responses, NK showed himself to be selectively impaired in experiencing disgust, as contrasted with fear or anger. NK also showed significant and selective impairment in disgust recognition (attribution), in both visual and auditory modalities. Similarly, Adolphs et al. (2003) had a patient B who suffered extensive damage to the anterior insula and was able to recognize the six basic emotions *except disgust* when observing dynamic displays of facial expressions. The inability of these two patients to undergo a normal disgust response in their anterior insula apparently prevented them from mindreading disgust in others, although their attribution of other basic emotions was preserved. It is reasonable to conclude that when *normal* individuals recognize disgust through facial expressions of a target, this is causally mediated by a mirrored experience of disgust (Goldman & Sripada, 2005; Goldman, 2006).

Low-level mindreading, then, can be viewed as an elaboration of a primitive tendency to engage in automatic mental mimicry. Both behavioral and mental mimicry are fundamental dimensions of social cognition. Meltzoff & Moore (1983) found facial mimicry in neonates less than an hour old. Among adults, unconscious mimicry in social situations occurs for facial expressions, hand gestures, body postures, speech patterns, and breathing patterns (Hatfield, Cacioppo, & Rapson, 1994; Bavelas et al., 1986; Dimberg, Thunberg, & Elmehed, 2000; Paccalin & Jeannerod, 2000). Chartrand & Bargh (1999) found that automatic mimicry occurs even between strangers, and that it leads to higher liking and rapport between interacting partners. Mirroring, of course, is mental mimicry usually unaccompanied by behavioral mimicry. The sparseness of behavioral imitation (relative to the amount of mental mimicry) seems to be the product of inhibition. Compulsive behavioral imitation has been found among patients with frontal lesions, who apparently suffer from an impairment of inhibitory control (Lhermitte et al., 1986; de Renzi et al., 1996). Without the usual inhibitory control, mental mimicry would produce an even larger amount of behavioral mimicry. Thus, mental mimicry is a deep-seated property of the social brain, and low-level mindreading builds on its foundation.

7. SIMULATION AND HIGH-LEVEL MINDREADING

The great bulk of mindreading, however, cannot be explained by mirroring. Can it be explained (in whole or part) by another form of simulation? The general idea of mental simulation is the re-experiencing or reenactment of a mental event or process, or an *attempt* to re-experience or reenact a mental event (Goldman, 2006: chap. 2). Where does the traditional version of simulation theory fit into the picture? It mainly fits into the second category (i.e., *attempted* interpersonal reenactment). This captures the idea of mental pretense, or what we call "enactment imagination" (E-imagination), which consists of trying to construct in oneself a mental state that isn't generated by the usual means (Goldman, 2006; Currie & Ravenscroft, 2002). *Simulating Minds* argues that E-imagination is an intensively used cognitive operation, one commonly used in reading others' minds.

Let us first illustrate E-imagination with intrapersonal applications, for example, imagining seeing something or launching a bodily action. The products of such applications constitute, respectively, visual and motor imagery. To visualize something is to (try to) construct a visual image that resembles the visual experience we would undergo if we were actually seeing what is visualized. To visualize the *Mona Lisa* is to (try to) produce a state that resembles a seeing of the *Mona Lisa*. Can visualizing really resemble vision? Cognitive science and neuroscience suggest an affirmative answer.

Kosslyn (1994) and others have shown how the processes and products of visual perception and visual imagery have substantial overlap. An imagined object "overflows" the visual field of imagination at about the same imagined distance from the object as it overflows the real visual field. This was shown in experiments where subjects actually walked toward rectangles mounted on a wall and when they merely visualized the rectangles while imagining a similar walk (Kosslyn, 1978). Neuroimaging reveals a notable overlap between parts of the brain active during vision and during imagery. A region of the occipitotemporal cortex known as the fusiform gyrus is activated both when we see faces and when we imagine them (Kanwisher et al., 1997). Lesions of the fusiform face area impair both face recognition and the ability to imagine faces (Damasio et al., 1990).

An equally (if not more) impressive story can be told for motor imagery. Motor imagery occurs when you are asked to imagine (from a motoric perspective) moving your effectors in a specified way, for example, playing a piano chord with your left hand or kicking a soccer ball. It has been convincingly shown that motor imagery corresponds closely, in neurological terms, to what transpires when one actually executes the relevant movements (Jeannerod, 2001).

At least in some modalities, then, E-imagination produces strikingly similar experiences to ones that are usually produced otherwise. Does the same hold for mental events like forming a belief or making a decision? This has not been established, but it is entirely consistent with existing evidence. Moreover, a core brain network has recently been proposed that might underpin high-level simulational mindreading as a special case. Buckner & Carroll (2007) propose a brain system that subserves at least three, and possibly four, forms of what they call "self-projection." Self-projection is the projection of the current self into one's personal past or one's personal future, and also the projection of oneself into other people's minds or other places (as in navigation). What all these mental activities share is projection of the self into alternative situations, involving a perspective shift from the immediate environment to an imagined environment (the past, the future, other places, other minds). Buckner and Carroll refer to the mental construction of an imagined alternative perspective as a "simulation."

So, E-imaginative simulation might be used successfully for reading other minds. But what specific evidence suggests that we deploy E-imaginative simulation in trying to mindread others most of the time or even much of the time? This is what simulation theory concerning high-level mindreading needs to establish. (This assumes that simulation theory no longer claims that each and every act of mindreading is executed by simulation.

Rather, it is prepared to accept a hybrid approach in which simulation plays a central but not exclusive role.)

Two lines of evidence will be presented here (for additional lines of argument, see Goldman, 2006: chap. 7). An important feature of the imagination-based simulation story is that successful mindreading requires a carefully pruned set of pretend inputs in the simulational exercise. The exercise must not only *include* pretend or surrogate states that correspond to those of the target, but must also *exclude* the mindreader's own genuine states that don't correspond to ones of the target. This implies the possibility of two kinds of error or failure: failure to include states possessed by the target and failure to exclude states lacked by the target. The second type of error will occur if a mindreader allows a genuine state of his own, which he "knows" that the target lacks, to creep into the simulation and contaminate it. This is called *quarantine failure*. There is strong evidence that quarantine failure is a serious problem for mental-state attributors. This supports ST because quarantine failure is a likely affliction if mindreading is executed by simulation, but should pose no comparable threat if mindreading is executed by theorizing.

Why is it a likely problem under the simulation story? If one tries to predict someone's decision via simulation, one sets oneself to make a decision (in pretend mode). In making this decision, one's own relevant desires and beliefs try to enter the field to "throw their weight around," because this is their normal job. It is difficult to monitor the states that don't belong there, however, and enforce their departure. Enforcement requires suppression or inhibition, which takes vigilance and effort. No analogous problem rears its head under a theorizing scenario. If theorizing is used to predict a target's decision, an attributor engages in purely factual reasoning, not in mock decision-making. So there's no reason why his genuine first-order desires or beliefs should intrude. What matters to the factual reasoning are the mindreader's beliefs *about* the target's desires and beliefs, and these second-order beliefs pose no comparable threat of intrusion.

Evidence shows that quarantine failure is in fact rampant, a phenomenon generally known as "egocentric bias." Egocentric biases have been found for knowledge, valuation, and feeling. In the case of knowledge, egocentric bias has been labeled "the curse of knowledge," and it's been found in both children (Birch & Bloom, 2003) and adults (Camerer et al., 1989). To illustrate the bias for valuations, Van Boven, Dunning, & Loewenstein (2000) gave subjects Cornell coffee mugs and then asked them to indicate the lowest price they would sell their mugs for, while others who didn't receive mugs were asked to indicate the highest price they would pay to purchase one. Because prices reflect valuations, the price estimates were, in effect, mental-state predictions. Both owners and sellers substantially underestimated the differences in valuations between themselves and their opposite numbers, apparently projecting their own valuations onto others. This gap proved very difficult to eliminate. To illustrate the case of feelings, Van Boven & Loewenstein (2003) asked subjects to predict the feelings of hikers lost in the woods with neither food nor water. What would bother them more, hunger or thirst? Predictions were elicited either before or after the subjects engaged in vigorous exercise, which would make one thirsty. Subjects who had just exercised were more likely to predict that the hikers would be more bothered by thirst than by hunger, apparently allowing their own thirst to contaminate their predictions.

Additional evidence that effective quarantine is crucial for successful third-person mindreading comes from neuropsychology. Samson et al. (2005) report the case of patient WBA, who suffered a lesion to the right inferior and middle frontal gyri. His brain lesion includes a region previously identified as sustaining the ability to inhibit one's own perspective. Indeed, WBA had great difficulty precisely in inhibiting his own perspective (his own knowledge, desires, emotions, etc.). In nonverbal false-belief tests, WBA made errors in eleven out of twelve trials where he had to inhibit his own knowledge of reality. Similarly, when asked questions about other people's emotions and desires, again requiring him to inhibit his own perspective, fifteen of twenty-seven responses involved egocentric errors. This again supports the simulationist approach to high-level mindreading. There is, of course, a great deal of other relevant evidence that requires considerable interpretation and analysis. But ST seems to fare well in light of recent evidence (for contrary assessments, see Saxe, 2005; Carruthers, 2006).

8. FIRST-PERSON MINDREADING

Our last topic is self-mentalization. Philosophers have long claimed that a special method—"introspection" or "inner sense"—is available for detecting one's own mental states, although this traditional view is the object of skepticism and even scorn among many scientifically minded philosophers and cognitive scientists. Most theory-theorists and rationality theorists would join these groups in rejecting so-called privileged access to one's own current mental states. Theory-theorists would say that self-ascription, like other-person ascription, proceeds by theoretical inference (Gopnik, 1993). Dennett holds that the intentional stance is applied even to oneself. But these positions can be challenged with simple thought experiments.

I am now going to predict my bodily action during the next twenty seconds. It will include, first, curling my right index finger, then wrinkling my nose, and finally removing my glasses. There, those predictions are verified! I did all three things. You could not have duplicated these predictions (with respect to *my* actions). How did I manage it? Well, I let certain intentions form, and then I detected (i.e., introspected) those intentions. The predictions were based on the introspections. No other clues were available to me, in particular, no behavioral or environmental cues. The predictions must have been based, then, on a distinctive form of access I possess vis-à-vis my current states of mind, in this case, states that were primed to cause the actions. I seem to have similar access to my own itches and memories.

In an important modification of a well-known paper that challenged the existence or reliability of introspective access (Nisbett & Wilson, 1977), the co-author Wilson subsequently provides a good example and a theoretical correction to the earlier paper:

> The fact that people make errors about the causes of their own responses does not mean that their inner worlds are a black box. I can bring to mind a great deal of information that is inaccessible to anyone but me. Unless you can read my mind, there is no way you could know that a specific memory just came to mind, namely an incident in high school in which I dropped my bag lunch out a third-floor window, narrowly missing a gym teacher.... Isn't this a case of my having privileged, "introspective access to higher order cognitive processes"? (Wilson, 2002: 105)

Nonetheless, developmentalists have adduced evidence that putatively supports a symmetry or parallelism between self and other. They deny the existence of a special method, or form of access, available only to the first-person. Nichols & Stich (2003: 168–192) provide a comprehensive analysis of this literature, with the clear conclusion that the putative parallelism doesn't hold up and fails precisely in ways that favor introspection or self-monitoring.

If there is such a special method, how exactly might it work? Nichols & Stich (2003: 160–161) present their own model of self-monitoring. To have beliefs about one's own beliefs, they say, all that is required is that there be a monitoring mechanism that, when activated, takes the representation *p* in the Belief Box as input and produces the representation *I believe that p* as output. To produce representations of one's own beliefs, the mechanism merely has to copy representations from the Belief Box, embed the copies in a representation schema of the form *I believe that* ___, and then place the new representations back into the Belief Box. The proposed mechanism would work in much the same way to produce representations of one's own desires, intentions, and imaginings.

One major lacuna in this account is its silence about an entire class of mental states: bodily feelings. They don't fit the model because, at least in the orthodox approach, sensations lack representational content, which is what the Nichols-Stich account relies upon. Their account is a syntactic theory, which says that the monitoring mechanism operates on the syntax of the mental representations monitored. A more general problem is what is meant by saying that the proposed mechanism would work in "much the same way" for attitude types other than belief. How does the proposed mechanism decide *which* attitude to ascribe? Which attitude verb should be inserted into the schema *I ATTITUDE that* ____? Should it be belief, desire, hope, fear, etc.? Each contentful mental state consists, at a minimum, of an attitude type plus a content. The Nichols-Stich theory deals only with contents, not types. In apparent recognition of the problem, Nichols and Stich make a parenthetical suggestion: Perhaps a distinct but parallel mechanism exists for each attitudes type. But what a profusion of mechanisms this would posit, each mechanism essentially "duplicating" the others! Where is nature's parsimony that they appeal to elsewhere in their book?

The Nichols-Stich model of monitoring belongs to a family of self-attribution models that can be called "redeployment" theories, because they try to explain self-attribution in terms of redeploying the content of a first-level mental state at a meta-representational level. Another such theory is that of Evans (1982), defended more recently by Gordon (1996), who calls it the "ascent-routine" theory. Gordon describes the ascent routine as follows: The way in which one determines whether or not one believes that *p* is simply to ask oneself the question whether or not *p*. The procedure is presumably to be completed as follows: If one answers the whether-*p* question in the affirmative, one then "ascends" a level and also gives an affirmative answer to the question, "Do I think/believe that *p*?"

The ascent-routine theory faces a problem previously encountered with the monitoring theory. The basic procedure is described only for belief and lacks a clear parallel for classifying other attitudes or sensations. How is it supposed to work with hope, for example? Another problem concerns the procedure's details. When it says that a mindreader "answers" the whether-*p* question, what exactly does this mean? It cannot mean *vocalizing* an affirmative answer, because this won't cover cases of self-ascription where the answer is only *thought*, not vocalized. What apparently is meant by saying that one gives the "answer" *p* is that one *judges* the answer to be *p*. But how is one supposed to *tell* whether or not one judges that *p*? Isn't this the same question of how one determines whether one (occurrently) believes that *p*? This is the same problem we started with, so no progress appears to have been made.

As we return to an introspectivist approach, notice that it is not committed to any strong view about introspection's reliability. Traditionally, introspection was associated with infallibility, but this is an easily detachable feature that few current proponents espouse. Introspectionism is often associated with a perceptual or quasi-perceptual model of self-knowledge, as the phrase "inner sense" suggests. Is that a viable direction? Shoemaker (1996) argues to the contrary. There are many disanalogies between outer sense and introspection, though not all of these should deter a theorist, says Shoemaker. Unlike standard perceptual modalities, inner sense has no proprietary phenomenology, but this shouldn't disqualify a quasi-perceptual analogy. A more serious disanalogy, according to Shoemaker, is the absence of any organ that orients introspection toward its cognitive objects (current mental states), in the manner in which the eyes or nose can be oriented toward their objects. Shoemaker considers but rejects attention as a candidate organ of introspection.

This rejection is premature, however. A new psychological technique called "descriptive experience sampling" has been devised by Hurlburt (Hurlburt & Heavey, 2001) for studying introspection. Subjects are cued at random times by a beeper, and they are supposed to pay immediate attention to their ongoing experience upon hearing the beep. This technique revealed thoughts that they hadn't initially been aware of, though they were not unconscious. Schooler et al. (2004) have made similar findings, indicating that attention is typically required to trigger reflective awareness via introspection. Actually, the term "introspection" is systematically ambiguous. It can refer to a process of inquiry, that is, inwardly directed attention, that chooses selected states for analysis. Or it can refer to the process of performing an analysis of the states and outputting some descriptions or classifications. In the first sense, introspection itself is a form of attention, not something that requires attention in order to do its job. In the latter sense, it is an operation that performs an analysis or description of a state once attention has picked out the object or objects to be analyzed or described.

If introspection is a perception-like operation, should it not include a transduction process? If so, this raises two questions: What are the inputs to the transduction process, and what are the outputs? Goldman (2006: 246–255) addresses these questions and proposes some answers. There has not yet been time for these proposals to receive critical attention, so it remains to be seen how this new quasi-perceptual account of introspection will be received. In any case, the problem of first-person mentalizing is as difficult and challenging as the problem of third-person mentalizing, though it has thus far received a much smaller dollop of attention, especially among cognitive scientists.

9. CONCLUSION

Like most topics at the cutting edge of either philosophy or cognitive science, mindreading is awash with competing theories and rival bodies of evidence. The landscape is especially difficult to negotiate because it involves investigations using a myriad of disparate methodologies, ranging from a priori reflection to the latest techniques of contemporary neuroscience. The resulting variety of evidential sources ensures that new and fascinating findings are always around the corner; but it also makes it likely that we will not see a settled resolution of the debate in the very near future. It would be misguided to conclude that the amount of research effort devoted to the subject is disproportionate to its importance. To the contrary, the target phenomenon is a key to human life and sociality. People's preoccupation with mindreading, arguably at multiple levels, is a fundamental facet of human nature, and a philosophico-scientific understanding of how we go about this task must rank as a preeminent intellectual desideratum for philosophy of mind and for cognitive science.

REFERENCES

Adolphs, R., Tranel, D., & Damasio, A. R. (2003). Dissociable neural systems for recognizing emotions. *Brain and Cognition* 52:61–69.

Baron-Cohen, S., Leslie, A., & Frith, U. (1985). Does the autistic child have a 'theory of mind'? *Cognition* 21:37–46.

——. (1986). Mechanical, behavioral, and intentional understanding of picture stories in autistic children. *British Journal of Developmental Psychology* 4:113–125.

Bavelas, J. B., Black, A., Lemery, C. R., & Mullett, J. (1986). "I show how you feel": Motor mimicry as a communicative act. *Journal of Personality and Social Psychology* 50:322–329.

Birch, S. A. J., & Bloom, P. (2003). Children are cursed: An asymmetric bias in mental-state attribution. *Psychological Science* 14:283–286.

Buccino, G., Binkofski, F., Fink, G. R., Fadiga, L., Fogassi, L., Gallese, V., Seitz, R. J., Zilles, K., Rizzolatti, G., & Freund, H.-J. (2001). Action observation activates premotor and parietal areas in a somatotopic manner: An fMRI study. *European Journal of Neuroscience* 13(2):400–404.

Buckner, R. L., & Carroll, D. C. (2007). Self-projection and the brain. *Trends in Cognitive Sciences* 11:49–57.

Calder, A. J., Keane, J., Manes, F., Antoun, N., & Young, A. W. (2000). Impaired recognition and experience of disgust following brain injury. *Nature Reviews Neuroscience* 3:1077–1078.

Camerer, C., Loewenstein, G., & Weber, M. (1989). The curse of knowledge in economic settings: An experimental analysis. *Journal of Political Economy* 97:1232–1254.

Carlson, S. M., & Moses, L. J. (2001). Individual differences in inhibitory control and children's theory of mind. *Child Development* 72:1032–1053.

Carruthers, P. (2006). *The Architecture of the Mind.* Oxford, UK: Oxford University Press.

Chartrand, T. L., & Bargh, J. A. (1999). The chameleon effect: The perception-behavior link and social interaction. *Journal of Personality and Social Psychology* 76:893–910.

Churchland, P. (1981). Eliminative materialism and the propositional attitudes. *Journal of Philosophy* 78:67–90.

Collingwood, R. G. (1946). *The Idea of History.* Oxford, UK: Clarendon Press.

Currie, G., & Ravenscroft, I. (2002). *Recreative Minds.* Oxford, UK: Oxford University Press.

Damasio, A. R., Tranel, D., & Damasio, H. (1990). Face agnosia and the neural substrates of memory. *Annual Review of Neuroscience* 13:89–109.

Dennett, D. C. (1987). *The Intentional Stance*. Cambridge, MA: MIT Press.

De Renzi, E., Cavalleri, F., & Facchini, S. (1996). Imitation and utilization behavior. *Journal of Neurology and Neurosurgical Psychiatry* 61:396–400.

Dimberg, U., Thunberg, M., & Elmehed, K. (2000). Unconscious facial reactions to emotional facial expressions. *Psychological Science* 11(1):86–8

Evans, G. (1982). *The Varieties of Reference.* Oxford, UK: Oxford University Press.

Fadiga, L., Fogassi, L, Pavesi, G., & Rizzolatti, G. (1995). Motor facilitation during action observation: A magnetic stimulation study. *Journal of Neurophysiology* 73:2608–2611.

Fodor, J. A. (1983). *The Modularity of Mind.* Cambridge, MA: MIT Press.

———. (1987). *Psychosemantics.* Cambridge, MA: MIT Press.

Fogassi, L., Ferrari, P. F., Gesierich, B., Rozzi, S., Chersi, F., & Rizzolatti, G. (2005). Parietal lobe: From action organization to intention understanding. *Science* 308:662–667.

Gallese, V., & Goldman, A. (1998). Mirror neurons and the simulation theory of mind-reading. *Trends in Cognitive Sciences* 2:493–501.

Gallese,V., Fadiga, L., Fogassi, L., & Rizzolatti, G. (1996). Action recognition in the premotor cortex. *Brain* 119:593–609.

Gergely, G., & Csibra, G. (2003). Teleological reasoning in infancy: The naïve theory of rational action. *Trends in Cognitive Sciences* 7(7):287–292.

Gergely, G., Nadasdy, Z., Csibra, G., & Biro, S. (1995). Taking the intentional stance at 12 months of age. *Cognition* 56:165–193.

Goldman, A. I. (1989). Interpretation psychologized. *Mind and Language* 4:161–185.

———. (2006). *Simulating Minds: The Philosophy, Psychology, and Neuroscience of Mindreading.* New York: Oxford University Press.

———. (2008). Mirroring, mindreading, and simulation. In J. Pineda, ed., *Mirror Neuron Systems: The Role of Mirroring Processes in Social Cognition,* 311–330. New York: Humana Press.

Goldman, A. I., & Sripada, C. S. (2005). Simulationist models of face-based emotion recognition. *Cognition* 94:193–213.

Gopnik, A. (1993). How we know our minds: The illusion of first-person knowledge of intentionality. *Behavioral and Brain Sciences* 16:1–14.

Gopnik, A., Glymour, C., Sobel, D. M., Schulz, L. E., Kushnir, T., & Danks, D. (2004). A theory of causal learning in children: Causal maps and Bayes nets. *Psychological Review* 111:3–32.

Gopnik, A., & Meltzoff, A. N. (1997). *Words, Thoughts and Theories*. Cambridge, MA: MIT Press.

Gopnik, A., & Wellman, H. (1992). Why the child's theory of mind really *is* a theory. *Mind and Language* 7:145–171.

Gordon, R. M. (1986). Folk psychology as simulation. *Mind and Language* 1:158–171.

———. (1996). 'Radical' simulationism. In P. Carruthers & P. Smith, eds., *Theories of Theories of Mind*, 11–21. Cambridge, UK: Cambridge University Press.

Hatfield, E., Cacioppo, J. T., & Rapson, R. L. (1994). *Emotional Contagion*. Cambridge, UK: Cambridge University Press.

Heal, J. (1986). Replication and functionalism. In J. Butterfield, ed., *Language, Mind, and Logic*, 135–150. Cambridge, UK: Cambridge University Press.

Hurlburt, R. T., & Heavey, C. L. (2001). Telling what we know: Describing inner experience. *Trends in Cognitive Sciences* 5:400–403.

Iacoboni, M., Molnar-Szakacs, I., Gallese, V., Buccino, G., Mazziotta, J. C., & Rizzolatti, G. (2005). Grasping the intentions of others with one's own mirror neuron system. *PLoS Biology* 3:529–535.

Jackson, P. L., Meltzoff, A. N., & Decety, J. (2004). How do we perceive the pain of others? A window into the neural processes involved in empathy. *NeuroImage* 24:771–779.

Jeannerod, M. (2001). Neural simulation of action: A unifying mechanism for motor cognition. *NeuroImage* 14:S103–S109.

Johnson, S. C. (2005). Reasoning about intentionality in preverbal infants. In P. Carruthers, S. Lawrence, & S. Stich, eds., *The Innate Mind: Structure and Contents*, 254–271. Oxford, UK: Oxford University Press.

Johnson, S. C., Slaughter, V., & Carey, S. (1998). Whose gaze will infants follow? Features that elicit gaze-following in 12-month-olds. *Developmental Science* 1:233–238.

Kanwisher, N., McDermott, J., & Chun, M. M. (1997). The fusiform face area: A module in human extrastriate cortex specialized for face perception. *Journal of Neuroscience* 17:4302–4311.

Keysers, C., Wicker, B., Gazzola, V., Anton, J.-L., Fogassi, L., & Gallese, V. (2004). A touching sight: SII/PV activation during the observation of touch. *Neuron* 42:335–346.

Kohler, E., Keysers, C., Umilta, M. A., Fogassi, L., Gallese, V., & Rizzolatti, G. (2002). Hearing sounds, understanding actions: Action representation in mirror neurons. *Science* 297:846–848.

Kosslyn, S. M. (1978). Measuring the visual angle of the mind's eye. *Cognitive Psychology* 7:341–370.

———. (1994). *Image and Brain: The Resolution of the Imagery Debate*. Cambridge, MA: MIT Press.

Leslie, A. M. (1994). Pretending and believing: Issues in the theory of ToMM. *Cognition* 50:211–238.

Leslie, A. M., & German, T. (1995). Knowledge and ability in 'theory of mind': One-eyed overview of a debate. In M. Davies & T. Stone, eds., *Mental Simulation*. 123–150, Oxford, UK: Blackwell.

Lhermitte, F., Pillon, B., & Serdaru, M. (1986). Human autonomy and the frontal lobes. Part I: Imitation and utilization behavior: a neuropsychological study of 75 patients. *Annals of Neurology* 19(4):326–334.
Meltzoff, A. N., & Moore, M. K. (1983). Newborn infants imitate adult facial gestures. *Child Development* 54:702–709.
Mitchell, P., & Lacohee, H. (1991). Children's early understanding of false belief. *Cognition* 39:107–127.
Morrison, I., Lloyd, D., de Pelligrino, G., & Roberts, N. (2004). Vicarious responses to pain in anterior cingulate cortex. Is empathy a multisensory issue? *Cognitive Affective Behavioral Neuroscience* 4:270–278.
Nichols, S., & Stich, S. P. (2003). *Mindreading*. Oxford, UK: Oxford University Press.
Nisbett, R. E., & Wilson, T. D. (1977). Telling more than we can know. *Psychological Review* 84:231–259.
Onishi, K. H., & Baillargeon, R. (2005). Do 15-month-old infants understand false beliefs? *Science* 308:255–258.
Paccalin, C., & Jeannerod, M. (2000). Changes in breathing during observation of effortful actions. *Brain Research* 862:194–200.
Perner, J. (1991). *Understanding the Representational Mind*. Cambridge, MA: MIT Press.
Rizzolatti, G., Fadiga, L., Gallese, V., & Foggasi, L. (1996). Premotor cortex and the recognition of motor actions. *Cognitive Brain Research* 3:131–141.
Samson, D., Apperly, I. A., Kathirgamanathan, U., & Humphreys, G. W. (2005). Seeing it my way: A case of a selective deficit in inhibiting self-perspective. *Brain* 128:1102–1111.
Saxe, R. (2005). Against simulation: The argument from error. *Trends in Cognitive Sciences* 9:174–179.
Scholl, B., and Leslie, A. M. (1999). Modularity, development and 'theory of mind'. *Mind and Language* 14:131–153.
———. (2001). Minds, modules and meta-analysis. *Child Development* 72:696–701.
Schooler, J., Reichle, E. D., & Halpern, D. V. (2004). Zoning-out during reading: Evidence for dissociations between experience and meta-consciousness. In D. Levin, ed., *Thinking and Seeing: Visual Meta-Cognition in Adults and Children*. Cambridge, MA: MIT Press.
Schulz, L. E., & Gopnik, A. (2004). Causal learning across domains. *Developmental Psychology* 40:162–176.
Sellars, W. (1956). Empiricism and the philosophy of mind. In H. Feigl & M. Scriven, eds., *Minnesota Studies in Philosophy of Science*, vol. 1, 253–329. Minneapolis: University of Minnesota Press.
Shoemaker, S. (1996). *The First-Person Perspective and Other Essays*. New York: Cambridge University Press.
Singer, T., Seymour, B., O'Doherty, J., Kaube, H., Dolan, R., & Frith, C. (2004). Empathy for pain involves the affective but not sensory components of pain. *Science* 303:1157–1162.
Spelke, E. (1994). Initial knowledge: Six suggestions. *Cognition* 50:431–445.
Stich, S. (1981). Dennett on intentional systems. *Philosophical Topics* 12:38–62.
Stueber, K. (2006). *Rediscovering Empathy*. Cambridge, MA: MIT Press.
Tager-Flusberg, H. (2000). Language and understanding minds: Connections in autism. In S. Baron-Cohen, H. Tager-Flusberg, & D. Cohen, eds., *Understanding Other Minds: Perspectives from Developmental Cognitive Neuroscience*, 2nd ed., 124–149. Oxford, UK: Oxford University Press.

Van Boven, L., & Loewenstein, G. (2003). Social projection of transient drive states. *Personality and Social Psychology Bulletin* 29(9):1159–1168.

Van Boven, L., Dunning, D., & Loewenstein, G. (2000). Egocentric empathy gaps between owners and buyers: Misperceptions of the endowment effect. *Journal of Personality and Social Psychology* 79:66–76.

Wellman, H. M., Cross, D., & Watson, J. (2001). Meta-analysis of theory-of-mind development: The truth about false belief. *Child Development* 72:655–684.

Wicker, B., Keysers, C., Plailly, J., Royet, J.-P., Gallese, V., & Rizzolatti, G. (2003). Both of us disgusted in *my* insula: The common neural basis of seeing and feeling disgust. *Neuron* 40:655–664.

Wilson, T. D. (2002). *Strangers to Ourselves: Discovering the Adaptive Unconscious.* Cambridge, MA: Harvard University Press.

Wimmer, H., & Perner, J. (1983). Beliefs about beliefs: Representation and constraining function of wrong beliefs in young children's understanding of deception. *Cognition* 13:103–128.

Woodward, A. L. (1998). Infants selectively encode the goal object of an actor's reach. *Cognition* 69:1–34.

Zaitchik, D. (1991). Is only seeing really believing? Sources of true belief in the false belief task. *Cognitive Development* 6:91–103.

CHAPTER 2

Mirror Neurons and the Simulation Theory of Mindreading

(WITH VITTORIO GALLESE)

How do we understand other people's behavior? How can we assign goals, intentions, or beliefs to the inhabitants of our social world? A possible way to answer these challenging questions is to adopt an evolutionary frame of reference, both in phylogenetical and ontogenetical terms, envisaging these mindreading capacities as rooted in antecedent, more "ancient" and simple mechanisms. This approach can capitalize on the results of different fields of investigation: Neurophysiology can investigate the neural correlates of precursors of these mechanisms in lower species of social primates, such as macaque monkeys. Developmental psychology can study how the capacity to attribute propositional attitudes to others develops.

In the present chapter, we propose that humans' mindreading abilities rely on the capacity to adopt a simulation routine. This capacity might have evolved from an action execution/observation matching system whose neural correlate is represented by a class of neurons recently discovered in the macaque monkey premotor cortex: mirror neurons (MNs).

1. THE MACAQUE MONKEY PREMOTOR AREA F5 AND MIRROR NEURONS

Converging anatomical evidence (see Matelli and Luppino[1] for review) supports the notion that the ventral premotor cortex (referred to also as inferior area 6) is composed of two distinct areas, designated as F4 and F5 (figure 2.1).[2] Area F5 occupies the most rostral part of inferior area 6,

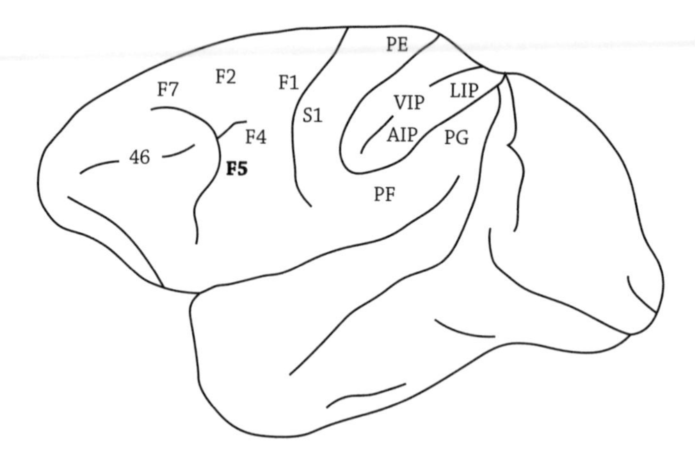

Figure 2.1: Anatomical location and functional properties of mirror neurons.
Lateral view of the macaque brain showing the cytoarchitectonic parcellation of the agranular frontal cortex and of the posterior parietal cortex. Motor and premotor areas, indicated by the letter F, are defined according to Matelli et al. Mirror neurons were all recorded from area F5 (shown in bold).

extending rostrally within the posterior bank of the inferior limb of the arcuate sulcus. Area F5 is reciprocally connected with the hand field of the primary motor cortex[3–5] and has direct, although limited, projections to the upper cervical segments of the spinal cord.[6] Microstimulation in F5 evokes hand and mouth movements at thresholds generally higher than in the primary motor cortex.[7,8] The functional properties of F5 neurons were assessed in a series of single unit recording experiments.[9–11] These experiments showed that the activity of F5 neurons is correlated with specific hand and mouth motor acts and not with the execution of individual movements like contractions of individual muscle groups. What makes a movement into a motor act is the presence of a goal. This distinction is very important because it allows one to interpret the role of the motor system not just in terms of the control of the dynamic variables of movement (like joint torques, etc.), but rather as a possible candidate for the instantiation of mental states such as purpose or intention. Using the effective motor act as the classification criterion, the following types of neurons were described: Grasping neurons, holding neurons, tearing neurons, and manipulation neurons. Grasping neurons discharge when the monkey performs movements aimed to take possession of objects with the hand (grasping-with-the-hand neurons), with the mouth (grasping-with-the-mouth neurons), or with both. Grasping-with-the-hand neurons form the largest class of F5 neurons. Most neurons of this class are selective for different types of grips. The role of these neurons has been conceptualized by Rizzolatti[12] as a "motor vocabulary" of actions related to prehension.

The study of F5 neurons' responsiveness to visual stimuli led to the discovery of two distinct classes of neurons: canonical neurons,[13] which are activated during observation of graspable objects, and MNs,[14,15] which discharge when the monkey observes another individual performing an action. We will describe in more detail the functional properties of this class of neurons.

MNs respond both when a particular action is performed by the recorded monkey and when the same action performed by another individual is observed. All MNs, as mentioned above, discharge during specific goal-related motor acts. Grasping, manipulating, and holding objects are by far the most effective actions triggering their motor response. About half of them discharge during a specific type of prehension, precision grip (prehension of small objects by opposing the thumb and the index finger) being the most common one. The most effective visual stimuli triggering MNs' visual responses are actions in which the experimenter or a second monkey interacts with objects with their hand or mouth. Neither the sight of the object alone nor the agent alone is effective in evoking the neuronal response. Mimicking the action without a target object or performing the action by using tools is similarly ineffective. In over 90 percent of MNs, a clear correlation between the most effective observed action and their motor response was observed. In many neurons, this correlation was strict both in terms of the general goal of the action (e.g., grasping) and in terms of the way in which it was executed (e.g., precision grip).

On the basis of their functional properties summarized here, MNs appear to form a cortical system that matches observation and execution of motor actions. What could be the possible functional role of this matching system? Before addressing this issue it is important to stress that the existence of an equivalent system has also been demonstrated in humans.

2. THE MIRROR SYSTEM IN HUMANS

Two lines of evidence strongly suggest that an action/observation matching system similar to that discovered in monkeys also exists in humans. The first refers to an elegant study by Fadiga et al.[16] in which the excitability of the motor cortex of normal human subjects was tested by using transcranial magnetic stimulation (TMS). The basic assumption underlying this experiment was the following: If the observation of actions activates the premotor cortex in humans, as it does in monkeys, this mirror effect should elicit an enhancement of the motor evoked potentials (MEPs) induced by TMS of the motor cortex, given its strong anatomical links to premotor areas. TMS was performed during four different conditions: observation of an experimenter grasping objects; observation of an experimenter making aimless movements in the air with his arm; observation of objects; and detection of the dimming of a small spot of light. The results of this study showed that during grasping observation, MEPS recorded from the hand muscles markedly increased with respect to the other conditions, including the attention-demanding dimming detection task. Even more intriguing was the finding that the increase of excitability was present only in those muscles that subjects would use when actively performing the observed movements. This study provided for the first time evidence that humans have a mirror system similar to that in monkeys. Every time we are looking at someone performing an action, the same motor circuits that are recruited when we ourselves perform that action are concurrently activated.

These results posed the question of the anatomical location of the mirror system within the human brain. This issue has been addressed by two brain-imaging experiments utilizing the technique of Positron Emission Tomography (PET).[17,18] These two experiments, although different in many respects, shared a condition in which normal human subjects observed the experimenter grasping 3-D objects. Both studies used the observation of objects as a control condition. The results showed that grasping observation significantly activates the cortex of the left superior temporal sulcus (Brodmann's area 21) of the left inferior parietal lobule (Brodmann's area 40) and of the anterior part of Broca's region (Brodmann's area 45). The activation, during action observation, of a cortical sector of the human brain traditionally linked with language raises the problem of the possible homologies between Broca's region and the premotor area F5 of the monkey, in which MNs have been discovered. This issue is outside the scope of the present chapter and will not be dealt with here (for discussion, see endnote).[19]

3. MIRROR NEURONS AND MINDREADING

What is the function of the mirror system? One possible function could be to promote learning by imitation. When new motor skills are learned, one often spends the first training phases trying to replicate the movements of an observed instructor. MNs could in principle facilitate that kind of learning. We do not favor this possible role of MNs, at least in nonhuman primates (see box 2.1). Here we explore another possibility: that MNs underlie the process of mindreading, or serve as precursors to such a process.

BOX 2.1 MIRROR NEURONS AND IMITATION

The ability of nonhuman primates to imitate the behavior of conspecifics is a highly controversial issue. Tomasello et al.[a] identify three strict criteria to delimit imitational learning:

(1) The imitated behavior should be novel for the imitator;
(2) it should reproduce the behavioral strategies of the model; and
(3) it should share with it the same final goal.

Behaviors not satisfying these criteria should not be considered as true imitational ones, and are rather to be explained by means of other mechanisms, such as stimulus enhancement, emulation, or response facilitation. By applying these strict criteria to the extant literature, Tomasello et al. exclude the possibility that wild animals may display true imitative behavior. A different perspective is offered by Byrne and Russon.[b] These authors start from the concept that behaviors display a hierarchical structure, and can therefore be described at several levels of increasing complexity. Manual skills represent a good example. Because complex behaviors are hierarchically structured, "... there exists a range of possibilities for how imitation might take place, beyond the simple dichotomy of imitation versus no imitation." Byrne and Russon single out an action-level imitation in which a detailed specification of the various motor sequences composing a complex action is made, and a program-level imitation in which the broader, more highly structured component of a complex skill is retained, with subjective solutions to the low-level specifications. Byrne and Russon conclude that imitational learning in nonhuman primates might have been overlooked by the exclusive application of the action-level strategy as the defining criterion.

What is the relevance of MNs for imitation in nonhuman primates? First of all, it should be stressed that imitation behavior has never been observed in association with MN activity. Furthermore, even adopting Byrne and Russon's criteria, we are not aware of any clear-cut evidence of imitation of grasping behavior among adult macaque monkeys, although this possibility is not precluded for young monkeys during development. On the basis of these considerations, we are inclined not to favor the hypothesis that MNs in area F5 promote grasping imitation learning in adult monkeys.

Notes

a Tomasello, M., Kruger, A. C., & Ratner, H. H. (1993). Cultural learning. *Behav. Brain Sci.* 16:495–511.

b Byrne, R. W., & Russon, A. E. Learning by imitation: A hierarchical approach. *Behav. Brain Sci.* (in press).

Mindreading is the activity of representing specific mental states of others, for example, their perceptions, goals, beliefs, expectations, and the like. It is now agreed that all normal humans develop the capacity to represent mental states in others, a system of representation often called folk psychology. Whether nonhuman primates also deploy folk psychology is more controversial (see the last section of this chapter), but it certainly has not been precluded. The hypothesis explored here is that MNs are part of—albeit perhaps a rudimentary part of—the folk psychologizing mechanism.

Like imitation learning, mindreading could make a contribution to inclusive fitness. Detecting another agent's goals and/or inner states can be useful to an observer because it helps him anticipate the agent's future actions, which might be cooperative, non-cooperative, or even threatening. Accurate understanding and anticipation enable the observer to adjust his responses appropriately. Our discussion of mindreading initially and primarily focuses on humans; later, we return to its possible realization in nonhuman primates.

4. TWO THEORIES OF MINDREADING

There is a large literature concerned with the nature of (human) mindreading. Two types of approaches have dominated recent discussion: theory-theory (TT) and simulation theory (ST).[20–22] The fundamental idea of TT is that ordinary people accomplish mindreading by acquiring and deploying a commonsense theory of mind (ToM), something akin to a scientific theory. Mental states attributed to other people are conceived of as unobservable, theoretical posits, invoked to explain and predict behavior in the same fashion that physicists appeal to electrons and quarks to predict and explain observable phenomena. On the standard presentation, the ToM possessed by ordinary people consists of a set of causal/explanatory laws that relate external stimuli to certain inner states (e.g., perceptions), certain inner states (e.g., desires and beliefs) to other inner states (e.g., decisions), and certain inner states (e.g., decisions) to behavior. This picture has been articulated by functionalist philosophers of mind,[23–26] as well as by developmental psychologists.[27,28] According to TT, attributing particular mental states to others arises from theoretical reasoning involving tacitly known causal laws.

Much on this subject has been done by developmentalists eager to determine how the mindreading capacity is acquired in childhood.[29] Many interpret children's changes in mindreading skills as evidence in favor of TT because the skill changes are construed as manifestations of changes in theory.[30,31] Theory-theorists differ among themselves as to whether ToM is acquired by a general-purpose scientizing algorithm[32] or by the maturation of a domain-specific module or set of modules.[33,34] This debate will not concern us here.

ST arose partly from doubts about whether folk psychologizers really represent, even tacitly, the sorts of causal/explanatory laws that TT typically posits. ST suggests that attributors use their own mental mechanisms to calculate and predict the mental processes of others. For example, Kahneman and Tversky[35] gave subjects a description of two travelers who shared the same limousine en route to the airport and were caught in a traffic jam. Their planes were scheduled to depart at the same time, but they arrived thirty minutes late. Mr. A was told that his flight left on time; Mr. B was told that his flight was delayed and left just five minutes ago. Who was more upset? Ninety-six percent of the experimental subjects said that Mr. B was more upset. How did they arrive at this answer? According to TT, there must be some psychological law they exploited to infer the travelers' relative upsetness. According to ST, on the other hand, each subject would have put himself in each of the imaginary traveler's shoes and imagined how he would have felt in his place.[36] Another example concerns the prediction of decisions. To predict White's next move in a chess match, ST suggests that you try to simulate White's thought processes and arrive at a decision that you then attribute to him.[37–38] First you create in yourself *pretend* desires, preferences, and beliefs of the sort you assume White to have (e.g., preferences among chess strategies). These pretend preferences and beliefs are fed into your own decision-making mechanism, which outputs a (pretend) decision (see figure 2.2). Instead of acting on that decision, it is taken "offline" and used to predict White's decision. According to this simulation account, you need not know or utilize any psychological laws.

If simulation is going to make accurate predictions of targets' decisions, pretend desires and beliefs must be sufficiently similar to genuine desires and beliefs that the decision-making system operates on them the same way as it operates on genuine desires and beliefs. Are pretend states really similar enough to the genuine articles that this will happen? Homologies between pretend and natural (i.e., non-pretend) mental states are well documented in the domains of visual and motor imagery.[39–43] (We assume here that visual and motor imaging consist, respectively, in pretending to see and pretending to do; see Currie and Ravenscroft.[44]) These visual and motor homologies do not show, of course, that other pretend mental states (e.g., desires and beliefs) also functionally resemble their natural counterparts, but informal evidence suggests this (see Goldman[45]).

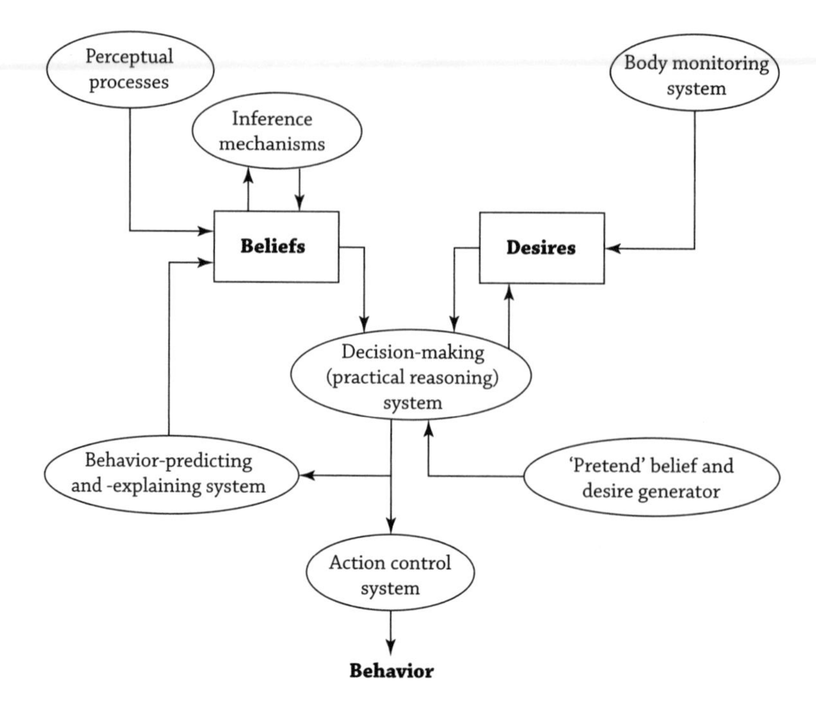

Figure 2.2: The basic elements of the simulation routine.
Cognitive steps in predicting or explaining someone's decision by means of simulation are shown in the lower part of the figure. A dedicated pretend-state generator generates pretend beliefs and desires suited to the target agent. These pretend beliefs and desires are fed into the attributor's decision-making system (the same system that normally operates on natural, non-pretend beliefs and desires). The output of the decision-making system is taken offline. That is, instead of being fed into the action control system, the output decision is sent to the behavior-predicting and -explaining system, which outputs a prediction that the target will make that very decision.[59]

5. THE DIFFERENCE BETWEEN TT AND ST

The core difference between TT and ST, in our view, is that TT depicts mindreading as a thoroughly "detached" theoretical activity, whereas ST depicts mindreading as incorporating an attempt to replicate, mimic, or impersonate the mental life of the target agent.[46] This difference can be highlighted diagrammatically, as shown in figure 2.3.

In the simulation scenario, there is a distinctive matching or "correspondence" between the mental activity of the simulator and the target. This is highlighted by the similar state-sequences the two undergo (figures 2.3A and 2.3B), the only exception being that the simulator uses pretend states rather than natural states. The attributor in the TT scenario (figure 2.3C) does not utilize any pretend states that mimic those of the target; nor does he utilize his own decision-making system to arrive at a prediction.

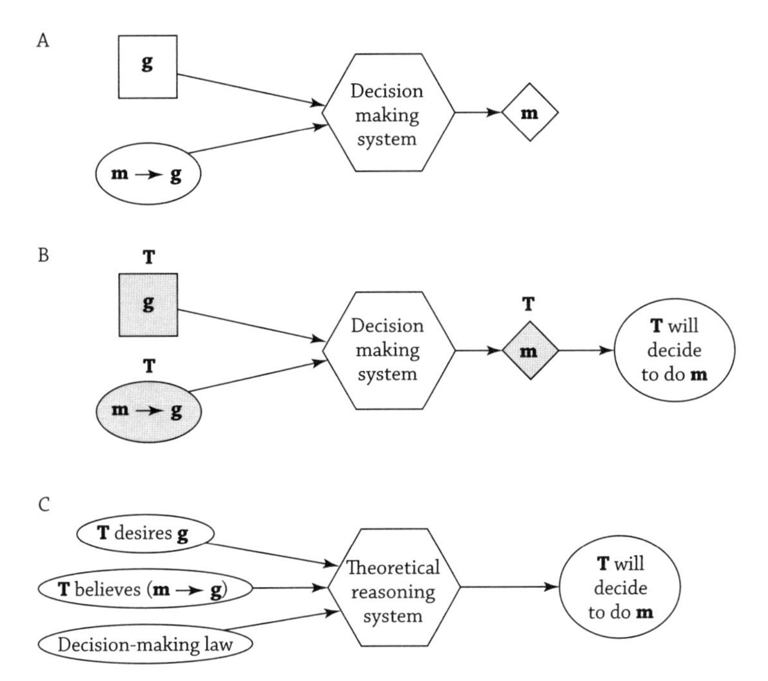

Figure 2.3: Two possible ways of predicting someone's decision.
(A) A simple decision by an agent. His desire for goal **g** and belief that action **m** would be a good means to **g** are fed into his decision-making system, which outputs a decision to perform **m**. **(B)** shows how an attributor can successfully predict this agent's decision using a simulation routine. After learning of the target's **(T)** desire and belief (e.g., from previous applications of the simulation routine), the attributor creates similar pretend states in himself. These states are "tagged" as belonging to the target and then fed into the attributor's decision-making system, which outputs a (pretend) decision to do **m**. The attributor takes this decision offline and predicts that the target will decide to do **m**. **(C)** represents the way an attributor might predict the target's decision using theoretical reasoning. The attributor starts with knowledge that the target has a desire for goal **g** and a belief that **m** would achieve **g**. He also believes some psychological law about human decision-making. These beliefs are all fed into his own theoretical-reasoning system, which outputs the belief that the target will decide to do **m**. Squares represent desires; ellipses represent beliefs; diamonds represent decisions; and hexagons represent cognitive mechanisms. Shading indicates that the mental state is a pretend state.

Thus, ST hypothesizes that a significant portion of mindreading episodes involves the process of mimicking (or trying to mimic) the mental activity of the target agent. TT predicts no such mimicking as part of the mindreading process. This contrast presents a potential basis for empirically discriminating between ST and TT. If there is evidence of mental mimicry in the mindreading process, that would comport nicely with ST and would not be predicted by TT.

Before turning to such evidence, however, we should note that simulation can be used to retrodict, as well as predict, mental states, that is, to

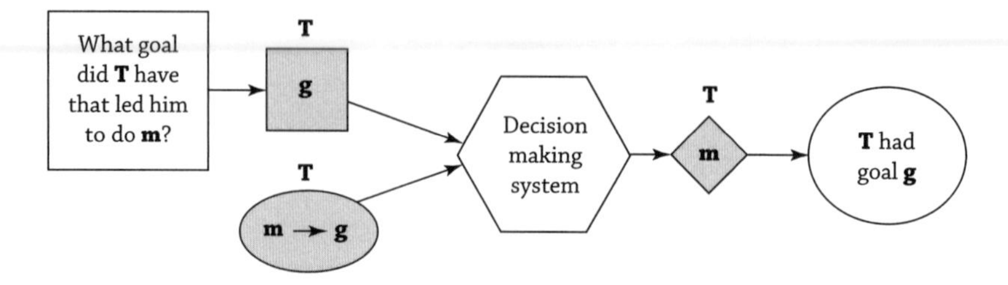

Figure 2.4: A retrodictive use of simulation.
After observing the target agent **(T)** perform action **m**, the attributor uses simulation to test whether goal **g** would have fitted with the choice of **m**. Goal **g** is recreated and fed into his decision-making system, which does output **m**.

determine what mental states of a target have *already* occurred. Figure 2.4 depicts a retrodictive use of simulation. The attributor starts with the question "What goal did the target have that led him to perform action **m**?" He conjectures that it was goal **g**, and tries out this conjecture by pretending to have **g**, as well as certain beliefs about the effectiveness or ineffectiveness of the action **m** vis-à-vis goal **g**. This simulation leads him to form a (pretend) decision to do **m**. He therefore uses this result to conclude that the target did indeed have goal **g**. In this fashion, the attributor ultimately makes a "backward" inference from the observed action to a hypothesized goal state.

6. MIRROR NEURONS AND SIMULATION

In a similar fashion, it is conceivable that externally generated MN activity serves the purpose of "retrodicting" the target's mental state, moving backward from the observed action. Let us interpret internally generated activation in MNs as constituting a plan to execute a certain action (e.g., the action of holding a certain object, grasping it, or manipulating it). When the same MNs are externally activated—by observing a target agent execute the same action—MN activation still constitutes a plan to execute this action. But in the latter case, the subject of the MN activity knows (visually) that the observed target is concurrently performing this very action. So we assume that he "tags" the plan in question as belonging to that target. In fact, externally generated MN activity does not normally produce motor execution of the plan in question. Externally generated plans are largely inhibited, or taken "offline," precisely as ST postulates. Thus, MN activity seems to be nature's way of getting the observer into the same "mental shoes" as the target—exactly what the conjectured simulation heuristic aims to do.

Although we compare externally generated MN activity with what transpires in figure 2.4, there clearly are differences. One difference is that the real attributor does not go back to a distal goal or set of beliefs. He only goes back to a motoric plan. Still, this seems to be a "primitive" use of simulation with the same structure as that depicted in figure 2.4. It also bears a resemblance to the motor theory of speech perception advocated by Liberman,[47] in which the common link between the sender and the receiver is not sound but the neural mechanism, shared by both, allowing the production of phonetic gestures.

A proponent of TT might say that TT also has ways of accounting for retrodictive attributions of mental states. Is it clear that anything similar to simulation occurs in externally generated MN activity? The point is that MN activity is not mere theoretical inference. It creates in the observer a state that matches that of the target. This is how it resembles the simulation heuristic. Nothing about TT leads us to expect this kind of matching. It should be emphasized that the hypothesis being advanced here is not that MNs themselves constitute a full-scale realization of the simulation heuristic. In particular, we do not make this conjecture for MNs in monkeys. Our conjecture is only that MNs represent a primitive version, or possibly a precursor in phylogeny, of a simulation heuristic that might underlie mindreading.

A further link between mirror neuron activity and simulation can be inferred from the fact that, as the TMS experiment by Fadiga et al. demonstrates, the human equivalent matching system facilitates in the observer the same muscle groups as those utilized by the target. This supports the idea that even when one is observing the action of another, one undergoes a neural event that is qualitatively the same as an event that triggers actual movement in the observed agent. It is as if the tracking process in the observer is not taken entirely offline. This might appear to be a violation of ST, but actually it is wholly within the spirit of ST, which postulates mental occurrences in the mindreader that are analogous to mental occurrences in the target, so it is not surprising that downstream motor activity is not entirely inhibited. If TT were correct, and an observer represents a target's behavior in purely theoretical fashion, it would not be predicted that the same muscle groups would be facilitated in the observer as in the target. But if ST were correct, and a mindreader represents an actor's behavior by recreating in himself the plans or movement intentions of the actor, then it is reasonable to predict that the same muscular activation will occur in the mindreader. As matching muscular activation is actually observed in the observer, this lends support to ST as opposed to TT.

Clinical evidence of a similar phenomenon is found in so-called imitation behavior. A group of patients with prefrontal lesions compulsively imitate gestures or even complex actions performed in front of them by an experimenter. This behavior is explained as an impairment of the inhibitory control normally governing motor schemas, or plans. It may be inferred from this that normal humans, when observing someone else perform an action, generate a plan to do the same action, or an image of doing it, themselves. Normally this plan is inhibited so that it does not yield motor output, but such inhibition is impaired in the patient population in question.[48]

7. NONHUMAN PRIMATES: BEHAVIORISTS OR MINDREADERS?

A mindreading capacity for nonhuman primates is a hotly debated issue among primatologists and behavioral scientists. In a recent paper, Heyes[49] argued that a survey of empirical studies of imitation, self-recognition, social relationships, deception, role-taking, and perspective-taking fails to support the ToM hypothesis over non-mentalist alternatives. Although, for the sake of concision, it is not possible here to address this issue thoroughly (for reviews, see note),[50] a few points are worth making.

Let us consider first the social nature of nonhuman primates. Social organization is by no means a distinctive feature of primates: Within the realm of insects, several species (ants are one example) are endowed with a clear social structure. The distinctive hallmark of the social organization of nonhuman primates is its sophisticated complexity. Nonhuman primates live in groups that can comprise as many as one hundred individuals. These groups are characterized by intense and diversified types of social interactions.[51] Within such a complex and hierarchically organized social structure, individuals are able to recognize kinship and hierarchical ranks, and distinguish allies from enemies. Stammbach[52] showed that dominant macaque monkeys modified their social relationships with lower-ranking individuals who had previously learned how to retrieve food by pressing a lever. Dominant individuals started grooming the low-ranking ones more often than before, once they "understood" that the newly acquired skills of the low-ranking individuals could be more easily triggered, and therefore exploited, by using this sort of social upgrading. All these examples, although not providing conclusive evidence of mindreading abilities, nevertheless, in our view, provide a strong argument supporting the hypothesis that nonhuman primates are endowed with cognitive abilities that cannot be easily dismissed as the result of simple stimulus—response operant conditioning.

Being a "cognizer," nevertheless, does not necessarily imply being a mindreader, or a possessor of the ability to detect intentional states in others.

The argument that seems to suggest the presence, in nonhuman primates, of elementary forms of mindreading abilities comes from the discovery of deceptive behavior. In a series of field experiments, Hauser[53,54] showed that rhesus monkeys can withhold information about food location in order to deceive conspecifics and obtain more food for themselves. Deception is particularly relevant here, because deceptive behavior calls for the existence of second-order intentionality, and therefore for the capability to attribute mental states to conspecifics.

8. OUTSTANDING QUESTIONS

- Is mirror-neuron activity innate or learned, and what is the relevance of this to the ST interpretation of mirror-neuron activity?
- Is the motor system involved in the semantic mode of internally coding actions?
- Can any evidence be found of "matching" events for observed agent's beliefs (as well as their plans or intentions) in nonhuman primates?

The relevance of the data on deceptive behavior has been questioned on the basis of two main arguments. First, field reports of ethologists are anecdotal and therefore intrinsically ambiguous. Second, alternative non-mentalistic explanations, such as chance behavior, associative learning, and inferences about observable features of the situation have been proposed as more parsimonious explanations of deceptive behavior (see Heyes).

However, according to Byrne,[55] who surveyed the literature thoroughly, there are at least eighteen independent reports of intentional deception in nonhuman primates, supporting the notion that they can represent the mental states of other conspecifics. On the basis of this evidence, Byrne and Whiten[56] suggested that primates act according to a manipulative strategy very similar to that put forward in the sixteenth century by Niccolò Machiavelli in his masterpiece *Il Principe*.[57]

Our speculative suggestion is that a "cognitive continuity" exists within the domain of intentional-state attribution from nonhuman primates to humans, and that MNs represent its neural correlate (see box 2.2). This continuity is grounded in the ability of both human and non-human primates to detect goals in the observed behavior of conspecifics. The capacity to understand action goals, already present in nonhuman primates, relies on a process that matches the observed behavior to the action plans of the observer. It is true, as pointed out by Meltzoff and Moore,[58] that the understanding of action goals does not imply a full grasp of mental states such as beliefs or desires. Action-goal understanding nevertheless constitutes a necessary phylogenetical stage within the evolutionary path leading to the fully developed mindreading abilities of human beings.

ACKNOWLEDGMENTS

This work was supported by a Human Frontier Scientific Program grant to V. G. The authors wish to thank Giacomo Rizzolatti, Elisabeth Pacherie, and the anonymous referees for their valuable comments and criticisms.

BOX 2.2 NEURAL CODING OF COMPLEX BIOLOGICAL STIMULI

Neurons responding to complex biological stimuli have been previously described in the macaque brain. A series of studies showed that in the inferior temporal cortex there are neurons that discharge selectively to the presentation of faces or hands.[a-c] More recently it has been shown that some of these neurons respond to specific features of these stimuli.[d] Neurons responding to complex biological visual stimuli, such as walking or climbing, were reported also in the amygdala.[e] Even more relevant to the issues addressed in this chapter is the work of Perrett and colleagues.[f,g] These authors showed that in the cortex buried within the superior temporal sulcus (STS), there are neurons selective to the observation of hand movements. These properties resemble the visual properties of F5 MNs very closely: Both populations of neurons code the same types of actions; they both generalize their responses to the different instances of the same action; and they both are not responsive to mimicked hand actions without the target object. However, the unique feature of F5 MNs resides in the fact that they also discharge during active movements of the observer. An observed action produces the same neural pattern of activation as does the action actively made by the observer.

The presence of two brain regions with neurons endowed with similar complex visual properties raises the question of their possible relationship. Two possibilities might be suggested. One is that F5 MNs and STS neurons have different functional roles: STS neurons would code the semantic properties, the meaning, of hand-object interactions, while F5 MNs would be engaged in the pragmatic coding of the same actions. A second possibility, which we favor, is that these two "action detector" systems could represent distinct stages of the same analysis. The STS neurons would provide an initial "pictorial" description of actions that would be then fed (most likely through an intermediate step in the posterior parietal cortex) to the F5 motor vocabulary, where it would acquire a meaning for the individual. The latter hypothesis stresses the role of action in providing meaning to what is perceived.

Notes

a Gross, C. G., et al. (1972). Visual properties of neurons in inferotemporal cortex of the monkey. *J. Neurophysiol.* 35:96–111.
b Perrett, D. I., Rolls, E. T., & Caan, W. (1982). Visual neurons responsive to faces in the monkey temporal cortex. *Exp. Brain Res.* 47:329–342.
c Gross, C. G., et al. (1985). Inferior temporal cortex and pattern recognition. In *Pattern Recognition Mechanisms* (Chagas, C., Gattass, R., & Gross, C., eds.), Berlin: Springer-Verlag.
d Tanaka, K., et al. (1991). Coding visual images of objects in the inferotemporal cortex of the macaque monkey. *J. Neurophysiol.* 66:170–189.
e Brothers, L., Ring, B., & Kling, A. (1990). Response of neurons in the macaque amygdala to complex social stimuli. *Behav. Brain Res.* 41:199–213.
f Perrett, D. I. (1989). Frameworks of analysis for the neural representation of animate objects and actions. *J. Exp. Biol.* 146:87–113.
g Perrett, D. I. (1990). Understanding the visual appearance and consequence of hand actions. In *Vision and Action: the Control of Grasping* (Goodale, M. A., ed.), 163–180, New York: Ablex.

REFERENCES

1. Matelli, M., & Luppino, G. (1997). Functional anatomy of human motor cortical areas. In *Handbook of Neuropsychology* (vol. 11) (Boiler, F., & Grafman, J., eds), 9–26. Amsterdam: Elsevier.
2. Matelli, M., Luppino, G., & Rizzolatti, G. (1985). Patterns of cytochrome oxidase activity in the frontal agranular cortex of macaque monkey. *Behav. Brain Res.* 18: 125–137.
3. Matsumura, M., & Kubota, K. (1979). Cortical projection of hand-arm motor area from postarcuate area in macaque monkey: A histological study of retrograde transport of horseradish peroxidase. *Neurosci. Lett.* 11:241–246.
4. Muakkassa, K. F., & Strick, P. L. (1979). Frontal lobe inputs to primate motor cortex: evidence for four somatotopically organized 'premotor' areas. *Brain Res.* 177:176–182.
5. Matelli, M., et al. (1986). Afferent and efferent projections of the inferior area 6 in the macaque monkey. *J. Comp. Neurol.* 251:281–298.
6. He, S. Q., Dum, R. P., & Strick, P. L. (1993). Topographic organization of corticospinal projections from the frontal lobe: Motor areas on the lateral surface of the hemisphere. *J. Neurosci.* 13:952–980.
7. Gentilucci, M., et al. (1988). Functional organization of inferior area 6 in the macaque monkey: I. Somatotopy and the control of proximal movements. *Exp. Brain Res.* 71: 475–490.
8. Hepp-Reymond, M-C., et al. (1994). Force-related neuronal activity in two regions of the primate ventral premotor cortex. *Can. J. Physiol. Pharmacol.* 72:571–579.

9. Rizzolatti, G., et al. (1981). Afferent properties of periarcuate neurons in macaque monkey. II. Visual responses. *Behav. Brain Res.* 2:147–163.
10. Okano, K., & Tanji, J. (1987). Neuronal activities in the primate motor fields of the agranular frontal cortex preceding visually triggered and self-paced movement. *Exp. Brain Res.* 66:155–166.
11. Rizzolatti, G., et al. (1988). Functional organization of inferior area 6 in the macaque monkey: II. Area F5 and the control of distal movements. *Exp. Brain Res.* 71:491–507.
12. Rizzolatti, G., & Gentilucci, M. (1988). Motor and visual-motor functions of the premotor cortex. In *Neurobiology of Neocortex* (Rakic, P., & Singer, W., eds.), 269–284. New York: John Wiley & Sons.
13. Murata, A., et al. (1997). Object representation in the ventral premotor cortex (Area F5) of the monkey. *J. Neurophysiol.* 78:2226–2230.
14. Gallese, V., et al. (1996). Action recognition in the premotor cortex. *Brain* 119:593–609.
15. Rizzolatti, G., et al. (1996). Premotor cortex and the recognition of motor actions. *Cognit. Brain Res.* 3:131–141.
16. Fadiga, L., et al. (1995). Motor facilitation during action observation: A magnetic stimulation study. *J. Neurophysiol.* 73:2608–2611.
17. Rizzolatti, G., et al. (1996). Localization of grasp representations in humans by PET: 1. Observation versus execution. *Exp. Brain Res.* 111:246–252.
18. Grafton, S. T., et al. (1996). Localization of grasp representations in humans by PET: II. Observation compared with imagination. *Exp. Brain Res.* 112:103–111.
19. Rizzolatti, G., & Arbib, M. A. (1998). Language within our grasp. *Trends Neurosci.* 21:188–194.
20. Davies, M., & Stone, T., eds. (1995). *Folk Psychology*. Oxford, UK: Blackwell.
21. Davies, M., & Stone, T., eds. (1995). *Mental Simulation*. Oxford, UK: Blackwell.
22. Carruthers, P., & Smith, P., eds. (1996). *Theories of Theories of Mind*. Cambridge, UK: Cambridge University Press.
23. Sellars, W. (1963). Empiricism and the philosophy of mind. In *Science, Perception and Reality*, 127–194. London: Routledge.
24. Lewis, D. (1972). Psychophysical and theoretical identifications. *Australas. J. Philos.* 50:249–258.
25. Fodor, J. A. (1987). *Psychosemantics*. Cambridge, MA: MIT Press.
26. Churchland, P. N. (1988). *Matter and Consciousness*. Cambridge, MA: MIT Press.
27. Wellman, H. (1990). *The Child's Theory of Mind*. Cambridge, MA: MIT Press.
28. Gopnik, A. (1993). How we know our minds: The illusion of first-person knowledge of intentionality. *Behav. Brain Sci.* 16:1–14.
29. Astington, J., Harris, P., & Olson, D., eds. (1988). *Developing Theories of Mind*. Cambridge, UK: Cambridge University Press.
30. Perner, J. (1991). *Understanding the Representational Mind*. Cambridge, MA: MIT Press.
31. Gopnik, A., & Wellman, H. (1992). Why the child's theory of mind really is a theory. *Mind Lang.* 7:145–171.
32. Gopnik, A., & Meltzoff, A. (1997). *Words, Thoughts, and Theories*. Cambridge, MA: MIT Press.
33. Leslie, A. (1994) Pretending and believing: Issues in the theory of TOMM. *Cognition* 50:211–238.

34. Baron-Cohen, S. (1995). *Mindblindness*. Cambridge, MA: MIT Press.
35. Kahneman, D., & Tversky, A. (1982). The simulation heuristic. In *Judgment Under Uncertainty* (Kahneman, D., Slovic, P., & Tversky, A., eds.), 201–208. Cambridge, UK: Cambridge University Press.
36. Goldman, A. (1989). Interpretation psychologized. *Mind Lang*. 4:161–185.
37. Gordon, R. (1986). Folk psychology as simulation. *Mind Lang*. 1:158–171.
38. Heal, J. (1986). Replication and functionalism. In *Language, Mind and Logic* (Butterfield, J., ed.), 135–150. Cambridge, UK: Cambridge University Press.
39. Kosslyn, S. (1978). Measuring the visual angle of the mind's eye. *Cognit. Psychol.* 10:356–389.
40. Decety, J., et al. (1991). Vegetative response during imagined movement is proportional to mental effort. *Behav. Brain Res*. 42:1–5.
41. Farah, M., Soso, M., & Dasheiff, R. (1992). Visual angle of the mind's eye before and after unilateral/occipital lobectomy. *J. Exp. Psychol. Hum. Percept. Perform.* 18:241–246.
42. Yue, G., & Cole, K. (1992). Strength increases from the motor program: Comparison of training with maximal voluntary and imagined muscle contractions. *J. Neurophysiol* 67:1114–1123.
43. Jeannerod, M. (1994). The representing brain: Neural correlates of motor intention and imagery. *Behav. Brain Sci*. 17:187–202.
44. Currie, G., & Ravenscroft, I. (1997). Mental simulation and motor imagery. *Philos. Sci*. 64:161–180.
45. Goldman, A. (1992). In defense of the simulation theory. *Mind Lang*. 7:104–119.
46. Goldman, A. (2000). The mentalizing folk. In *Metarepresentations* (Sperber, D., ed.), 171–196, Oxford, UK: Oxford University Press.
47. Liberman, A. (1996). *Speech: A Special Code*. Cambridge, MA: MIT Press.
48. Lhermitte, F., Pillon, B., & Serdaru, M. (1986). Human autonomy and the frontal lobes: I. Imitation and utilization behavior: A neuropsychological study of 75 patients *Ann. Neurol*. 19:326–334.
49. Heyes, C. M. (1998). Theory of mind in nonhuman primates. *Behav. Brain Sci.* 21:101–148.
50. Barresi, J., & Moore, C. (1996). Intentional relations and social understanding. *Behav. Brain Sci*. 19:107–154.
51. Cheney, D. L., & Seyfarth, R. M., eds. (1990). *How Monkeys See the World*. Chicago: Chicago University Press.
52. Stammbach, E. (1988). Group responses to specially skilled individuals in *Macaca fascicularis*. *Behaviour* 107:241–266.
53. Hauser, M. D. (1992). Costs of deception: Cheaters are punished in rhesus monkeys. *Proc. Natl. Acad. Sci. U. S. A.* 89:12137–12139.
54. Hauser, M. D. & Marier, P. (1993). Food-associated calls in rhesus macaques (*Macaca mulatta*): II. Costs and benefits of call production and suppression. *Behav. Ecology* 4:206–212.
55. Byrne, R. W. (1995). *The Thinking Ape. Evolutionary Origins of Intelligence*. Oxford, UK: Oxford University Press.
56. Byrne, R. W., & Whiten, A. (1988). *Machiavellian Intelligence: Social Expertise and the Evolution of Intellect in Monkeys, Apes and Humans*. Oxford, UK: Clarendon Press.
57. Machiavelli, N. (1532). *Il Principe* (English trans., 1979, *The Prince*). New York: Penguin Books.

58. Meltzoff, A., & Moore, M. K. (1995). Infants' understanding of people and things: from body imitation to folk psychology. In *The Body and the Self* (Bermudez, J. L. et al., eds.), 43–69. Cambridge, MA: MIT Press.
59. Stich, S., & Nichols, S. (1992). Folk psychology: Simulation versus tacit theory. *Mind Lang*. 7:29–65.

CHAPTER 3

Simulationist Models of Face-Based Emotion Recognition

(WITH CHANDRA SEKHAR SRIPADA)

1. INTRODUCTION

Mindreading is the capacity to identify the mental states of others (e.g., their beliefs, desires, intentions, goals, experiences, sensations, and also emotion states). One approach to mindreading holds that mental-state attributors deploy a naïve psychological theory to infer mental states in others from their behavior, the environment, and/or their other mental states. According to different versions of this theory-theory (TT), the naïve psychological theory is either a component of an innate, dedicated module or is acquired by domain-general learning. A second approach holds that people typically execute mindreading by a different sort of process, a simulation process. Roughly, according to simulation theory (ST), an attributor arrives at a mental attribution by simulating, replicating, or reproducing in his own mind the same state as the target's, or by attempting to do so. For example, the attributor would pretend to be in initial states thought to correspond to those of the target, and feed these states into parts of his own cognitive equipment (e.g., a decision-making mechanism), which would operate on them to produce an output state that is imputed to the target.

Mindreading has been studied in many disciplines, and both TT and ST have had proponents in each of them. In developmental psychology, TT has been endorsed by Gopnik & Meltzoff (1997), Gopnik & Wellman (1992, 1994), Leslie (1994), Perner (1991), Premack & Woodruff (1978), and Wellman (1990), whereas ST has been defended by Harris (1991, 1992). In philosophy, ST has been endorsed by Currie & Ravenscroft (2002),

Goldman (1989, 1992a, 1992b, 2000), Gordon (1986, 1992, 1996), and Heal (1986, 1996, 1998), whereas TT has been defended as an explicit approach to the execution of mindreading by Fodor (1992), Nichols, Stich, Leslie, & Klein (1996), and Stich & Nichols (1992), or as a theory of how the folk conceptualize mental states by Armstrong (1968), Lewis (1972), Sellars (1956), and Shoemaker (1975). Studies of mentalizing in neuroscience (e.g., Fletcher et al., 1995; Frith & Frith, 1999; McCabe, Houser, Ryan, Smith, & Trouard, 2001) typically ignore the TT-ST controversy but work with theory of mind (ToM) terminology (which is suggestive of TT) and cite the TT-leaning literature. On the other hand, much recent neuroscientific work is quite receptive to simulationist ideas (Blakemore & Decety, 2001; Carr, Iacoboni, Dubeau, Mazziotta, & Lenzi, 2003; Chaminade, Meary, Orliaguet, & Decety, 2001; Gallese, 2001, 2003; Gallese & Goldman, 1998; Iacoboni et al., 1999; Jeannerod, 2001), although the majority of this research is addressed less to mindreading per se than to related topics, such as simulation of action, imitation, or empathy. In recent years, a number of researchers have moved away from pure forms of TT or ST in the direction of some sort of TT/ST hybrid (Adolphs, 2002; Goldman, 2006; Nichols & Stich, 2003; Perner, 1996), though the exact nature of the hybrid is rather fluid. In light of this continuing controversy, any research that provides substantial evidence for either ST or TT, even in a single subdomain of mindreading, deserves close attention.

In this chapter, we review a body of neuropsychological research that, we shall argue, supports ST for a certain circumscribed mindreading task. This is the task of attributing emotion states to others based on their facial expressions. This task is different from those usually studied in the mindreading literature, in part because the attributed mental states differ from the usual ones. The vast majority of the literature is devoted to propositional attitudes, such as desires and beliefs, almost entirely ignoring emotion states like fear, anger, disgust, or happiness. There is no good reason to exclude these mental states, which are routinely attributed to others in daily life. So it is time to extend research and theory into this subdomain of the mental. At the same time, it cannot be assumed that the style of mindreading in this subdomain is the same as the style that characterizes other subdomains. So we make no attempt to generalize from the type of mental state ascriptions studied here to mindreading tout court.

There are at least two reasons why the properties of face-based emotion recognition (FaBER) might not be shared by methods of mindreading in other subdomains. First, the recognition or classification of propositional attitude contents may introduce a level of complexity that goes beyond the task of classifying emotion types. Second, the reading of emotions, especially basic emotions, may have unique survival value, so it is conceivable that specialized programs have evolved for the recognition of emotions, and these specialized programs may not operate in other mindreading tasks. Because of these differences between FaBER and other types of mindreading, it cannot be assumed that the processes characteristic of FaBER can be extrapolated to other types of mindreading.

We begin by reviewing existing findings, some clinical and some experimental, that display a striking pattern of paired deficits between emotion production and face-based recognition (attribution). These findings have not previously been brought together with the explicit intent of examining them in the context of the TT-ST controversy. Next, we argue that this pattern readily lends itself to a simulationist explanation, whereas existing data do not fit with a theory-based explanation. Finally, the core project of the chapter is to formulate and evaluate four specific models of how normal mindreaders could use simulation to arrive at emotion classifications.

2. PAIRED DEFICITS IN EMOTION PRODUCTION AND FACE-BASED RECOGNITION

In early studies, Ralph Adolphs and colleagues investigated whether damage to the amygdala affects face-based emotion recognition (Adolphs, 1995; Adolphs, Tranel, Damasio, & Damasio, 1994). These studies were motivated by the well-known role of the amygdala in mediating fear, including its prominent role in fear-conditioning and the storage of fear-related emotion memories (LeDoux, 1993, 2000). One patient studied by Adolphs et al. was SM, a thirty-year-old woman with Urbach-Wiethe disease, which resulted in bilateral destruction of her amygdalae with sparing of adjacent hippocampus and other neocortical structures. Consistent with the important role of the amygdala in mediating fear, SM was indeed abnormal in her experience of fear. Antonio Damasio notes that SM "approaches people and situations with a predominantly, indeed excessively, positive attitude":

> S[M] does not experience fear in the same way you or I would in a situation that would normally induce it. At a purely intellectual level she knows what fear is supposed to be, what should cause it, and even what one may do in situations of fear, but little or none of that intellectual baggage, so to speak, is of any use to her in the real world. The fearlessness of her nature, which is the result of the bilateral damage to her amygdalae, has prevented her from learning, throughout her young life, the significance of the unpleasant situations that all of us have lived through. (Damasio, 1999: 66)

Other lines of evidence also suggest that SM is abnormal in her experience of fear. For example, in one experiment, SM was presented with a conditioned stimulus repeatedly paired with a startle-inducing unconditioned stimulus, a boat horn delivered at 100 decibels. However, she failed to demonstrate a conditioned autonomic reaction to the conditioned stimulus, indicating she had an abnormality in acquiring or expressing conditioned emotion responses (Bechara et al., 1995).

Adolphs et al. (1994) tested SM against a number of brain-damaged controls in various FaBER tasks. Subjects were presented with photographs or video slides depicting facial expressions and asked to identify the emotion states to which the expressions corresponded. SM was abnormal in face-based recognition of the emotion fear; her ratings of fearful faces correlated less with normal ratings than did those of any of twelve brain-damaged control subjects, and fell 2–5 standard deviations below the mean of the controls when the data were converted to a normal distribution. Subsequent studies have both confirmed and qualified these findings regarding co-occurring deficits in the production of fear and the ability to recognize expressions of fear in others. Sprengelmeyer et al. (1999) studied NM, another patient with bilateral amygdala damage. Like SM, NM was abnormal in his experience of fear. He was prone to dangerous activities (such as hunting jaguars in the Amazon river basin and deer in Siberia while dangling from a helicopter) and tested as abnormal on a self-assessment questionnaire measuring experience of the emotion fear. NM also exhibited a severe and selective impairment in face-based recognition for fear.

Adolphs et al. (1999) conducted an additional study of patients with bilateral amygdala damage with a larger-sized sample (nine patients, including SM). They again found that face-based fear recognition is abnormal among these patients. Although deficits were most severe for fear recognition, recognition of other emotions, anger in particular, was also somewhat abnormal. Other neuropsychological studies are also broadly consistent with these findings (see Adolphs, 2002; Lawrence & Calder, 2004 for reviews).

The pattern noted here (i.e., a *paired deficit* in the production and face-based recognition of an emotion) is not unique to fear. A similar pattern emerges with respect to at least two other emotions, disgust and anger, to which we now turn.

Paul Rozin and colleagues (Rozin, Haidt, & McCauley, 2000) conceptualized the emotion of disgust as an elaboration of a phylogenetically more primitive distaste response. Many aspects of taste processing are known from animal studies to be localized in the anterior insula region, known as the "gustatory cortex" (Rolls, 1995). Functional neuroimaging studies confirm that the anterior insula plays a similar role in taste processing in humans (Small et al., 1999, 2003).

What neural structures are implicated in the *recognition* of facial expressions of disgust? Sprengelmeyer and colleagues (Sprengelmeyer et al., 1996, 1997), using standard face-based emotion recognition tasks, found that patients with Huntington's disease display selective deficits in face-based recognition of disgust. In light of these findings, Phillips and colleagues undertook an fMRI study to see which brain areas are activated when subjects observe facial expressions of disgust (Phillips et al., 1997, 1998). The most striking finding for perception of facial expressions of disgust was activation in the right insula (adjacent regions such as the amygdala were not activated). They concluded that "appreciation of visual stimuli depicting other's disgust is closely linked to the perception [i.e., experience] of unpleasant tastes and smells" (Phillips et al., 1997: 496).

Lesion studies have also found paired deficits in the experience and facial recognition of disgust. Calder and colleagues found this pairing in NK, who suffered insula and basal ganglia damage (Calder, Keane, Manes, Antoun, & Young, 2000). On a questionnaire, NK's overall score for disgust was significantly lower than the scores of controls, whereas his scores for anger and fear did not significantly differ from the controls' mean scores. In tests of his ability to recognize emotions in faces, NK showed significant and selective impairment in disgust recognition. Adolphs, Tranel, & Damasio (2003) similarly found pronounced deficits in the experience and face-based recognition of disgust in a patient with bilateral insular and temporal lobe damage.

Wicker et al. (2003) did an fMRI study of disgust to determine whether the same neural regions are activated in normal subjects both during the experience of disgust and during the observation of the facial expression of disgust. In two "visual" runs, participants passively viewed movies of individuals smelling the contents of a glass (disgusting, pleasant, or neutral) and expressing the facial expressions of the respective emotions. In two "olfactory" runs, the same participants inhaled disgusting or pleasant odorants through a mask over their nose and mouth. The core finding of Wicker et al. was that the left anterior insula and the right anterior cingulate cortex are preferentially activated during the experience of the emotion of disgust evoked by disgusting odorants (compared to activation levels during pleasant and neutral odors), and this same region is preferentially activated during the observation of disgust facial expressions (compared to activation levels during pleasure-expressive and neutral faces). In other words, observation of disgust-expressive faces automatically activates the same neural substrates implicated in the experience of the same emotion.[1]

Anger is a third emotion system for which a paired deficit in emotion production and face-based recognition is found. Social agonistic encounters represent a distinct and phylogenetically recurrent adaptive problem for many animals. Various lines of evidence, reviewed in Lawrence and Calder (2004), suggest that the dopamine system has evolved as a neural subsystem involved in the processing of aggression in social-agonistic encounters in a wide variety of species, and this system plays an important role in mediating the experience of the emotion anger. They note that dopamine levels in rats and a number of other species are elevated in social-agonistic encounters. Conversely, administration of dopamine antagonists, such as the D_2 receptor antagonist *sulpiride*, selectively impairs responses to agonistic encounters. Lawrence and colleagues hypothesized that dopaminergic blockade by the administration of sulpiride would lead to selective disruption of face-based recognition of anger, while sparing the recognition of other emotions (Lawrence, Calder, McGowan, & Grasby, 2002). This was indeed found. Following sulpiride administration, subjects were significantly worse at recognizing angry faces, though there were no such impairments in recognizing facial expressions of other emotions.

Based on the studies reviewed above, there is substantial evidence that *deficits in the production of an emotion and deficits in the face-based recognition of that emotion reliably co-occur*. How is this evidence relevant to the question of whether (face-based) emotion mindreading proceeds by tacit theorizing or by simulation?

3. EMOTION MINDREADING BY THEORY VERSUS SIMULATION

Let us further clarify and expand upon the two basic theoretical positions toward mindreading, which have loomed large in the literature. There are numerous ways of developing the TT idea, but the main idea is that the mindreader selects a mental state for attribution to a target based on *inference* from other information about the target. According to one popular version of TT, such an inference is guided by folk psychological generalizations concerning relationships or transitions between psychological states and/or behavior of the target (Gopnik & Wellman, 1992; Wellman, 1990). But we shall not insist on law-like generalizations. The fundamental feature of TT is that it is an *information-based* approach. It says that attributors engage in mindreading by deploying folk psychological information. What they don't do, as a means to reading a target's mental state, is (try to) *model* or *instantiate* the very mental process that the target herself undergoes.

The core idea of ST is that the attributor selects a mental state for attribution after reproducing or "enacting" within herself the very state in question, or a relevantly similar state. In other words, she tries to replicate a target's mental state by undergoing (what she takes to be) the same or a similar mental process to one the target undergoes. If, in her own case, the process yields mental state M as an output, she attributes M to the target. For example, if she wants to attribute a future decision to a target, she might try to replicate the target's decision-making process in her own mind and use the output of this process as the decision to assign to the target. Alternatively, she may test a hypothesized state by simulating it in her own mind and seeing whether its upshots match those of the target. In either scenario, the attributor must recognize her own state as being of type M in order to select M as the state type occupied by the target. This presumably requires some sort of "information" about states of type M, so simulation isn't entirely information-free (as some proponents of simulationism maintain, e.g., Gordon, 1996). However, in contrast to TT, ST says that the relevant information about M is applied to something like a token or facsimile of a mental state in her own mind, not simply to information about the target from which she infers that the target instantiates M. There is, of course, much more to be said about the TT/ST contrast, but these points should suffice for present purposes (for additional details, see Gallese & Goldman, 1998; Goldman, 2000, 2006.).

How would TT and ST be applied to the task of face-based emotion recognition? First, how would TT explain the capacity to attribute emotions from facial expressions? We are not aware of any *specific* TT-based proposal in the literature. However, a general outline of what a TT-based account would look like is not hard to provide. It would propose, at a minimum, that people have a mentally represented body of generalizations for mapping representations of particular facial configurations to names for emotion states. When a target is observed displaying a particular facial expression, the attributor utilizes this body of information, coupled with ordinary capacities for factual reasoning, to infer and attribute an emotion state to the target. Of course, this account presupposes that there is enough information in the facial expression itself to uniquely select an appropriate corresponding emotion state. This supposition seems plausible, as it has been shown that facial expressions exhibit rich geometric and configural properties sufficient for the purposes of inferring a corresponding emotion state (Calder, Burton, Miller, Young, & Akamatsu, 2001). Thus a TT-based account, elaborated along the lines we have suggested, is one legitimate contender for explaining how face-based emotion recognition occurs.

ST would approach this question in a different way. It would propose that an attributor selects an emotion category to assign to a target by producing an emotion in herself, or running her own emotional "equipment," and seeing which emotion has an appropriate link to the observed facial expression. Exactly how this simulational story would go is a matter to be addressed in detail below. In outline, however, the distinctive characteristic of the simulationist approach is to hypothesize that (normal) attributors execute face-based emotion attribution by means that somehow involve the production of that very emotion (at least in cases of accurate emotion detection).

4. EXPLAINING THE EMOTION RECOGNITION DATA BY TT VERSUS ST

The central claim presented in section 2 was that there is substantial evidence concerning three emotions indicating that deficits in the production (experience) of an emotion and deficits in the face-based recognition of that emotion reliably co-occur. This strongly suggests that the same neural mechanisms subserve both the experience and the recognition of an emotion. In addition, the Wicker et al. (2003) study found that, in normals, the same neural regions were implicated in both the experience of disgust and the observation of disgust-expressive faces. Putting the Wicker et al. data together with the paired-deficit data, the same neural substrate is implicated in normals when they both experience and observe disgust, and when this same substrate is damaged, subjects fail to experience or recognize disgust at normal levels.

How does this evidence bear on the choice between theorizing and simulating as the explanation of face-based emotion recognition? On the surface, it strongly favors ST. If one (successfully) mindreads via simulation, one undergoes the same, or a relevantly similar, process to the one the target undergoes in using or arriving at the target state. Someone impaired in experiencing a given emotion will be unable to simulate a process that includes that emotion. Thus, ST predicts that someone damaged in experiencing fear, or in the neural substrate of fear experience, would have trouble mindreading fear. Hence, the phenomenon observed in SM—a paired deficit in fear experience and recognition—is straightforwardly predictable based on the simulationist hypothesis. And it is similarly so for the other paired deficits. By contrast, there is no reason to expect a paired deficit under TT. Why should conceptual representations of fear occur in the same region that underlies fear experience? That is, why should the processing of theoretical information *about* fear occur in the same region (or one of the regions) as the region subserving fear itself? TT predicts no such finding. Perhaps TT could be supplemented with auxiliary assumptions to make it consistent with the finding. We will discuss possible auxiliary assumptions the theory-theorist might deploy in a moment, but here we merely note that, in the absence of such assumptions, TT would not lead one to expect the identified paired deficits. Thus, there is a prima facie case for ST over TT as an explanation of the phenomena.

Skeptics might reply that paired deficits could be due to a merely contingent relationship, for example, the contingent co-localization of emotion experience and face-based emotion recognition (or classification) capacities, rather than the functional codependence that follows from a simulationist account. For example, lesions to the fusiform gyrus of the right occipital cortex produce both prosopagnosia, an impaired ability to recognize faces (see Kanwisher, 2000), and achromatopsia, an impaired perception of color (see Bartels & Zeki, 2000). But these two deficits have no interesting functional relationship to one another. It just so happens that the impaired capacities are at least partially co-localized in the fusiform gyrus, leading to the paired deficit. Isn't it possible that such a purely happenstantial story also applies to the paired deficits found in fear, disgust, and anger recognition?

In theory this is possible. But the fact that paired deficits in emotion production and emotion recognition occur for three distinct emotions makes the pairings seem far from contingent or accidental. They seem to reflect a *systematic* relationship between emotion experience and FaBER. So while a theory-based account can appeal to various auxiliary assumptions or hypotheses to account for the paired deficit data (e.g., the contingent co-localization hypothesis), these assumptions and hypotheses appear ad hoc relative to the simulationist explanation, which predicts the paired deficits in a more principled way.

Let us be more specific about how a TT explanation of the paired deficits might go. Three types of declarative knowledge might be used in the normal execution of FaBER tasks according to TT: (1) visually obtained knowledge of the facial configuration of the target, (2) semantic knowledge concerning these configurations, in particular, knowledge that facial configuration C is paired with emotion E, and (3) general knowledge concerning a given emotion (i.e., its typical elicitors or behavioral effects). In order to account for a paired deficit in one emotion, TT must say that one or more of these types of knowledge concerning the emotion in question are selectively damaged, while similar types of knowledge are preserved for other emotions. As it happens, there is specific evidence from the paired-deficit literature that paired-deficit patients do *not* suffer from reduced knowledge of types (1) or (3). A deficit in knowledge of type (2) is not specifically contravened by the evidence, but this proposal suffers from other problems.

The evidence about knowledge of types (1) and (3) is as follows: First, the paired- deficit studies present evidence that subjects have no difficulty with perceptual processing of faces. In most of these studies, subjects performed normally on measures designed to identify any such impairments. The most common measure used was the Benton face-matching task, in which different views of unfamiliar faces must be categorized as belonging to the same face (Benton, Hamsher, Varney, & Spreen, 1983). Additionally, subjects were often found to be able to recognize other high-level properties of faces, including age, gender, and identity. SM's ability to recognize facial identity, for example, was fully preserved.

An informational deficit of type (3) is also disconfirmed by existing evidence. In the cited studies, deficits in FaBER routinely occur with preservation of subjects' general declarative knowledge regarding emotions. For example, subjects can readily cite situations in which a person might experience the emotion of which face-based recognition is impaired. Recall that in describing SM, Damasio (1999) noted that "at a purely intellectual level," SM "knows what fear is supposed to be, what should cause it, and even what one may do in situations of fear." Similarly, Calder, Lawrence, & Young (2001) reported that "...patients with disgust recognition impairments are able to provide plausible situations in which a person might feel disgusted and do not show impaired knowledge of the concept of disgust." Finally, in most cases, subjects' lesions occurred relatively late in life. So it cannot be plausibly argued that they lacked declarative knowledge about emotions because of deficits in their own emotional experiences. They did suffer from experience deficits at the time of examination, but most had ample opportunities in earlier life to undergo relevant experiences and build normal declarative knowledge from those experiences.

So in order to account for the paired-deficit data, the theory-theorist is likely to appeal to deficits in information of type (2), information consisting of semantic labels paired with representations of facial configurations. The theory-theorist will need to propose that labeling information of this kind for fear, disgust, and anger depends on the integrity of the amygdala, anterior insula, and dopaminergic system, respectively. Moreover, the theory-theorist must claim that it is possible to damage this labeling information quite selectively, in at least two ways. First, it must be possible to damage this labeling information for one emotion while leaving this information preserved for other emotions. Second, it must be possible to damage the labeling information in a way such that the label is inaccessible for *visual* representations of faces specifically, because, as just reported, impaired subjects have command of the label when verbally discussing general knowledge of the impaired emotion type. Although these postulations are certainly possible, in the absence of any independent reason to believe that naming information is stored in this way, such postulations seem quite ad hoc. Thus, the kinds of deficits to which theory-theorists might appeal are either specifically contravened by the evidence or are quite improvised.

5. POSSIBLE SIMULATIONIST MODELS

Although the foregoing case for a simulational approach to face-based emotion recognition strikes us as compelling, it leaves open the question of how the simulational process proceeds. Those skeptical of our case for simulation may remain skeptical as long as no plausible, sufficiently detailed story of the simulation process in these cases is forthcoming. We get little help on this question from the existing literature. Articles describing paired deficits often contain conclusions hinting at a simulational explanation, but few pursue any details about the computational mechanisms. Exploring the options for a simulational process is the next task we tackle.

There are several ways a simulation heuristic might be used to attribute a mental state, depending on the nature of the causal link between the evidence events in the target (known to the attributor) and the sought-after state of the target. The causal link might be of two general kinds: (A) the evidence events cause the state, or (B) the state causes the evidence events. When an attributor has knowledge of prior states of the target (e.g., specific desires and beliefs) and wants to predict a mental effect of those states (e.g., a decision), we have an instance of type (A). Here the evidence events cause (or will cause) the sought-after mental state. When an attributor witnesses a target's piece of behavior (including a facial expression) and seeks to identify a mental state responsible for that behavior, we have an instance of type (B). Here the sought-after mental state is what causes the observed evidence. In face-based emotion recognition, the relevant connection is presumably of type (B). The target's emotional state causes her facial expression, and this expression is the evidence used by the attributor to identify the antecedent emotion state. How might simulation be used to exploit this kind of evidence?

5.1. Generate-and-test model

One possibility is a *generate-and-test* heuristic. As shown in figure 3.1, the attributor starts by hypothesizing a certain emotion as the possible cause of the target's facial display and proceeds to "enact" that emotion (i.e., produce a facsimile of it in her own system). She lets this facsimile (or pretend) emotion run its typical course, which includes the production of its natural facial expression, or at least a neural instruction to the facial musculature to construct the relevant expression. If the resulting facial expression, or the instruction to construct such an expression, matches the expression observed in the target, then the hypothesized emotion is confirmed, and the attributor imputes that emotion to the target. The simulation interpretation of the paired-deficit findings would say that this is the sort of thing that happens in emotion interpreters who are normal with respect to the emotion in question. Someone impaired in the relevant emotion area, however, cannot "enact" that emotion or produce a facsimile of it. So she cannot generate the relevant face-related downstream activity necessary to recognize the emotion. Hence, a recognition impairment specific to that emotion arises.

Several issues about this model must be addressed. One question concerns the final phase of the postulated process, in which the system tries to "match" a constructed facial expression with the expression observed in the target. The representation of one's own facial expression is presumably a proprioceptive representation, whereas the representation of the target's expression is visual. How can one "match" the other? One possible answer is that the system has acquired an association between proprioceptive and visual representations of the same facial configuration through some type of learning. Alternatively, there might be an innate cross-modal matching of the sort postulated by Meltzoff & Moore (1997) to account for neonate facial imitation.

Second, there is a problem of how the generation process works. If candidate emotions are generated randomly, say, from the six basic emotions, the observer will have to covertly generate on average three facial expressions before hitting upon the right one. This would be too slow to account for actual covert mimicry of displayed facial expressions, which occurs as early as 300 ms after stimulus onset (Dimberg & Thunberg, 1998; Lundquist & Dimberg, 1995). An alternative is to say that "theoretical" information is used to guide the generation process—though it isn't clear what theoretical information it would be. However, this proposal seems to turn the generate-and-test model into more of a theory-simulation hybrid, rather than a pure simulationist model. Does this undercut the thrust of

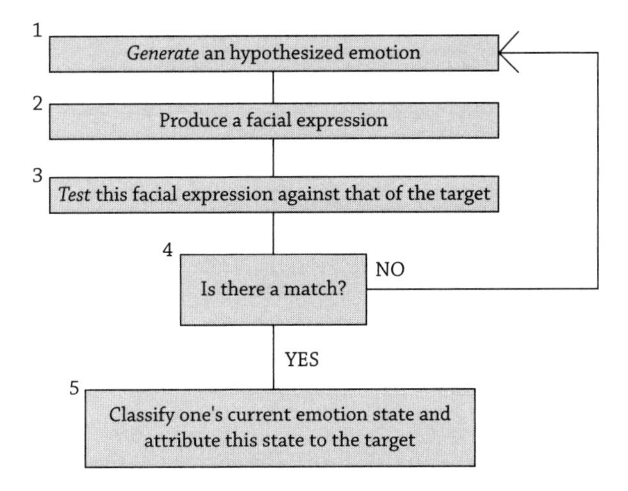

Figure 3.1: Generate-and-test simulation.

our simulationist argument? No. First, the simulational "test" phase of the generate-and-test heuristic is crucial, because without it, the model cannot explain the paired-deficit data. Second, the timing problems make this first model the least promising of the four we shall offer, and all of the other three are more purely simulationist in character.

5.2. Reverse simulation model

A second possibility, which seems to be implicitly endorsed by a number of theorists in the literature, is a *reverse simulation heuristic*. The idea in reverse simulation is that the attributor engages one of her own mental processes in the reverse direction, in order to attribute to the target a mental state that is temporally prior to the state that serves as the evidence for the attribution. In most cases in which simulation is deployed, reverse simulation is not an option: The standard forward directionality of mental processes precludes the possibility that these processes can be utilized in a reverse direction for the purposes of evidence-posterior interpretation tasks. However, there may be an important exception in the case of FaBER.

Under conditions of normal operation, the induction of an emotion episode causes a coordinated suite of cognitive and physiological changes, including, at least in the case of the so-called basic emotions, a characteristic facial expression (Ekman, 1992). Interestingly, this causal relationship appears bidirectional. There is substantial evidence that manipulation of the facial musculature, either voluntarily or involuntarily, has a causal effect in

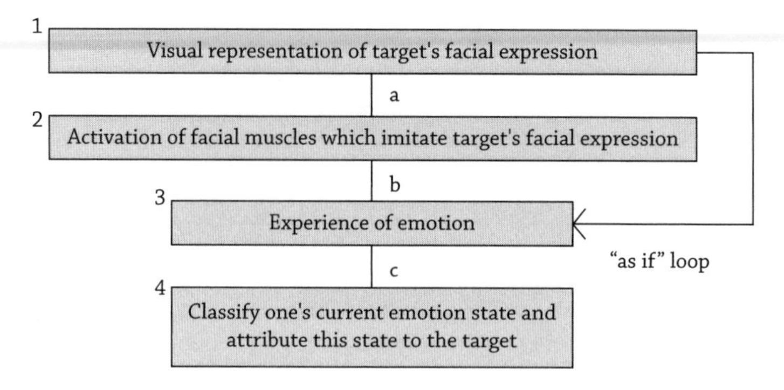

Figure 3.2: Reverse simulation.

generating, at least in attenuated form, the corresponding emotional state and its cognitive and physiological correlates (see the discussion below). Thus, the relationship between emotion states and their facial expressions exhibits a kind of rough one-to-one correspondence in both directions. For this reason, the standard mode of operation in which emotion states causally produce a characteristic facial expression could potentially be utilized in a backward direction for the purposes of reverse simulation. The underlying idea in reverse simulation has also been invoked by others in the mindreading literature. For example, Blakemore & Decety (2001) propose that the cerebellum houses a "forward model" that maps motor instructions into sensory expectations that would ensue, given that an action is performed. They propose that this database can be queried in the reverse direction for the purposes of mindreading.

How would reverse simulation for the purposes of face-based emotion recognition operate? As shown in figure 3.2, a potential attributor who sees an emotion-expressive face starts by mimicking the facial expression she observes, presumably in an attenuated and largely covert manner. As noted above, actual facial exertions appropriate to a certain emotion commonly produce the experience of traces of that very emotion. An experiencer of that mild emotion would then classify her own emotion state, and then, in keeping with the common core of all simulational heuristics, would classify the observed face as being expressive of the same state produced in herself. Of course, all this might happen at a subthreshold level.

The reverse simulation model provides a plausible explanation for the paired-deficit studies cited earlier. Someone impaired in the emotion in question would be unable to produce that emotion, or even significant traces thereof, in her own system. The requisite facial exertions would not arouse the appropriate neural activity for emotion production. Hence, such a person would be impaired in recognizing the corresponding emotion in a target. But in addition to explaining the paired-deficit results, several independent lines of evidence support the reverse simulation model.

As depicted in figure 3.2, reverse simulation begins with a visual representation of the target's facial expression, which serves to activate facial musculature imitating the expression of the target. That such imitation capacities exist is well established. Meltzoff & Moore (1983) found that infants as young as one hour old imitate tongue protrusion and a range of other facial displays that they see modeled before them. In addition to the finding that humans *can* imitate the facial expressions of others, there is further evidence that humans *do* in fact spontaneously and rapidly activate facial musculature corresponding to visually presented facial expressions. In a series of studies, Dimberg and colleagues found that presentation of pictures of facial expressions produces rapid, covert activation of one's own facial musculature, which mimics the presented faces (Dimberg & Thunberg, 1998; Lundquist & Dimberg, 1995). Such muscular activation is often subtle, but is detectable by electromyography and (as noted above) occurs extremely rapidly.

The finding that subjects spontaneously, rapidly, and covertly imitate visually presented facial expressions is consistent with the reverse simulation model (it supports link "a" in figure 3.2), but it is also consistent with a model in which these self-generated facial expressions are the *consequence* of an antecedently generated emotion state. Support for the claim that the muscle movements are primary and in fact give rise to a subsequent emotion state comes from two lines of evidence.

The first line of evidence suggesting that facial movements might occur prior to the emotion experience is that there are reasons to believe that these facial movements are an instance of a more general "mirroring" phenomenon. An action mirroring system, in which internal action representations activated in the production of an action are also activated when the same action-type is observed in others, is known to exist in the monkey and human ventral premotor cortex and neighboring regions (Gallese, Fadiga, Fogassi, & Rizzolatti, 1996; Iacoboni et al., 2001; Rizzolatti, Fadiga, Gallese, & Fogassi, 1996). Furthermore, the operation of the action mirroring system is known to generate covert activation of distal musculature. In an early experiment that helped establish an action mirroring system in humans, Fadiga and colleagues used transcranial magnetic stimulation (TMS) to enhance distal electromyographic recordings (Fadiga, Fogassi, Pavesi, & Rizzolatti, 1995). They found that observation of actions (e.g., grasping an object, tracing a figure in the air) modeled by a target reliably produced electromyographically detectable activation in the corresponding muscle groups of the observer.

Recent evidence indicates that the action mirroring system may also operate during the observation of facial expressions. An fMRI study by Carr et al. (2003) found that subjects passively observing emotion-expressive faces display neural activation in the premotor cortex and neighboring regions that are normally activated in the production of facial movements and that are in the region thought to house the action mirroring system. If there is indeed an action mirroring system that operates during the observation of facial expressions, it may help explain the covert activation of facial musculature discussed earlier. In other words, the phenomenon found by Dimberg and colleagues—covert activation of musculature that imitates the muscular activation patterns presented by a target—may be an instance of a mirroring phenomenon that obtains for somatic musculature more generally.[2] And if the facial movements found by Dimberg and colleagues are indeed an instance of a more general mirroring phenomenon, they need not be explained as the product of an antecedent emotion experience.

A second line of evidence suggesting that facial movements might occur prior to the emotion experience is the substantial accumulated data that suggest there is a causal pathway that links manipulations of facial expressions with corresponding changes in emotion states. Subjects made to produce facial expressions voluntarily or involuntarily (e.g., by holding a pencil in their mouth or saying "cheese") are found to exhibit cognitive and physiological correlates of emotion experience (Adelman & Zajonc, 1989). A number of theorists endorsing the "facial-feedback hypothesis" view this causal pathway by which facial expressions produce emotion states to be an important mechanism in mediating the experience of emotion generally (Laird & Bressler, 1992; Tomkins, 1962).

Facial feedback has also been implicated in the interpersonal communication of emotion states. A number of theorists have noted the interesting phenomenon of "primitive emotion contagion" in which motoric mimicry (mimicking facial expressions, prosody, posture, and movements) is the causal basis for convergence in emotion states between interacting individuals (Hatfield, Cacioppo, & Rapson, 1994). These theorists marshal substantial evidence in favor of the existence of primitive emotion contagion, and speculate that this phenomenon may play an important role in facilitating the interpretation of others' mental states (i.e., mindreading). The primitive emotion contagion hypothesis is broadly consistent with simulationist approaches to mindreading and with the reverse simulation model in particular.

Theorists endorsing a facial-feedback hypothesis (link "b" in figure 3.2) have generally assumed that the mechanism by which self-generated facial expressions produce corresponding emotion states is mediated by proprioceptive sensation of the self-generated facial expression (Tomkins, 1981). Thus link "b" in figure 3.2 of the reverse simulation model appears to require proprioceptive mediation, which in turn implicates proprioceptive centers in the brain, in particular the somatosensory regions of the parietal cortices, in the process of emotion recognition. Adolphs and colleagues studied a large number of patients (N = 108 subjects) with cortical lesions, and found a significant association between right parietal lesions and impaired face-based emotion recognition (Adolphs, Damasio, Tranel, Cooper, & Damasio, 2000). A link between somatosensory impairment and face-based emotion recognition is predicted by the reverse simulation model, but it is harder to make sense of under the generate-and-test simulation model. Additionally, it is not predicted at all under a theory-based model.

At least some data, however, are inconsistent with the reverse simulation model. Hess & Blairy (2001) used a more challenging FaBER task and found that, although spontaneous facial mimicry did occur, the occurrence of successful mimicry did not correlate with accuracy in facial recognition, suggesting that facial mimicry may accompany but not actually facilitate recognition. Additionally, a study by Calder and colleagues found that three patients with Mobius syndrome, a congenital syndrome the most prominent symptom of which is complete bilateral facial paralysis, performed normally on FaBER tasks (Calder, Keane, Cole, Campbell, & Young, 2000). Keillor and colleagues report a similar finding in which a patient with bilateral facial paralysis performed normally on FaBER tasks (Keillor, Barrett, Crucian, Kortenkamp, & Heilman, 2002). These findings need to be interpreted with caution, however, because given the longstanding nature of their impairments, these subjects' normal performances may reflect the operation of a compensatory strategy.

5.3. Reverse simulation with "as if" loop

Adolphs et al. (2000) note that, while facial feedback along the lines suggested by the reverse simulation model may be utilized in face-based emotion recognition, they speculate that there may be an alternative pathway. In keeping with earlier work (Damasio, 1994), Adolphs and colleagues propose that there may be direct links between a visual representation of a target's facial expression and a somatosensory representation of "what it would feel like" were the observer to make that expression. It is speculated that these linked visual-somatosensory representations are the basis of an independent pathway (labeled the "as if" pathway in figure 3.2) that bypasses the facial musculature and allows the observer to directly produce an emotion state that corresponds to the facial expression displayed by the target. Although details of the "as if" pathway are not fully in place yet, the "as if" model may count as a third simulationist model for face-based emotion recognition that differs from the two models that we've already proposed. In this third model, actual reverse simulation involving the facial musculature does not occur. Instead, the observer sees the facial expression displayed by the target and jumps directly to somatosensory representations of what it would feel like to have made the requisite facial exertions (thus the name "as if"), which in turn serve to bring about the corresponding emotion state as in standard reverse simulation. This "as if" model is superior to the reverse simulation model to the extent that it does not postulate a causal role for facial musculature in the recognition process, and is therefore unthreatened by the results of Calder et al. (2000), Hess & Blairy (2001), and Keillor et al. (2002).

5.4. Unmediated resonance model

A fourth possible model would also accommodate the findings by Calder et al. (2000) and Keillor et al. (2002) of preserved facial recognition in patients with facial paralysis. Unlike the third model, it would not appeal to somatosensory-based feelings normally associated with making a face of the same sort as the one visually presented by the target. This fourth model is what we shall call the *unmediated resonance* model. The idea here is that observation of the target's face directly, without any mediation of the sorts posited by any of the first three models, triggers (subthreshold) activation of the same neural substrate associated with the emotion in question. This is the idea behind Gallese's (2001, 2003) "shared manifold hypothesis," and is suggested by Wicker et al. when they speak of an automatic sharing, by the observer, of the displayed emotion (Wicker et al., 2003: 661). This would parallel findings of mirror-neuron matching systems found in monkeys and humans, in which internal action representations, normally associated with producing actions, are triggered during the observation of, or listening to, someone else's corresponding actions (Gallese et al., 1996; Kohler et al., 2002; Rizzolatti, Foggasi, & Gallese, 2001).[3] Finally, the fourth model must of course assume that the occurrence, or production, of the relevant emotion in an observer is transmitted to some cognitive center that "recognizes" the experienced emotion, leading to its overt (usually verbal) classification in the experimental set-up *as* that type of emotion.

But this assumption would be common to all of the models, not distinctive to the fourth model.

Does this fourth model really fit the pattern of ST? Because the model posits unmediated resonance, it does not fit the usual examples of simulation in which pretend states are created and then operated upon by the attributor's own cognitive equipment (e.g., a decision-making mechanism), yielding an output that gets attributed to the target. However, we do not regard the creation of pretend states, or the deployment of cognitive equipment to process such states, as essential to the generic idea of simulation. The general idea of simulation is that the simulating process should be similar, in relevant respects, to the simulated process (Goldman, 2006.). Applied to mindreading, a minimally necessary condition is that the state ascribed to the target is ascribed as a result of the attributor's instantiating, undergoing, or experiencing, that very state. In the case of successful simulation, the experienced state matches that of the target. This minimal condition for simulation is satisfied in the fourth model.

Finally, it should be emphasized that we make no attempt here to choose the "best" of the four simulationist models, or even to express a *rank ordering* among them (apart from the previously indicated doubts about the first model). Additional research is required before there is adequate evidence to select among them. Our sole aim is to show that several simulationist models are available with substantial surface plausibility and consistency with the evidence; this lends further credence to our initial conclusion that *some* sort of simulationist account of FaBER is highly probable.

6. WHY SIMULATION?

In the preceding sections, we marshaled evidence that FaBER proceeds by simulation rather than theory-based mechanisms. At any rate, simulation is the fundamental or primitive method of recognizing emotion from faces, although theorizing might also be used, for example, as a compensatory strategy. The evidence is much less clear-cut, however, with regard to distinguishing possible simulationist models, and much further study is warranted. Our hope is that the specific models we have formulated will lend structure to future discussion and investigation. We conclude by briefly addressing three topics. First, we reply to a reasonable worry that may have arisen about simulational models. Second, we speculate about the evolutionary origins of the simulational characteristics of FaBER. Third, we suggest one possible experiment to help select among the competing models.

Three of the simulationist models we proposed involve many disparate systems, including, for example, specific emotion production systems, and the facial musculature and somatosensory centers. In contrast, a theory-based mechanism seems to require little more than the mental representation of a number of generalizations linking facial configurations with emotion names. Given desiderata such as simplicity, efficiency, and elegance, it is hard to see why simulational mechanisms would be advantageous relative to theory-based mechanisms for the FaBER task. Our answer to this challenge is straightforward. Simulation might be (somewhat) complex from a functional perspective, but it might be simpler from an evolutionary perspective. Simulation relies upon running the same emotional apparatus (possibly in reverse) that is already used to generate or experience the emotion. As a consequence, simulation routines do not require an organism to be outfitted with entirely new processes in order to confer an ability to recognize emotions in others. Natural selection is a tinkerer, not a planner, and she frequently builds new capacities from existing ones. For this reason, simulation routines may have been favored in the course of evolution.

Our second remark features a bolder speculation—not original to us—about the evolutionary origins of a simulation mechanism for FaBER. As many writers point out (e.g., Wicker et al., 2003), there are many reasons why it would be adaptive to have mechanisms of emotion contagion. Consider the case of disgust, for example. Disgust is frequently experienced in response to a food item that should not be eaten. If an individual observes a conspecific having such a response to a food item, it would be adaptive for that individual to have the same disgust response vis-à-vis that food item, in order to induce avoidance. Thus, mechanisms of emotion contagion or resonance can be explained in terms of this kind of adaptive advantage. Once in place, the resonance mechanism could be transmuted into a *simulational* method of recognition. By contrast, there are no obvious steps whereby emotion contagion would be transmuted into *theory-based* recognition.

Finally, we turn to a suggestion for future experimental work prompted by our four models. D. Osherson (pers. comm.) asks what would happen if subjects were given the task of recognizing facially expressed emotions while doing (unrelated) face exercises, in particular exercises that prevent them from engaging their facial musculature in an emotion-expressive way. Previous studies have found that manipulation of the facial musculature in an emotion-expressive manner induces emotion experiences (see section 5.2), and also produces interference on various emotion-relevant tasks (see, e.g., Laird, Wagener, Halal, & Szegda, 1982). But no studies have examined the effects on FaBER of exercises incompatible with making emotion-expressive faces. Would such exercises make FaBER more difficult? Would they induce errors? Both the generate-and-test model and the reverse simulation model seem to predict interference and hence reduced recognition. On the other hand, neither the "as if" loop variant of the reverse simulation model nor the unmediated resonance model makes this prediction, because neither of them posits use of the attributor's own facial musculature in emotion recognition. So the first two models would predict a positive result (reduced recognition) in such a test, and would be undercut by a negative result. But a negative result would not undercut either the third or fourth model. This is one experimental means, then, by which to test the rival models.

ACKNOWLEDGMENTS

The authors wish to thank Andrew Lawrence, Vittorio Gallese, Giacomo Rizzolatti, Christian Keysers, Ralph Adolphs, Daniel Osherson, and three anonymous reviewers for valuable comments on earlier drafts of this chapter.

NOTES

1. Previous studies have indicated that the insula, among other structures, is activated during the experience of disgusting odors and tastes (Fulbright et al., 1998; Small et al., 2003). Additionally, previous studies have also established that the insula is preferentially activated during the observation of disgust-expressive faces (Krolak-Salmon et al., 2003; Phillips et al., 1998, 1997; Sprengelmeyer et al., 1998). However, the Wicker et al. (2003) study is the first to demonstrate within a single experiment, using a single mode of investigation and the same pool of subjects, that the same neural substrate subserves both the experience and recognition of disgust.
2. It is worth noting that the time after stimulus onset at which Fadiga and colleagues detected covert activation of somatic musculature, 360 ms, is consistent with the time at which Dimberg and colleagues found that differential activation of facial musculature reached significance, 300–400 ms.
3. Of course, the role of mirroring in recognizing actions is not wholly clear. For example, Buxbaum, Sirigu, Schwartz, & Klatzky (2003) and Halsband et al. (2001) found action production deficits alongside preserved action recognition. However, interpretation of these studies is difficult for several reasons,

including most prominently the fact that it is unclear in these studies whether the production deficits are due to damage in the region of the premotor cortex and neighboring areas thought to house the action mirroring system in humans.

REFERENCES

Adelman, P., & Zajonc, R. (1989). Facial efference and the experience of emotion. *Annual Review of Psychology* 40:249–280.

Adolphs, R. (1995). Fear and the human amygdala. *Journal of Neuroscience* 15:5879–5891.

Adolphs, R. (2002). Recognizing emotion from facial expressions: Psychological and neurological mechanisms. *Behavioral and Cognitive Neuroscience Reviews* 1(1):21–62.

Adolphs, R., Damasio, H., Tranel, D., Cooper, G., & Damasio, A. (2000). A role for the somatosensory cortices in the visual recognition of emotion as revealed by three-dimensional lesion mapping. *Journal of Neuroscience* 20(7):2683–2690.

Adolphs, R., Tranel, D., & Damasio, A. (2003). Dissociable neural systems for recognizing emotions. *Brain and Cognition* 52:61–69.

Adolphs, R., Tranel, D., Damasio, H., & Damasio, A. (1994). Impaired recognition of emotion in facial expressions following bilateral damage to the amygdala. *Nature* 372:669–672.

Adolphs, R., Tranel, D., Hamann, S., Young, A. W., Calder, A. J., Phelps, E. A., Anderson, A., Lee, G. P., & Damasio, A. R. (1999). Recognition of facial emotion in nine individuals with bilateral amygdala damage. *Neuropsychologia* 37:1111–1117.

Armstrong, D. M. (1968). *A materialist theory of the mind*. New York: Humanities Press.

Bartels, A., & Zeki, S. (2000). The architecture of the colour centre in the human visual brain: New results and a review. *European Journal of Neuroscience* 12:172–193.

Bechara, A., Tranel, D., Damasio, H., Adolphs, R., Rockland, C., & Damasio, A. R. (1995). Double dissociation of conditioning and declarative knowledge relative to the amygdala and hippocampus in humans. *Science* 269:1115–1118.

Benton, A. L., Hamsher, K., Varney, N. R., & Spreen, O. (1983). *Contributions to neuropsychological assessment*. New York: Oxford University Press.

Blakemore, S., & Decety, J. (2001). From the perception of action to the understanding of intention. *Nature Reviews Neuroscience* 2:561–567.

Buxbaum, L. J., Sirigu, A., Schwartz, M., & Klatzky, R. (2003). Cognitive representations of hand posture in ideomotor apraxia. *Neuropsychologia* 41:1091–1113.

Calder, A. J., Burton, A. M., Miller, P., Young, A. W., & Akamatsu, S. (2001). A principal component analysis of facial expressions. *Vision Research* 41:1179–1208.

Calder, A. J., Keane, J., Cole, J., Campbell, R., & Young, A. W. (2000). Facial expression recognition by people with Mobius syndrome. *Cognitive Neuropsychology* 17 (1/2/3): 73–87.

Calder, A. J., Keane, J., Manes, F., Antoun, N., & Young, A. W. (2000). Impaired recognition and experience of disgust following brain injury. *Nature Reviews Neuroscience* 3:1077–1078.

Calder, A. J., Lawrence, A. D., & Young, A. W. (2001). Neuropsychology of fear and loathing. *Nature Reviews Neuroscience* 2:352–363.

Carr, L., Iacoboni, M., Dubeau, M.-C., Mazziotta, J. C., & Lenzi, G. L. (2003). Neural mechanisms of empathy in humans: A relay from neural systems for imitation to limbic areas. *Proceedings of the National Academy of Science U.S.A.* 100(9):5497–5502.

Chaminade, T., Meary, D., Orliaguet, J.-P., & Decety, J. (2001). Is perceptual anticipation a motor simulation? A PET study. *NeuroReport* 12(17):3669–3674.

Currie, G., & Ravenscroft, I. (2002). *Recreative minds*. Oxford, UK: Oxford University Press.

Damasio, A. (1994). *Descartes' error*. New York: Grosset.

Damasio, A. (1999). *The feeling of what happens*. New York: Harcourt Brace.

Dimberg, U., & Thunberg, M. (1998). Rapid facial reactions to emotional facial expressions. *Scandinavian Journal of Psychology* 39:39–45.

Ekman, P. (1992). Are there basic emotions? *Psychological Review* 99(3):550–553.

Fadiga, L., Fogassi, L., Pavesi, G., & Rizzolatti, G. (1995). Motor facilitation during action observation: A magnetic stimulation study. *Journal of Neurophysiology* 73(6):2608–2611.

Fletcher, P. C., Happe, F., Frith, U., Baker, S. C., Dolan, R. J., Frackowiak, R. S. J., & Frith, C. D. (1995). Other minds in the brain: A functional imaging study of 'theory of mind' in story comprehension. *Cognition* 57:109–128.

Fodor, J. A. (1992). A theory of the child's theory of mind. *Cognition* 44:283–296.

Frith, C. D., & Frith, U. (1999). Interacting minds—a biological basis. *Science* 286:1692–1695.

Fulbright, R. K., Skudlarski, P., Lacadie, C. M., Warrenburg, S., Bowers, A. A., Gore, J. C., & Wexler, B. E. (1998). MR functional imaging of regional brain responses to pleasant and unpleasant odors. *American Journal of Neuroradiology* 19(9):1721–1726.

Gallese, V. (2001). The 'shared manifold' hypothesis: From mirror neurons to empathy. *Journal of Consciousness Studies* 8(5–7):33–50.

Gallese, V. (2003). The manifold nature of interpersonal relations: The quest for a common mechanism. *Philosophical Transactions of the Royal Society of London*, B, 358:517–528.

Gallese, V., Fadiga, L., Fogassi, L., & Rizzolatti, G. (1996). Action recognition in the premotor cortex. *Brain* 119:593–609.

Gallese, V., & Goldman, A. I. (1998). Mirror neurons and the simulation theory of mind-reading. *Trends in Cognitive Sciences* 2(12):493–501.

Goldman, A. I. (1989). Interpretation psychologized. *Mind and Language* 4:161–185.

Goldman, A. I. (1992a). In defense of the simulation theory. *Mind and Language* 7(1–2):104–119.

Goldman, A. I. (1992b). Empathy, mind and morals. *Proceedings and Addresses of the American Philosophical Association* 66/3:17–41.

Goldman, A. I. (2000). The mentalizing folk. In D. Sperber (ed.), *Metarepresentations: An interdisciplinary approach* (171–196). New York: Oxford University Press.

Goldman, A. I. (2006). *Simulating Minds*. New York: Oxford University Press.

Gopnik, A., & Meltzoff, A. (1997). *Words, thoughts, and theories*. Cambridge, MA: MIT Press.

Gopnik, A., & Wellman, H. M. (1992). Why the child's theory of mind really *is* a theory. *Mind and Language* 7(1–2):145–171.

Gopnik, A., & Wellman, H. M. (1994). The theory theory. In L. Hirschfeld & S. Gelman (eds.), *Mapping the mind: Domain specificity in cognition and culture* (257–293). New York: Cambridge University Press.

Gordon, R. M. (1986). Folk psychology as simulation. *Mind and Language* 1:158–171.

Gordon, R. M. (1992). The simulation theory: Objections and misconceptions. *Mind and Language* 7(1–2):11–34.

Gordon, R. M. (1996). 'Radical' simulationism. In P. Carruthers & P. K. Smith (eds.), *Theories of theories of mind* (11–21). New York: Cambridge University Press.

Halsband, U., Schmitt, J., Weyers, M., Binkofski, F., Grutzner, G., & Freund, H.-J. (2001). Recognition and imitation of pantomimed motor acts after unilateral parietal and premotor lesions: A perspective on apraxia. *Neuropsychologia* 39:200–216.

Harris, P. L. (1991). The work of the imagination. In A. Whiten (ed.), *Natural theories of mind* (283–304). Oxford, UK: Blackwell.

Harris, P. L. (1992). From simulation to folk psychology: The case for development. *Mind and Language* 7(1–2):120–144.

Hatfield, E., Cacioppo, J., & Rapson, R. (1994). *Emotional contagion*. New York: Cambridge University Press.

Heal, J. (1986). Replication and functionalism. In J. Butterfield (ed.), *Language, mind and logic* (135–150). Cambridge, UK: Cambridge University Press.

Heal, J. (1996). Simulation and cognitive penetrability. *Mind and Language* 11:44–67.

Heal, J. (1998). Co-cognition and off-line simulation. *Mind and Language* 13:477–498.

Hess, U., & Blairy, S. (2001). Facial mimicry and emotional contagion to dynamic facial expressions and their influence on decoding accuracy. *International Journal of Psychophysiology* 40:129–141.

Iacoboni, M., Koski, L. M., Brass, M., Bekkering, H., Woods, R. P., Dubeau, M., Mazziotta, J. C., & Rizzolatti, G. (2001). Reafferent copies of imitated actions in the right superior temporal cortex. *Proceedings of the National Academy of Science U.S.A.* 98(24):13995–13999.

Iacoboni, M., Woods, R. P., Brass, M., Bekkering, H., Mazziotta, J. C., & Rizzolatti, G. (1999). Cortical mechanisms of human imitation. *Science* 286:2526–2528.

Jeannerod, M. (2001). Neural simulation of action. A unifying mechanism for motor cognition. *NeuroImage* 14:S103–S109.

Kanwisher, N. (2000). Domain specificity in face perception. *Nature Neuroscience* 3(8):759–763.

Keillor, J. M., Barrett, A. M., Crucian, G. P., Kortenkamp, S., & Heilman, K. M. (2002). Emotional experience and perception in the absence of facial feedback. *Journal of the International Neuropsychological Society* 8(1):130–135.

Kohler, E., Keysers, C., Umilta, M. A., Fogassi, L., Gallese, V., & Rizzolatti, G. (2002). Hearing sounds, understanding actions: Action representation in mirror neurons. *Science* 297:846–848.

Krolak-Salmon, P., Henaff, M. A., Isnard, J., Tallon-Baudry, C., Geunot, M., Vighetto, A., Bertrand, O., & Mauguiere, F. (2003). An attention modulated response to disgust in human ventral anterior insula. *Annals of Neurology* 53:446–453.

Laird, J. D., & Bressler, C. (1992). The process of emotion experience: A self-perception theory. In M. Clark (ed.), *Review of personality and social psychology (vol. 13)* (213–234). New York: Sage.

Laird, J. D., Wagener, J. J., Halal, M., & Szegda, M. (1982). Remembering what you feel: Effects of emotion and memory. *Journal of Personality and Social Psychology* 42:646–675.

Lawrence, A. D., & Calder, A. J. (2004). Homologizing human emotions. In D. Evans & P. Cruse (eds.), *Emotions, evolution and* rationality, 15–47. New York: Oxford University Press.

Lawrence, A. D., Calder, A. J., McGowan, S. M., & Grasby, P. M. (2002). Selective disruption of the recognition of facial expressions of anger. *NeuroReport* 13(6):881–884.

LeDoux, J. E. (1993). Emotion memory systems in the brain. *Behavioral Brain Research* 58: 69–79.

LeDoux, J. E. (2000). Emotion circuits in the brain. *Annual Review of Neuroscience*, 23: 155–184.

Leslie, A. M. (1994). Pretending and believing: Issues in the theory of ToMM. *Cognition* 50: 211–238.

Lewis, D. K. (1972). Psychophysical and theoretical identifications. *Australasian Journal of Philosophy* 50:249–258.

Lundquist, L., & Dimberg, U. (1995). Facial expressions are contagious. *Journal of Psychophysiology* 9:203–211.

McCabe, K., Houser, D., Ryan, L., Smith, V., & Trouard, T. (2001). A functional imaging study of cooperation in two-person reciprocal exchange. *Proceedings of the National Academy of Science U.S.A.* 98:11832–11835.

Meltzoff, A. N., & Moore, M. K. (1983). Newborn infants imitate adult facial gestures. *Child Development* 54:702–709.

Meltzoff, A. N., & Moore, M. K. (1997). Explaining facial imitation: A theoretical model. *Early Development and Parenting* 6:179–192.

Nichols, S., & Stich, S. (2003). *Mindreading*. Oxford, UK: Oxford University Press.

Nichols, S., Stich, S., Leslie, A., & Klein, D. (1996). Varieties of off-line simulation. In P. Carruthers & P. Smith (eds.), *Theories of theories of mind* (39–74). Cambridge, UK: Cambridge University Press.

Perner, J. (1991). *Understanding the representational mind*. Cambridge, MA: MIT Press.

Perner, J. (1996). Simulation as explicitation of predication-implicit knowledge about the mind: Arguments for a simulation-theory mix. In P. Carruthers & P. Smith (eds.), *Theories of theories of mind* (90–104). Cambridge, UK: Cambridge University Press.

Phillips, M. L., Young, A. W., Scott, S. K., Calder, A. J., Andrew, C., Giampietro, V., Williams, S. C., Bullmore, E. T., Brammer, M., & Gray, J. A. (1998). Neural response to facial and vocal expressions of fear and disgust. *Proceedings of the Royal Society of London*, B, 265:1809–1817.

Phillips, M. L., Young, A. W., Senior, C., Brammer, M., Andrew, C., Calder, A. J., Bullmore, E. T., Perrett, D. I., Rowland, D., Williams, S. C. R., Gray, J. A., & David, S. (1997). A specific neural substrate for perceiving facial expressions of disgust. *Nature* 389: 495–498.

Premack, D., & Woodruff, G. (1978). Does the chimpanzee have a theory of mind? *Behavioral and Brain Sciences* 4:515–526.

Rizzolatti, G., Fadiga, L., Gallese, V., & Foggasi, L. (1996). Premotor cortex and the recognition of motor actions. *Cognitive Brain Research* 3:131–141.

Rizzolatti, G., Foggasi, L., & Gallese, V. (2001). Neurophysiological mechanisms underlying the understanding and imitation of action. *Nature Reviews, Neuroscience* 2:661–670.

Rolls, E. T. (1995). Central taste anatomy and neurophysiology. In R. L. Doty (ed.), *Handbook of clinical olfaction and gustation* (549–573). New York: Dekker.

Rozin, P., Haidt, J., & McCauley, C. (2000). Disgust. In M. Lewis & J. Haviland (eds.), *Handbook of emotions* (637–653). New York: The Guilford Press.

Sellars, W. (1956). Empiricism and the philosophy of mind. In H. Feigl & M. Scriven (eds.), *The foundations of science and the concepts of psychology and psychoanalysis*

(253–329). *Minnesota studies in the philosophy of science 1*. Minneapolis, MN: University of Minnesota Press.

Shoemaker, S. (1975). Functionalism and qualia. *Philosophical Studies* 27:291–315.

Small, D. M., Gregory, M., Mak, R., Gitelman, D., Mesulam, M. M., & Parrish, T. (2003). Dissociation of neural representation of intensity and affective valuation in human gustation. *Neuron* 39:701–711.

Small, D. M., Zald, D. H., Jones-Gotman, M., Zatorre, R. J., Pardo, J. V., Frey, S., & Petrides, M. (1999). Brain imaging: Human cortical gustatory areas: A review of functional neuroimaging data. *NeuroReport* 10:7–14.

Sprengelmeyer, R., Rausch, M., Eysel, U. T., & Przuntek, H. (1998). Neural structures associated with recognition of facial expressions of basic emotions. *Proceedings of the Royal Society of London (Series B: Biology)* 265:1927–1931.

Sprengelmeyer, R., Young, A. W., Calder, A. J., Karnat, A., Lange, H., Homberg, V., Perrett, D. I., & Rowland, D. (1996). Loss of disgust: Perception of faces and emotions in Huntington's disease. *Brain* 119:1647–1665.

Sprengelmeyer, R., Young, A. W., Schroeder, U., Grossenbacher, P. G., Federlein, J., Buttner, T., & Przuntek, H. (1999). Knowing no fear. *Proceedings of the Royal Society (Series B: Biology)* 266:2451–2456.

Sprengelmeyer, R., Young, A. W., Sprengelmeyer, A., Calder, A. J., Rowland, D., Perrett, D., Hornberg, V., & Lange, H. (1997). Recognition of facial expressions: Selective impairment of specific emotions in Huntington's disease. *Cognitive Neuropsychology* 14(6):839–879.

Stich, S., & Nichols, S. (1992). Folk psychology: Simulation or tacit theory? *Mind and Language* 7(1–2):35–71.

Tomkins, S. (1962). *Affect, imagery, consciousness: The positive affects* (vol. 1). New York: Springer.

Tomkins, S. (1981). The role of facial response in the experience of emotion: A reply to Tourangeau and Ellsworth. *Journal of Personality and Social Psychology* 45:355–357.

Wellman, H. M. (1990). *The child's theory of mind*. Cambridge, MA: MIT Press.

Wicker, B., Keysers, C., Plailly, J., Royet, J.-P., Gallese, V., & Rizzolatti, G. (2003). Both of us disgusted in my insula: The common neural basis of seeing and feeling disgust. *Neuron* 40:655–664.

CHAPTER 4

Mirroring, Mindreading, and Simulation

1. INTRODUCTION

Mirror systems are well established as a highly robust feature of the human brain (Rizzolatti, Fogassi, & Gallese, 2004; Gallese, Keysers, & Rizzolatti, 2004; Iacoboni et al., 1999; Rizzolatti & Craighero, 2004). Mirror systems and mirroring processes are found in many domains, including action planning, sensation, and emotion (for reviews, see Keysers & Gazzola, 2006; Gallese, Keysers, & Rizzolatti, 2004; Goldman, 2006). Because mirroring commonly features an interpersonal matching or replication of a cognitive or mental event, it is a social interaction. It involves two people sharing the same mental-state type, although activations in observers are usually at a lower level than endogenous ones, commonly below the threshold of consciousness. There is strong evidence that mirror systems play pivotal roles in empathy and imitation (Iacoboni et al., 1999; Rizzolatti, 2005; Iacoboni, 2005; Gallese, 2005; Decety & Chaminade, 2005). Indeed, in a minimal sense of the term, "empathy" might simply *mean* the occurrence of a mirroring process. In this chapter, however, I shall focus on the connection between mirroring processes and another category of social cognition, viz., mindreading or mentalizing.

By "mindreading" I mean the attribution of a mental state to self or other. In other words, to mindread is to form a judgment, belief, or representation that a designated person occupies or undergoes (in the past, present, or future) a specified mental state or experience. This judgment may or may not be verbally expressed. Clearly, not all judgments about other people are acts of mindreading. To judge that someone makes a certain facial expression, or performs a certain action, or utters a certain sound is not to engage in mindreading, because these are not attributions of mental states. To attribute a mental state, the judgment must deploy a mental concept or category. Thus, if "empathize" simply means "echo the emotional state of another," empathizing isn't sufficient for mindreading. A person who merely echoes another's emotional state may not *represent* the second person at all, and may not represent her *as* undergoing that emotional state (a species of mental state). It is certainly possible that mirroring processes are responsible for acts of mindreading, as well as for imitation or empathy, but this involvement in mindreading does not logically follow from their role in either imitation or empathy. A connection between mirroring and mindreading must be considered separately.

2. DEFINITIONAL ISSUES

The phrase "mirroring process" can have either a wide sense or a narrow sense. In the wide sense, it refers to an interpersonal process that spans both sender and receiver. In a narrower sense, it refers to an intrapersonal process that includes only a receiver (a single individual). Unless otherwise specified, I shall understand mirroring process in the narrow sense. To define it in the narrow sense, we first need a definition of "mirror neuron" or "mirror system." Rizzolatti, Fogassi, & Gallese (2004) offer the following definition of mirror neurons:

> Mirror neurons are a specific class of neurons that discharge both when the monkey performs an action and when it observes a similar action done by another monkey or the experimenter. (2004: 431)

This is a good definition of *action* mirror neurons, but not mirror neurons in general. We don't want to restrict mirror neurons or mirroring processes to action-related events; they should equally be allowed in the domains of touch, pain, and emotion, for example. So let me propose a more general definition:

> Mirror neurons are a class of neurons that discharge both when an individual (monkey, human, etc.) undergoes a certain mental or cognitive event *endogenously* and when it observes a *sign* that another individual undergoes or is about to undergo the same type of mental or cognitive event.

One type of "sign" to which an observer's mirror neuron might respond is a behavioral manifestation of the mental event in question; another example is a facial expression. A third type of sign is a stimulus that can be expected to produce the mental event in question. Thus, an observer's pain mirror neurons discharge when he sees a sharp knife being applied to someone else's body.

The above definition of mirror neurons can be extended essentially unchanged to mirror *systems* or *circuits*:

> Mirror systems are neural systems that get activated both when an individual undergoes a certain mental or cognitive event endogenously and when he observes a sign that another individual is undergoing, or is about to undergo, the same type of mental or cognitive event.

What counts as an "endogenous" occurrence varies from one type of event to another. For present purposes I won't try to characterize endogenousness any further.

Even given the definitions of mirror neurons and mirror systems, it is not trivial to produce a definition of a *mirroring process*. One cannot say there is a mirroring process whenever a mirror neuron or mirror system discharges or is activated. When a mirror neuron or system is endogenously activated, it is not a mirroring event. Only when a mirror neuron or system is activated in the *observation* mode is there a mirroring process. What exactly do we mean by "observation mode"? Must the observer perceive a genuine behavioral manifestation or expression (etc.) of an endogenous mirror event? I think not. If a good imitation of a pained expression triggers an observer's pain mirror neuron to fire, the process in the observer is still a mirroring process, even though the imitation is not a genuine expression, or sign, of an endogenous mirror event.

It would also be incorrect to say, as a simple definition, that every non-endogenous activation of a mirror neuron or mirror system is a mirroring process. An important non-endogenous mode of activation is imagination-generated activation. Motor imagery, for instance, is the result of imagining the execution of a motor act, and the generation of motor imagery uses largely the same mirror circuits as used in the endogenous generation of an action (M. Jeannerod, pers. comm.). However, we would not consider the process of creating motor imagery a type of mirroring process. In light of these points, let us define a mirroring process as follows:

> Neural process N is a mirroring process if and only if (1) N is an activation of a mirror neuron or mirror system, and (2) N results from observing something that is normally a behavioral or expressive manifestation (or a predictive sign) of a matching mirror event in another individual.

3. FOUR THESES ABOUT MIRRORING PROCESSES AND MINDREADING

With these clarifications and definitions in hand, let me now present the main theses I wish to advance and defend in this chapter:

(1) Mirroring processes in themselves do not constitute mindreading.
(2) *Some* acts of mindreading (low-level mindreading) are caused by, or based on, mirroring processes.
(3) Not *all* acts of mindreading (in particular, not high-level mindreading) are based on mirroring.
(4) Simulation-based mindreading is broader than mirroring-based mindreading; some simulation-based mindreading (the low-level type) involves mirroring and some of it (the high-level type) does not.

In this section I'll defend thesis 1, and in succeeding sections theses 2, 3, and 4.

An act of mindreading consists of a belief or judgment about a mental state. So, if a mirroring process in itself were to *constitute* mindreading (as opposed to merely *cause* it), the "receiving" mirroring event would itself have to *be*, or *include*, a judgment or attribution of a mental state. In particular, it would have to be an attribution to a third person, presumably the originator of the mirroring process. Is there reason to suppose that belief "constitution" of this kind generally holds for mirroring processes?

Our definition of mirror neurons and mirror systems says that they are neural units that serve as substrates of one and the same cognitive event type, whether activated endogenously or observationally. This presumably implies that tokens of a mirror unit have substantially the same functional properties in whichever mode they are activated. If they had sharply different functional properties under different modes of activation, nobody would regard them as tokens of a mirroring type.[1] Having the same neuroanatomical location is not sufficient, because there are well-known cases in which the same neural region underpins more than one functionally distinct activity.[2] If that were the case in multimodal cells or circuits, I doubt that anyone would speak of mirroring. So what are the cognitive or mental units that mirror neurons or mirror circuits underpin? They are units like "planning to grasp an object," "planning to tear an object," "feeling touch in bodily area X," "feeling pain in bodily area X," "feeling disgust," and so forth.

Now if the "receiving" mirror events are tokens of the same event types (i.e., they co-instantiate the same event types), then they too will be units like "planning to grasp an object," "planning to tear an object," "feeling touch in bodily area X," and so forth. They won't also be beliefs, judgments, or attributions to the effect that the observed agent is planning to grasp an object, planning to tear an object, feeling touch in bodily area X, and so forth. If they were beliefs, judgments, or attributions of these sorts (in addition to being plannings, feelings, etc.), then, because they are mirroring events, the original endogenous occurrences would also have to be beliefs, judgments, or attributions with the same contents. But nobody has ever proposed that the sending mirror events are, or include, beliefs, judgments, or attributions. These are strong considerations in favor of thesis 1.

The truth of thesis 1 does not spell doom for the idea that mirroring is pivotally involved in mindreading. Thesis 1 denies that a mirroring process in itself *constitutes* a mindreading event, but it allows a mirroring process to *cause* or *generate* a mindreading event. Lightning doesn't constitute thunder, but lightning can certainly cause thunder. There is strong evidence for causal links between mirroring processes and selected mindreading events. This is thesis 2, which is addressed in the next several sections.

4. MIRRORING AND INTENTION ATTRIBUTION

Since mirror systems were initially discovered in the domain of motor planning, one would reasonably expect this to be the domain for which mirror-based mindreading is best supported. Also, because the first proposal of a possible link between mirroring processes and mindreading (Gallese & Goldman, 1998) focused on the motoric domain, one might expect this domain to be favored as a locus of evidence for this connection. This is especially so in light of two recent studies concerning mirror-based intention attribution, one pertaining to monkeys and the other to humans. I don't agree entirely with the researchers' own interpretations of their findings, but the second paper does provide plausible evidence to endorse their principal conclusion with respect to humans.

Fogassi et al. (2005) studied the discharge of monkey parietal mirror neurons during the viewing of a grasping act that would be followed by one of two subsequent acts: either bringing the object to the mouth or placing it in a container. Different grasping neurons of the viewer coded one or the other of the subsequent acts (or neither). In other words, for most of these grasping neurons, the level of discharge was influenced by the subsequent motor act, although the discharge occurred in the observing monkey before the subsequent act began. According to Fogassi et al.,"Thus, these [mirror] neurons not only code the observed motor act but also allow the observer to understand the agent's intentions" (2005: 662). A similar conclusion is drawn from an experiment on humans by Iacoboni et al. (2005), namely that the motor mirror system infers an agent's intention. Now, to attribute an intention is to engage in mindreading. Do such intention attributions occur in virtue of the mirroring processes in and of themselves?

This is possible, but it isn't definitely implied by the experimental evidence. There are two rival, comparatively deflationary, interpretations of the main findings that would not warrant this conclusion. The first rival interpretation would say that the parietal mirror neuron activity didn't constitute the attribution of an *intention*, only the prediction of an *action*. The prediction of an action—because it is not the attribution of a mental state—would not qualify as mindreading. The second rival interpretation would say that the parietal mirror neuron activity in the observer constituted a simulation, or mimicking, of the agent's intention by the observer,[3] but not an intention attribution. *Possessing* an intention (or "tokening" an intention, as philosophers would say) should not be confused with attributing such an intention to the agent. Only such an attribution would be a belief or judgment *about* an intention.

Under either of these rival interpretations, the reported scanning results do not, in isolation, definitively imply mindreading. The observing monkey or human who underwent mirroring processes during the experiment underwent *some* sort of mental events related to the conditions that discriminated between the different intentions of the agent. But on one interpretation, the relevant mirror events in the observer were merely *action* predictions (hence not *mind*reading events); and on the other interpretation, the mirror events were intention tokenings—again not mindreading events, this time because they were intentions rather than beliefs or attributions.[4]

If this were as far as the evidence goes, it would not firmly establish intention mindreading. But in fact there is additional evidence in the Iacoboni et al. study with human subjects, which had a special feature that lends support to intention attribution. After being scanned, participants were debriefed about the grasping actions they had witnessed. They all reported that they associated the intention of drinking with the grasping action in the "during tea" condition and the intention of cleaning up with the grasping in the "after tea" condition. These verbal reports did not depend on the instructions they had been given, that is, whether or not they had been instructed to pay attention to the agent's intention. So here there is independent evidence of intention attribution (at least in humans). One highly probable scenario, then, is that human observers mirrored the agent's intention by undergoing a matching intention themselves, and, in addition, these intentions were the causal bases for intention attributions to the agent. On this interpretation there is no suggestion that the observers' intentions *constituted* attributions; but it is agreed that intention attributions occurred.

Admittedly, the participants' reports during the debriefing session do not show that their intention ascriptions were *caused* by mirroring processes. But that is a reasonable inference, fully consistent with all findings. Even more open is the question of where their intention ascriptions occurred, whether in a motor mirroring area or elsewhere. Certainly the observers' (mirroring) intentions occurred in a motor mirroring area. But whether the attributions also occurred there is undetermined. However, an act of mindreading produced by a mirroring process need not have a mirroring process as its substrate. An attribution does not have to be *part* of a mirroring process; it only has to be caused by such a process. So the Iacoboni et al. study does provide support, if not conclusive support, for thesis 2.

5. MIRROR-BASED ATTRIBUTION OF EMOTION

Support for thesis 2 finds additional fertile ground in the mindreading of emotion (commonly labeled emotion recognition in the studies to be reviewed). The best way to assemble the relevant evidence is to conjoin two sorts of studies: (1) studies of normal participants that establish the existence of mirroring processes for emotions, and (2) neuropsychological studies of emotion-specific brain damage showing that such damage is accompanied by selective impairment in attributing the same emotion. The two types of studies together yield convincing evidence that the substrate underpinning experience of an emotion is causally implicated in normal attribution of that emotion to an observed other. Failure to (fully) mirror an emotion in oneself while observing its expression in another prevents one from reliably attributing that emotion to the other.

The best example of this twofold pattern of evidence pertains to disgust. Wicker et al. (2003) conducted an fMRI study of normal participants who were scanned both during the experience of disgust and during the observation of disgust-expressive faces. Participants viewed movies of individuals smelling the contents of a glass (disgusting, pleasant, or neutral) and spontaneously expressing the respective emotions. Then the same participants inhaled disgusting or pleasant odorants through a mask. The left anterior insula and the right anterior cingulate cortex were preferentially activated both during the experience evoked by inhaling disgusting odorants and during the observation of disgust-expressive faces. This establishes the existence of a mirroring process. However, the Wicker et al. study did not feature any emotion recognition tasks, so the study did not address the question of disgust attribution.

There are lesion studies, however, that contain relevant evidence about disgust attribution. Calder et al. (2000) studied patient NK who suffered insula and basal ganglia damage. In questionnaire responses, NK showed himself to be selectively impaired in the experience of disgust (as contrasted with fear or anger). NK also showed significant and selective impairment in disgust recognition or attribution, which was established in two modalities: visual and auditory. Similarly, Adolphs et al. (2003) had a patient B who suffered extensive damage to the anterior insula (among other regions) and was able to recognize the six basic emotions *except disgust* when shown dynamic displays of facial expressions or told stories about actions. Apparently, an inability to mirror disgust because of damage to the anterior insula prevented these patients from attributing disgust, though their ability to attribute other basic emotions remained intact. It is a reasonable inference that when normal individuals recognize disgust when viewing the facial expression of disgust, this recognition is causally based on the production in the viewer of a (mirrored) experience of disgust.[5]

Analogous findings have been made in the case of fear, though here the results are a bit more qualified (Adolphs et al., 1994; Sprengelmeyer et al., 1999; Goldman & Sripada, 2005; Goldman, 2006: 115–116, 119–124; Keysers & Gazzola, 2006). Turning to the emotion of anger, Lawrence et al. (2002) reported selective anger recognition impairment as a result of damage to one of its substrates. Previous studies indicated that the neurotransmitter dopamine is involved in the experience of anger. Lawrence et al. therefore hypothesized that a temporary, drug-induced suppression of the dopamine system would also result in impairment of the recognition of angry faces while sparing recognition of other emotions. This is indeed what they found, though this has not been replicated in other studies.

Another finding in the emotion category concerns the secondary emotion guilt. According to the 1991 revised psychopathy checklist (PCL-R), psychopaths lack remorse or guilt. Blair et al. (1995) examined the ability of psychopaths and nonpsychopathic controls to attribute emotions to others, using a story understanding task. Responses of psychopaths and controls to happiness, sadness, and embarrassment stories did not significantly differ. But psychopaths were significantly less likely than controls to attribute guilt to others. This is indirect evidence, once again, that possessing the substrate of an emotion is critical to accurate attribution of that emotion to others, implicating a mirroring process as critical to normal attribution.

6. PAIN AND TOUCH

Continuing with thesis 2, is there evidence that mirroring plays a causal role in the (third-person) attribution of sensations like touch or pain? We begin with touch. Keysers et al. (2004) showed that large extents of the secondary somatosensory cortex that respond to a subject's own legs being touched also respond to the sight of someone else's legs being touched. This is a clear demonstration of empathy for touch (at least in a minimal sense of "empathy"). However, there haven't been tests to determine if observation-mediated somatosensory activity also causes attributions, or judgments, to the effect that another is undergoing such sensations.

Subsequent experiments do provide dramatic support for the mirroring-of-touch phenomenon, and even show that mirroring events can rise above the threshold of consciousness. Blakemore et al. (2005) described a subject C for whom the observation of another person being touched is experienced as tactile stimulation on the equivalent part of C's own body. They call this vision-touch synaesthesia. fMRI experiments also reveal that, in C, the mirror system for touch (in both SI and SII) is hyperactive, above the threshold for conscious tactile perception. Banissy & Ward (2007) followed up on this study and confirmed that synaesthetic touch feels like real touch. However, neither of these studies specifically addressed the question of whether synaesthetic touch leads the subject to attribute the felt touch to the observed person, which would be interpersonal mindreading. Their findings are entirely consistent with this claim, but their experimental manipulations did not specifically address the question.

There is more evidence for mirroring-based pain attribution. Mirror cells for pain were initially discovered serendipitously by Hutchison et al. (1999) while preparing a neurological patient for cingulotomy. More recently, Singer et al. (2004), Jackson et al. (2004), and Morrison et al. (2004) all reported pain resonance or mirroring. All three of these reports were restricted to the affective portion of the pain system, but subsequent transcranial magnetic stimulation (TMS) studies by Avenanti et al. (2005, 2006) highlighted the sensorimotor side of empathy for pain.

On the question of whether mirrored pain can cause pain attribution to others, results from both Jackson et al. (2004) and Avenanti et al. (2005) are especially pertinent. Jackson et al. had subjects watch depictions of hands and feet in painful or neutral conditions and were asked to rate the intensity of pain they thought the target was feeling. This intensity rating is a third-person attribution task. There was a strong correlation between the ratings (attributions) of pain intensity and the level of activity within the posterior ACC (a crucial component of the affective portion of the pain network). This confirms the idea that a mirror-induced feeling can serve as the causal basis of third-person pain attribution.

Avenanti et al. (2005; for a review, see Singer & Frith, 2005) found that there is sharing of pain between self and others not only in the affective portion of the pain system but also in the fine-grained somatomotor representations. When a participant experiences pain, motor evoked potentials (MEPs) elicited by TMS indicate a marked reduction of corticospinal excitability. Avenanti and colleagues found a similar reduction of corticospinal excitability when participants saw someone else receiving a painful stimulus (e.g., when participants watched a video showing a sharp needle being pushed into someone's hand). No change in excitability occurred when they saw a Q-tip pressing the hand or a needle being pushed into a tomato. The neural effects were quite precise. Corticospinal excitability measured from hand muscles was not affected by seeing a needle being thrust into someone's foot. Thus, there appears to be a pain resonance system that extracts basic sensory qualities of another person's painful experience and maps these onto the observers' own sensorimotor system in a somatotopically organized manner. Avenanti et al. also analyzed subjective judgments about the sensory and affective qualities of the pain ascribed to the model during needle penetration. These judgments were obtained by means of the McGill pain questionnaire (MPQ) and visual analogue scales, one for pain intensity and one for pain unpleasantness. The amplitude changes of MEPs recorded from the first dorsal interosseus (FDI) muscle were negatively correlated with sensory aspects of the pain purportedly felt by the model during the "needle in FDI" condition, both for the sensory scale of MPQ and for pain intensity on the visual analogue scale. Thus, judgments of sensory pain to the model seemed to be based on the mirroring process in the sensorimotor pain system. Finally, in a follow-up study, Avenanti et al. (2006) again found a significant reduction in amplitudes of MEPs correlated with the intensity of the pain being attributed to the model, and no MEPs modulation contingent upon different task instructions was found. In particular, specific sensorimotor neural responses did not depend on observers being explicitly asked to mentally simulate sensory qualities of others' sensations.

To sum up, there is adequate evidence in the case of pain to conclude that mirroring states or processes are often causally responsible for third-person mental attributions of the mirrored state, in further support of thesis 2.

7. THE LIMITS OF MIRROR-BASED MINDREADING

There is clear evidence, then, that mirroring plays a causal role in certain types of mindreading. How wide a range of mindreading is open to a mirroring explanation? At present the range appears fairly narrow, because the types of mental activity known to participate in mirroring—viz., motoric activity, sensation, and emotion—appear to be circumscribed, although their ramifications for other phenomena are quite pervasive. Is it possible, then, that massive amounts of other mindreading are also based on mirroring? Or are there principled reasons to think that other types of mindreading differ from mirror-based mindreading?

Thesis 3 denies that there is a universal connection between mindreading and mirroring. A salient reason for being dubious of a universal connection comes from extensive fMRI studies of ToM that identify different brain regions associated with desire and belief attribution, perhaps a dedicated mentalizing network. These regions are disjoint from both the well-known motoric mirror areas and the areas involving pain and emotion that are cited above as loci of mirroring-based mindreading. The so-called ToM regions are sometimes called "cortical midline structures," and consist in the medial frontal cortex (MFC), perhaps subsuming the anterior cingulate cortex, the temporo-parietal junction, the superior temporal sulcus, and the temporal poles. Because these structures are strongly associated with mentalizing—at least certain types of mentalizing—there is a neuroanatomical challenge to the thesis that *all* mindreading is the product of mirroring. This is the "argument from neuroanatomy" for thesis 3.

There are two challenges to this neuroanatomy argument for doubting the universality of mirroring in mindreading. The first challenge comes from the study of a stroke patient, GT, with extensive damage to the medial frontal lobes bilaterally, including regions identified as critical for ToM. Bird et al. (2004) carried out a thorough assessment of GT's cognitive profile and found no significant impairment in ToM tasks. They concluded that the extensive medial frontal regions destroyed by her stroke are not necessary for mindreading.

It is not clear, however, that medial prefrontal cortex should have been identified in the first place as the region dedicated to (high-level) mentalizing. Saxe & Wexler (2005) argue that the critical region is the right temporo-parietal junction (RTPJ). If they are correct, then GT has no substantial bearing on the challenge to a universal role for mirroring in mindreading.

A more dramatic response to the neuroanatomy argument comes from the recent discovery of mirror neurons in the medial frontal lobe by Iacoboni's group (Mukamel et al., 2007). Using single-cell recordings in humans, they found mirror cells for grasping and for facial emotional expressions in the medial frontal cortex, the sites being the preSMA/SMA proper complex, the dorsal sector of ACC, and the ventral sector of ACC. Iacoboni (pers. comm.) suggests that the findings may show that even the higher forms of mindreading are based on some mechanism of neural mirroring. Obviously, confirmation of the latter theory would undermine the argument from neuroanatomy. It remains to be seen, however, exactly which types of mindreading, if any, might be subserved by this new group of mirror cells.

Putting aside the argument from neuroanatomy, let us consider a second type of argument for the non-universality of mirror-based mindreading: a theoretical argument that I will call the "argument from error." This argument says that some forms of mindreading are susceptible to a form of error to which mirror-based mindreading is not susceptible.[6] Therefore, not all mindreading is mirror-based. Let us spell this out.

Mirror-based mindreading is comparatively immune to error. In the first stage of mirror-based mindreading, the observer sees a behavioral or expressive sign in the agent that produces a matching mirror event in him. True, there might be a misfire here if the sign does not genuinely manifest a mental event of which it is typical. But this is not the kind of error I am thinking of in the case of "other" forms of mindreading. In the second stage of mirror-based mindreading, the mindreader classifies the mental event "received" from the agent and attributes it to the agent. If the classification process is normal, as well as the mirror-matching, the resulting act of mindreading will be accurate.

There are other types of mindreading, however, that are susceptible to a different kind of error. In particular, mindreading is prone to "egocentric" errors, largely from failures of perspective-taking. Children are especially prone to this kind of error, but it is also found in adults. One form of perspective-taking failure is the failure to inhibit self-perspective (for a review, see Goldman, 2006: 164–175). There is no place for this kind of error in mirror-based mindreading. Thus, there must be a kind of mindreading that doesn't fit the mirroring mold.[7]

Notice that my point does not rest on the claim that mirror-based mindreading leaves no room at all for error. Recent evidence suggests that mirroring does not always guarantee matching, because it can be modulated by other information or preferences. Singer et al. (2006) found that empathic responses to pain are modulated by learned preferences. Participants played an economic game in which two confederates played fairly or unfairly, and participants then underwent functional imaging while observing the confederates receiving pain stimuli. Participants of both sexes exhibited mirrored pain responses, but in males the responses were significantly reduced when observing an unfair person receiving pain. If these mirror responses also generated pain attributions of varying levels, the indicated modulation would tend to produce errors. This is one way that mirror-based mindreading is open to error, but it is quite different from patterns of error found in other cases of mindreading.

A third argument for the non-universality of mirror-based mindreading is more straightforward than the first two. It is the simple point that a great deal of mindreading is initiated by imagination, and according to our definition of mirroring processes, imagination-driven events do not qualify as mirroring processes. Thus, if a person attempts to determine somebody else's mental state, not by observing their behavior or their facial or postural expression, but by learning about their situation from an informant's description, this act of mindreading will not involve a mirroring process. It may proceed by inference, by imagination, or by "putting oneself in the target's shoes," but none of these qualifies as a mirroring process.

A fourth argument pertains to the types of mental states known to possess mirror properties. Most of these states are not states with propositional contents, like beliefs or desires; or if they do have propositional contents, these contents are of a bodily sort, pertaining to bodily location or movement. Thus, states of pain and touch have mirror properties and the mirroring extends to their felt bodily locations. Intentions to act have mirror properties, and these intentions have contents concerning the types of effectors used (hand, foot, mouth) and the types of actions intended (coded in rich motoric terms). But there is no evidence that beliefs, for example, have mirror properties, especially beliefs with abstract contents. Observing another person grasp or manipulate an object with his hands elicits in the observer a covert intention to grasp or manipulate an object. But observing someone else who is reflecting on the problem of global warming does not elicit a similar thought in one's own mind (except by sheer coincidence). Beliefs and other reflective states do not elicit matching contentful states by a mirroring process, nor do desires that go unexpressed in a distinctive motoric signature. Thus, there is a large class of mental states that aren't mirrored. As they are surely the targets of mindreading, they must be read in a different fashion. This establishes thesis 3.

8. HIGH-LEVEL SIMULATION-BASED MINDREADING

I turn finally to thesis 4. Many writers equate mirroring with simulation. So if a given mental state cannot be read by a mirroring process, it cannot be read by simulation. I take a different view (Goldman, 2006). Simulation and mirroring are not equivalent; mirroring is just one species of simulation. Hence, if a type of mental state is not readable by mirroring, it is still possible that it can be read by simulating, just a different form of simulating. It is also possible, of course, that it can be read by theorizing, and I do not wish to deny that some acts of mindreading, partly or wholly, consist of theorizing (Goldman, 2006: 43–46). Here I shall focus on the second form of simulational mindreading.

The basic idea of simulation of this second kind is to reenact or recreate a scenario in one's mind that differs from what one currently experiences in an endogenous fashion. It is to imagine a scenario, not merely in the sense of "supposing" that it has occurred or will occur, but to imagine being immersed in, or witnessing, the scenario. In other words, it involves engaging in mental pretense in which one tries to construct the scenario as one would experience or undergo it if it were currently happening. This is what philosopher-simulationists had in mind originally by "simulation" (Gordon, 1986; Heal, 1986; Goldman, 1989, 2006; Currie & Ravenscroft, 2002), not mirroring, which is a more recent entrant onto the scene (Gallese & Goldman, 1998). Mirroring features an automatic recreation in an observer's mind of an episode that initially occurs in another's mind. In "enactment simulation," by contrast, one *attempts* to create such a matching event without currently observing another person who undergoes it. One tries to construct the event with the help of experience or knowledge that, it is hoped, will facilitate the construction. Successful reenactment or recreation is more problematic than accurate mirroring. Reenactment must typically be guided by knowledge stored in memory, the quality of which is quite variable. However, any attempt at reenactment can be called "simulation," whether or not there is successful, accurate matching (Goldman, 2006: 38).

Enactment simulation as sketched here approximates the notion of simulation evoked by the neuroscientists Buckner & Carroll (2007), who discuss it under the heading of "self-projection." They conceive self-projection as the mental exploration and construction of alternative perspectives to one's current actual perspective, including perspectives on one's own future (prospection), one's own past (autobiographical memory), the viewpoint of others (ToM), and navigation. Buckner and Carroll refer to imagining an alternative perspective as "simulation." They also argue that all these forms of self-projection involve a shared neural network involving frontal and medial temporo-parietal lobe systems that are traditionally linked to planning and episodic memory.

Buckner and Carroll cite a variety of evidence in support of their view, beginning with the fact that among the deficits created by frontal lobe lesions are deficits in planning and structuring events in an appropriate temporal sequence. Patients with frontal lesions often perform normally in well-established routines and can show high intellectual function, but when confronted with challenging situations and new environments, reveal an inability to plan. They are unable to order sequences temporally, plan actions on tasks requiring foresight, and adjust behaviors flexibly as rules change. Mesulam (2002) noted that the prefrontal cortex might have a pivotal role in the ability to "transpose the effective reference point [of perception] from the self to other, from here to there, and from now to then."

Other evidence concerns the medial temporal lobe, damage to which often causes amnesia. A lesser-studied aspect of the amnesic syndrome is the inability to conceive the personal future. In his seminal description of amnesia in Korsakoff's syndrome, Talland (1965) noted that his amnesic patients could say little about their future plans. The same was true of the amnesic patient HM. Similarly, Klein, Loffus, & Kihlstrom (2002) observed that their amnesic patient DB, when questioned about his future, either confabulated or did not know what he would be doing. Although DB had general knowledge of the future—he knew there was a threat of weather change—he lacked the capacity to consider himself in the future.

A propos of theory of mind, Buckner and Carroll draw on Gallagher & Frith's (2003) account of the role of frontopolar cortex. They suggest that the paracingulate cortex, the anterior-most portion of the frontal midline, is recruited in executive components of simulating others' perspectives. This region is contiguous with but distinct from those involved in episodic remembering. Gallagher and Frith also conclude that this region helps to "determine [another's] mental state, such as a belief, that is decoupled from reality, and to handle simultaneously these two perspectives on the world." Obviously, this is the kind of ability crucial in solving false-belief tasks in mindreading.

If Buckner and Carroll are right that a substantial sector of mentalizing activities are simulations that conform to the foregoing description, and if they are right that such activities take place (roughly) in the brain systems they identify, then it appears that these are not *mirroring* activities. Nonetheless, they are *simulation* activities, in the sense intended. Thus, a substantial chunk of mindreading is simulationist in character without being the product of mirroring. In *Simulating Minds* (Goldman, 2006), I distinguish two types of simulation for mindreading: low-level and high-level. Low-level simulation features mirroring, and high-level simulation does not. *Simulating Minds* does not try to pinpoint precisely all the brain regions associated with high-level mindreading, and that is not essential here either. What is interesting about Buckner and Carroll's contribution is that it identifies a certain network or circuit of brain regions that accomplish a certain general type of task (adopting an alternative perspective), which is instantiated in other domains, as well as mindreading. This tends to substantiate thesis 4.

9. INTERACTIONS BETWEEN CORTICAL MIDLINE STRUCTURES AND MIRROR SYSTEMS?

Uddin, Iacoboni, Lange, & Keenan (2007) propose a unifying model to account for data on self and social cognition by sketching links between cortical midline structures (CMS) and the (motor) mirror-neuron system (MNS). The former is taken to consist of the medial prefrontal cortex, the anterior cingulate cortex, and the precuneus, and the latter is composed of the inferior frontal cortex and the rostral part of the inferior parietal lobule. They argue that a right-lateralized frontoparietal network that overlaps with mirror-neuron areas seems to be involved with self-recognition and social understanding. Because both MNS and CMS are involved in self-other representations, it seems only natural, they propose, that the two systems interact. One pathway by which this might occur is a direct connection between the precuneus (which they regard as a major node of the CMS) and the inferior parietal lobule (the posterior component of the MNS). Also there are direct connections between mesial frontal areas and the inferior frontal gyrus. Thus, the anterior and posterior nodes of the CMS and MNS are in direct communication.

However, it seems that MNS and CMS perform quite different functions vis-à-vis self-understanding. Both the self-face and the self-body activate the right frontoparietal network (Uddin et al., 2005; Sigiura et al., 2006). So the right-lateralized system, associated with the mirror-neuron system, seems to be related to representations of the *physical* self rather than the *mental* self. CMS, on the other hand, seem to be more involved in internal aspects of representing self and others, including mentalizing, as Uddin et al. (2007) themselves concede.

Uddin et al. (2007) propose a division of labor in which the CMS might support "evaluative simulation" in the same way that the MNS supports "motor simulation." This division of labor between the two networks would yield specializations for two related processes that are crucial to navigating the social world: understanding physical actions of intentional agents and understanding the attitudes of others. It is unclear, however, exactly what they mean by "evaluative simulation." Not all the mentalizing work done by the CMS involves evaluation in any straightforward sense. Attributing beliefs to other people (including false beliefs) is a principal mentalizing activity executed by portions of the CMS. But there is nothing "evaluative" (as opposed to "descriptive") about a belief attribution; nor are beliefs themselves evaluative states. It is also unclear what Uddin et al. (2007) mean by "simulation" in this context; they offer no explanation of this (somewhat slippery) notion. However, it appears that we agree on two important points: that simulation plays a central role in different sectors of mentalizing, and that mirror-neuron systems perform only a portion, albeit a very fundamental portion, of the mentalizing work that the human mind undertakes.

10. CONCLUSION

Mirroring per se does not constitute mindreading. Nonetheless, there is evidence of mirror-based mindreading in several domains, including action intention, emotion, and pain. Mirror-based mindreading is what I call low-level mindreading. There are also many reasons, however, to doubt that all mindreading is based on mirroring. How does this bear on the simulation theory of mindreading? Mirroring is one kind of simulational process, but not the only one. Attempting to take another person's perspective, or put oneself in their shoes, is another type of simulational process, and this kind of process is extensively used in mindreading. Thus, simulation figures importantly in high-level, as well as low-level, mindreading.

ACKNOWLEDGMENTS

The author thanks Vittorio Gallese, Marco Iacoboni, and Frederique de Vignemont for detailed comments that resulted in many helpful changes in the manuscript. Other helpful comments were provided by Holly M. Smith.

NOTES

1. The functional properties of two mirror tokenings need not be identical, however. First, it is taken for granted that mirror discharges in execution and observation mode are not perfectly identical (for a review, see Csibra, 2007). In observation mode, the frequency or amplitude of firing may not coincide with that of the execution mode. Thus, the "strength" of two tokenings may

diverge slightly, with implications of slight differences in functional properties. Second, the Parma group from the beginning has distinguished between "strictly" and "broadly" congruent mirror neurons (Gallese et al., 1996). In the case of broad congruence, functional properties are not identical. For present purposes, however, we can ignore this issue. Our approach focuses, for simplicity, on strictly congruent mirror neurons (or their analogue in mirror systems or circuits).

2. For example, lesions to the fusiform gyrus of the right occipital lobe produce both prosopagnosia and achromatopsia (Bartels & Zeki, 2000). But these two deficits have no interesting functional relationship to one another. It just so happens that the impaired capacities are at least partially co-localized in the fusiform gyrus.
3. I usually speak of the simulation relation as holding between *processes* rather than *states* (including intentions). However, as I use the term, a process is a series of causally related states; so, as a limiting case, we may consider a state to be a process with a single member. Hence, we may also speak of states, such as intentions, as items that figure in simulation relations.
4. Vignemont & Haggard (2008) make a strong case for the claim that the best candidate for what is shared in a pair of mirroring events is an "intention in action." If this is right, it argues against the intention-*prediction* interpretation of the Iacoboni et al. (2005) imaging results per se.
5. It is assumed in all of these studies that the participant not only "recognizes" the emotion in the sense of classifying or categorizing it, but also views the emotion as occurring *in the observed target* (whose facial expression is shown or depicted). This implies that the participant is not merely categorizing the emotion but also attributing it *to* the target. If the categorization results from the mirroring process—which includes the observation of the target—it is hardly open to question that the attribution also results from the mirroring process. Thanks to F. de Vignemont for emphasizing this point.
6. Saxe (2005) uses a somewhat analogous argument from error to criticize the general simulation theory of mindreading. Here an argument from error is being used to resist the claim that all mindreading takes a specific simulationist form, namely mirroring-based mindreading. Many errors associated with non-mirror-based mindreading are readily accommodated by a second form of simulation, discussed in section 8. More generally, see Goldman (2006: chap. 7).
7. It might be replied that mirror-based mindreading *is* susceptible to egocentric error. F. de Vignemont (pers. comm.) suggests that if I myself have a terrible back pain and I see you carrying a heavy box, I would feel pain and ascribe this feeling to you. This might be an error because you are perfectly fine with a box that heavy; you are not in pain. Is this not an egocentric error? No doubt, it is an egocentric error. The question is whether it is a case of mirroring, or at least a *pure* case of mirroring. It is not a case in which I see you exhibiting a behavioral or expressive manifestation of pain. And it is questionable whether the perceived heaviness of the box is a "sign" of pain comparable to a knife or needle penetrating a body. It might be a case of inference- or imagination-caused pain, rather than mirror-produced pain. Admittedly, the case puts pressure on our definition of mirroring, but this is not a problem only for me. It is a problem for anyone seeking to be precise about what counts as mirroring. In any case, there are other arguments offered here in favor of thesis 3. It does not rest exclusively on the lesser-liability-to-error argument.

REFERENCES

Adolphs, R., Tranel, D., & Damasio, A. R. (2003). Dissociable neural systems for recognizing emotions. *Brain and Cognition* 52:61–69.

Adolphs, R., Tranel, D., Damasio, H., & Damasio, A. (1994). Impaired recognition of emotion in facial expressions following bilateral damage to the amygdala. *Nature* 372:669–672.

Avenanti, A., Bueti, D., Galati, G., & Aglioti, S. M. (2005). Transcranial magnetic stimulation highlights the sensorimotor side of empathy for pain. *Nature Neuroscience* 8:955–960.

Avenanti, A., Paluello, I. M., Bufalari, I., & Aglioti, S. M. (2006). Stimulus-driven modulation of motor-evoked potentials during observation of others' pain. *NeuroImage* 32:316–324.

Bartels, A., & Zeki, S. (2000). The architecture of the color centre in the human visual brain: New results and a review. *European Journal of Neuroscience* 12:172–193.

Bird, C. M., Castelli, F., Malik, O., Frith, U., & Husain, M. (2004). The impact of extensive medial frontal lobe damage on 'Theory of Mind' and cognition. *Brain* 127:914–928.

Banissy, M. J., & Ward, J. (2007). Mirror-touch synaesthesia is linked with empathy. *Nature Neuroscience* 10:815–816.

Blair, R. J. R., Sellars, C., Strickland, I., Clark, F., Williams, A. O., Smith, M., & Jones, L. (1995). Emotion attributions in the psychopath. *Personality and Individual Differences* 19:431–437.

Blakemore, S.-J., Bristow, D., Bird, G., Frith, C., & Ward, J. (2005). Somatosensory activations during the observation of touch and a case of vision-touch synaesthesia. *Brain* 128:1571–1583.

Buckner, R. L., & Carroll, D. C. (2007). Self-projection and the brain. *Trends in Cognitive Sciences* 11(2):49–57.

Calder, A. J., Keane, J., Manes, F., Antoun, N., & Young, A.W. (2000). Impaired recognition and experience of disgust following brain injury. *Nature Reviews Neuroscience* 3:1077–1078.

Currie, G., & Ravenscroft, I. (2002). *Recreative Minds, Imagination in Philosophy and Psychology*. Oxford, UK: Oxford University Press.

Csibra, G. (2007). Action mirroring and action understanding: An alternative account. In P. Haggard, Y. Rosetti, & M. Kawato (eds.), *Sensorimotor Foundations of Higher Cognition: Attention and Performance XXII* (435–459). Oxford, UK: Oxford University Press.

Decety, J., & Chaminade, T. (2005). The neurophysiology of imitation and intersubjectivity. In S. Hurley & N. Chater, eds., *Perspectives on Imitation: From Neuroscience to Social Science* (119–140). Cambridge, MA: MIT Press.

Fogassi, L., Ferrari, P. F., Gesierich, B., Rozzi, S., Chersi, F., & Rizzolatti, G. (2005). Parietal lobe: From action organization to intention understanding. *Science* 308: 662–667.

Gallagher, H. L., & Frith, C. D. (2003). Functional imaging of 'theory of mind.' *Trends in Cognitive Sciences* 7:77–83.

Gallese, V. (2005). "Being-like-me": Self-other identity, mirror neurons, and empathy. In S. Hurley & N. Chater, eds., *Perspectives on Imitation*, vol. 1 (101–118). Cambridge, MA: MIT Press.

Gallese, V., Fadiga, L., Fogassi, L., & Rizzolatti, G. (1996). Action recognition in the premotor cortex. *Brain* 119:593–609.

Gallese, V., & Goldman, A. I. (1998). Mirror neurons and the simulation theory of mind-reading. *Trends in Cognitive Sciences* 2:493–501.

Gallese, V., Keysers, C., and Rizzolatti, G. (2004). A unifying view of the basis of social cognition. *Trends in Cognitive Sciences* 8:396–403.

Goldman, A. I. (1989). Interpretation psychologized. *Mind and Language* 4:161–185.

Goldman, A. I. (2006). *Simulating Minds: The Philosophy, Psychology, and Neuroscience of Mindreading*. New York: Oxford University Press.

Goldman, A. I., & Sripada, C. S. (2005). Simulationist models of face-based emotion recognition. *Cognition* 94:193–213.

Gordon, R. M. (1986). Folk psychology as simulation. *Mind and Language* 1:158–171.

Heal, J. (1986). Replication and functionalism. In J. Butterfield, ed., *Language, Mind and Logic* (135–150). Cambridge, UK: Cambridge University Press.

Hutchison, W. D., Davis, K. D., Lozano, A. M., Tasker, R. R., & Dostrovsky, J. O. (1999). Pain-related neurons in the human cingulate cortex. *Nature Neuroscience* 2(5):403–405.

Iacoboni, M., et al. (1999). Cortical mechanisms of human imitation. *Science* 286:2526–2528.

Iacoboni, M. (2005). Understanding others: Imitation, language, and empathy. In S. Hurley & N. Chater, eds., *Perspectives on Imitation: From Neuroscience to Social Science* (76–100). Cambridge, MA: MIT Press.

Iacoboni, M., Molnar-Szakacs, I., Gallese, V., Buccino, G., Mazziotta, J. C., & Rizzolatti, G. (2005). Grasping the intentions of others with one's own mirror neuron system. *PLoS Biology* 3:529–535.

Jackson, P. L., Meltzoff, A. N., & Decety, J. (2004). How do we perceive the pain of others? A window into the neural processes involved in empathy. *NeuroImage* 24: 771–779.

Keysers, C., &Gazzola, V. (2006). Towards a unifying neural theory of social cognition. *Progress in Brain Research* 156:383–406.

Keysers, C., Wicker, B., Gazzola, V., Anton, J-L., Fogassi, L., & Gallese, V. (2004). A touching sight: SII/PV activation during the observation of touch. *Neuron* 42: 335–346.

Klein, S. B., Loffus, J., & Kihlstrom, J. F. (2002). Memory and temporal experience: The effect of episodic memory loss on an amnesic patient's ability to remember the past and imagine the future. *Social Cognition* 20:353–379.

Lawrence, A. D., Calder, A. J., McGowan, S. M., & Grasby, P. M. (2002). Selective disruption of the recognition of facial expressions of anger. *NeuroReport* 13(6), 881–884.

Mesulam, M. M. (2002). The human frontal lobes: Transcending the default mode through contingent encoding. In D. T. Stuss & R. T. Knight (eds.), *Principles of Frontal Lobe Function* (8–30). Oxford, UK: Oxford University Press.

Morrison, I., Lloyd, D., de Pelligrino, G., & Roberts, N. (2004). Vicarious responses to pain in anterior cingulate cortex. Is empathy a multisensory issue? *Cognitive Affective Behavioral Neuroscience* 4:270–278.

Mukamel, R., Ekstrom, A. D., Kaplan, J., Iacoboni, M., & Fried, I. (2007). Mirror properties of single cells in human medial frontal cortex. *Social Neuroscience Abstracts*.

Rizzolatti, G. (2005). The mirror neuron system and imitation. In S. Hurley & N. Chater (eds.), *Perspectives on Imitation*, vol. 1 (55–76). Cambridge, MA: MIT Press.

Rizzolatti, G., & Craighero, L. (2004). The mirror-neuron system. *Annual Review of Neuroscience* 27:169–192.

Rizzolatti, G., Fogassi, L., & Gallese, V. (2004). Cortical mechanisms subserving object grasping, action understanding, and imitation. In M. Gazzaniga (ed.), *The Cognitive Neurosciences III* (427–440). Cambridge, MA: MIT Press.

Saxe, R. (2005). Against simulation: The argument from error. *Trends in Cognitive Sciences* 9:174–179.

Saxe, R., & Wexler, A. (2005). Making sense of another mind: The role of the right temporo-parietal junction. *Neuropsychologia* 43:1391–1399.

Sigiura, M., Sassa, Y., Jeong, H., Miura, N., Akitsaki, Horie, K., Sato, S., & Kawashima, R. (2006). Multiple brain networks for visual self-recognition with different sensitivity for motion and body part. *Neuroimage* 32:1905–1917

Singer, T., & Frith, C. (2005). The painful side of empathy. *Nature Neuroscience* 8: 845–846.

Singer, T., Seymour, B., O'Doherty, J., Kaube, H., Dolan, R., & Frith, C. (2004). Empathy for pain involves the affective but not sensory components of pain. *Science* 303:1157–1162.

Singer, T., Seymour, B., O'Doherty, J. Stephan, K. E., Dolan, R. J., & Frith, C. D. (2006). Empathic neural responses are modulated by the perceived fairness of others. *Nature* 439:466–469.

Sprengelmeyer, R., Young, A. W., Schroeder, U., Grossenbacher, P. G., Federlein, J., Buttner, T., & Przuntek, H. (1999). Knowing no fear. *Proceedings of the Royal Society, Series B: Biology* 266:2451–2456.

Talland, G. A. (1965). *Deranged Memory: A Psychonomic Study of the Amnesic Syndrome*. New York: Academic Press.

Uddin, L. Q., Kaplan, J.T., Molnar-Szakacs, I., Zaidel, E., & Iacoboni, M. (2005). Self-face recognition activates a frontoparietal 'mirror' network in the right hemisphere: An event-related fMRI study. *Neuroimage* 25:926–935.

Uddin, L. Q., Iacoboni, M., Lange, C., & Keenan, J. P. (2007). The self and social cognition: The role of cortical midline structures and mirror neurons. *Trends in Cognitive Sciences* 11(4):153–157.

Vignemont, F., & Haggard P. (2008). What is shared? *Social Neuroscience* 3:421–433.

Wicker, B., Keysers, C., Plailly, J., Royet, J-P., Gallese, V., & Rizzolatti, G. (2003). Both of us disgusted in *my* insula: The common neural basis of seeing and feeling disgust. *Neuron* 40:655–664.

CHAPTER 5

Mindreading by Simulation

The Roles of Imagination and Mirroring

(WITH LUCY JORDAN)

1. CRITERIA OF ADEQUACY FOR A THEORY OF "THEORY OF MIND"

There is consensus in cognitive science that ordinary people are robust mindreaders and that mindreading begins early in life. Many other questions concerning mindreading, however, remain in dispute, including the following:

(1) By what method(s) do cognizers read other people's minds—that is, attribute mental states to them? Which cognitive capacities, mechanisms, or processes play pivotal roles in mindreading?[1] Call this the *task-execution* question.
(2) How did the human species, and how do individuals, acquire mindreading capacities? What are the phylogenetic and ontogenetic parts of the story? Call this the *acquisition* question.
(3) What neural substrates underlie mindreading? In other words, does the proposed story pass neuroscientific muster? Call this the *neural plausibility* question.
(4) How does the proffered story of mindreading mesh with the general story of human cognition? Is mindreading a typical example of cognition, or is it a singularity—a one-off piece of cognitive hardware? Call this the question of *mesh*.

Any theory or approach to mindreading must answer these questions, or most of them. It should provide systematic answers that address the entire scope of mindreading: the full range of states that get attributed and the full range of contexts or cues on which mental attributions are based. The mental states imputed to others include at least three types: emotions (e.g., fear, anger, disgust), sensations (pain, touch, tickle), and propositional attitudes (belief, desire, intention). Attribution of all such states needs to be covered by an adequate theory. Our own approach will offer plausible answers to all of the foregoing questions. In that sense, it constitutes a "full scope" approach. Some of its rivals, by contrast, don't pass this test of adequacy. The rationality or teleological approach, for example (Dennett, 1987; Gergely et al., 1995; Csibra et al., 2003), seems to lack the resources to explain attributions of sensations or emotions.[2]

2. LEVELS OF MINDREADING

The general contours of the simulation approach have been laid down by a number of contributors. In the 1980s, philosophers advocated simulation (or "replication") as an alternative to the dominant functionalist, or theory-theory, approach to folk psychology (Gordon, 1986; Heal, 1986; Goldman, 1989). In the 1990s, a developmental slant on simulation theory was presented by Harris (1992). Later in the 1990s and 2000s, neuroscientific findings steered much of the impetus for simulation theory (Gallese & Goldman, 1998; Currie & Ravenscroft, 2002; Decety & Greze, 2006; Goldman & Sripada, 2005; Gallese, 2007). The present chapter begins by reviewing the original model that focused on the mindreading of propositional attitudes. It then moves to a later variant directed at the attribution of emotions, sensations, and intentional motion. The second half of the chapter examines more recent findings that could play pivotal roles in the ongoing debate.

Many treatments of theory of mind (ToM) postulate two or more *levels*, *components*, or *systems* of mindreading, and we too offer a bi-level approach (see Goldman, 2006). But not all duplex theories draw the same partitions or have the same rationale. An early two-component architecture similar to one we favor is that of Tager-Flusberg & Sullivan (2000). They distinguish between a "social cognitive" component and a "social perceptual" component. Their social-cognitive component features a conceptual understanding of the mind as a representational system, and is highly interactive with other domains such as language. The social-perceptual component is involved in the perception of biological and intentional motion and the recognition of emotion via facial expression. This distinction maps well onto our distinction between high-level mindreading of the attitudes versus low-level, or mirror-based, mindreading of non-propositional states. Essentially the same distinction is carved out neurologically by Waytz & Mitchell (2011). Our two methods of mindreading exemplify many of the contrasts that typify the popular dual-systems, or dual-processes, approach in contemporary cognitive science. Low-level mindreading is comparatively fast, stimulus-driven, and automatic; high-level mindreading is comparatively slow, reflective, and controlled.

Apperly advocates a different two-system approach (Apperly & Butterfill, 2009; Apperly, 2011). Both of his systems concern the propositional attitudes, but two systems are posited by analogy with numerical cognition. One system is characterized as efficient but inflexible, the other as flexible but effortful. It is hard to get a firm grip on his systems, however, partly because the account changes significantly between the two publications. The 2009 publication distinguishes between two types of *states* that are mindread—"registrations" and "beliefs"—but registrations disappear in the 2011 publication.

3. HIGH-LEVEL SIMULATIONAL MINDREADING

Early formulations of the simulation theory (ST) correspond to what we call high-level mindreading. To pinpoint its most significant features, we contrast it with its perennial foil, theory-theory (TT). We begin with an example:

> Shaun just left the house and drove away. I ask you where he is going, and you reply: "He didn't say. But I know he wants an espresso and thinks that the best espresso is at Sergio's café. So he probably decided to go to Sergio's."

You have executed a mindreading process, the upshot being the attribution of a certain decision to Shaun. How did you arrive at this? TT would reconstruct your mental process as in figure 5.1. You start with three beliefs, two specifically about Shaun and one about human psychology in general. All the beliefs are depicted as ovals on the left of figure 5.1. You believe of Shaun that he wants an espresso and thinks that Sergio's is the best (nearby) espresso place. Your general belief about human psychology is the "theoretical" proposition that people generally choose actions most likely (by their lights) to satisfy their desires. These three "premise" beliefs are

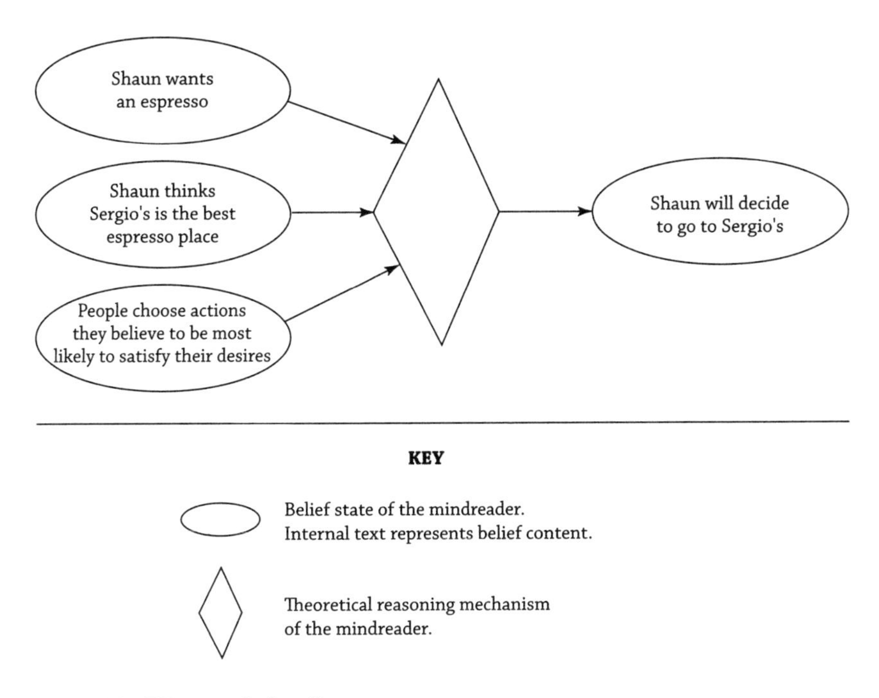

Figure 5.1: TT-type mindreading process.

fed into your reasoning mechanism, which outputs the conclusion that Shaun decided to go to Sergio's.

Several things about this simple TT story are noteworthy. First, the mindreader's states that do all the "work" in the TT story are *belief* states, and the only *processes* used are applications of a theoretical reasoning mechanism. The same would hold of somebody trying to understand and explain the workings of a physical system. Nothing transpires under the theory-oriented account like putting oneself in the target system's shoes. Second, the belief states that do the work are structurally rather complex. They are all *metarepresentational* states, which refer to states of the target that are themselves representational (have *content*). Shaun is portrayed as having a desire and a belief, each of which is a representational state. Third, under TT, mindreading's aptness for success critically depends on the content of the mindreader's naïve psychological theory. If this theory is ample enough in detail and (approximately) descriptively correct, it may tend to supply fairly accurate mental attributions. But if it is meager or misguided, it will frequently lead the mindreader astray. This will happen especially when the target's mental processes are sophisticated and complex.

One form of TT exploits the adequacy or inadequacy of the mindreader's psychological theory to explain influential patterns of error found in early childhood mindreading, especially errors in (verbal) false-belief tasks. Proponents of this form of TT—so-called child scientist theory-theorists—contend that children gradually refine and improve their ToM during their early years, much as adult scientists refine their theories over time. One such refinement is the replacement of a nonrepresentational ToM by a representational theory. The later-developing representational theory allows them to conceptualize the possibility of false belief and thereby improve their performance in false-belief tasks between three and four years of age.

A second form of TT, the *modularity theory,* denies that children develop a ToM mind by a science-like process. Rather, a ToM is an innate endowment of one or more dedicated modules (Baron-Cohen, 1995; Leslie, 1994). How, then, might this sort of theory explain comparatively poor performance by three-year-olds on false-belief tasks? Leslie et al. (2005) introduce an additional, non-modular mechanism, the *selection processor,* to account for this phenomenon. The selection processor selects among candidate belief contents in a target agent's mind by inhibiting the default content—namely the content true of the world—and instead selects an alternative content (which is false of the world). Three-year-olds are weak at this task because their selection processor includes an inhibitory mechanism that hasn't fully matured at three years. Thus, three-year-olds have a *performance* problem with false-belief tasks, not a *competence* problem, as child scientist theory-theorists claim. Despite this difference, both types of TT hold that mindreading is executed by reliance on a ToM, whether an innate theory (embedded in one or more modules) or a gradually developing one.

Could the same tasks be executed in a less informationally demanding manner? Specifically, could they be done with less reliance on refined generalizations about causal connections among mental states? ST takes this tack. It conjectures that mindreaders exploit their own mind as a prototype, or model, of the target's mind. If different minds have the same fundamental processing characteristics, and if the attributor puts her own mind in the same "starting state" as the target's and lets it be guided by her own cognitive mechanisms, mental mimicry may allow her to determine what the target is going to do. ST embraces this alternative hypothesis, depicted in figure 5.2.

A distinctive feature of figure 5.2 is the presence of *shaded* shapes, which represent *pretend* states (products of pretense or imagination). Imagination is assumed to be a faculty that, when you wish to be in a specified mental state M, proceeds to construct an M-*like* state in you. By an "M-like state," we mean a state that is (at least functionally) very similar to a genuine state M, but would normally be produced by cognitive mechanisms other than imagination (e.g., perception, reasoning, emotion-generation). One crucial similarity between a genuine M state and an M-like state is that they

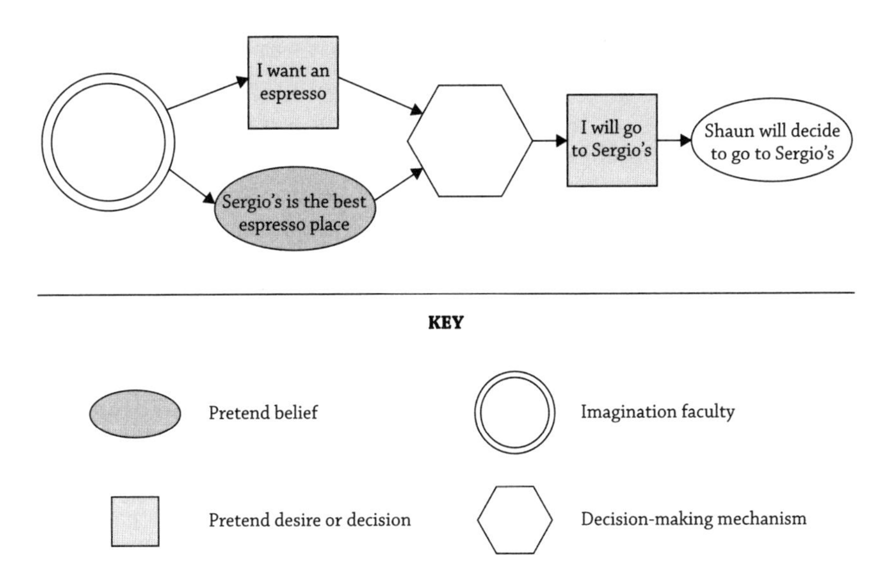

Figure 5.2: High-level ST mindreading process.

produce similar output states when fed into a cognitive mechanism (e.g., a choice or decision-making mechanism). Figure 5.2 depicts the mindreader as constructing a *pretend* desire (intended to match Shaun's relevant desire or goal) and a *pretend* belief (intended to match Shaun's relevant belief). These pretend states are fed into a decision-making mechanism, which operates over these inputs and generates an output state: a decision. This output state is also depicted by shading because it is still under the control of the imagination. Notice that figure 5.2, unlike figure 5.1, makes no reference to a *factual reasoning* mechanism or to a psychological theory of mental processes. The need for theoretical reasoning is replaced in ST by a simulation process, which in this case employs a decision-making mechanism that helps to replicate Shaun's decision-making process. The simulation routine terminates when the decision-making mechanism outputs a decision. This decision is attributed to Shaun (as shown at the far right of the diagram), the attribution being a *genuine* (hence unshaded) belief of the mindreader.

Do any interesting predictions flow from the simulationist model? One prediction is that if the mindreader's imagination performs poorly in constructing the target's starting state(s), the mindreading routine is not likely to succeed (be accurate). A second prediction requires more ground-laying. Mindreading by simulation runs the risk of letting the mindreader's own mental states get entangled with the pretend ones. A mindreader, after all, will always have her own "genuine" desires, beliefs, and intentions alongside the pretend ones. These genuine desires and thoughts must be segregated from the pretend ones, an activity that may not be trivial. There is a danger that genuine states will interfere with pretend ones, causing confusion and error. To avoid such entanglement, genuine states must be "quarantined" or "inhibited" to avoid confusion with mimicked states of the target. Thus, intensive use of simulation predicts a high incidence of mindreading error—specifically, egocentric error, reflecting the penetration of the mindreader's own genuine desires, beliefs, and emotions into the interpersonal tracking process.

Would egocentric errors be equally predicted by TT? Since a theorizing mindreader would also have her own thoughts running on a parallel track with those of the target, does she not face an equal danger of interference? If so, egocentric mindreading errors will not constitute a discriminating test of the rival theories. We argue that the likelihood of interference is higher under ST than TT. Why? Because it is easier to confuse genuine (first-order) cognitive states with the types of pretend states posited by ST than it is to confuse them with the higher-order states posited by TT. States that do the principal work of mindreading under TT are *metarepresentational* states, whereas those deployed during simulation are first-order states. Hence, it should be easier for normal (i.e., genuine as opposed to pretend) thoughts and plans to encroach or insinuate themselves into simulational thinking as compared with theory-guided mindreading.

Now, the mindreading literature is replete with reports of egocentric errors, or biases. Much of it goes under the heading of "curse of knowledge." This phrase was originally introduced in a study of adults who were forewarned that their own knowledge of the task domain exceeded that of their targets whose decisions they were charged with predicting. Despite this warning and related payoff incentives, they allowed their own knowledge states to seep into attributions to their targets, generating poor task performance (Camerer, Loewenstein, & Weber, 1989; Nickerson, 1999). The same leakage phenomenon is found in children (Birch & Bloom, 2003, 2004). For the reasons indicated, this lends greater support to ST as compared with TT.[3]

Although ST easily comports with the observed pattern of egocentric biases, should it not predict much more error than is actually observed? Shouldn't simulation lead to rampant error in view of the fact that pretend beliefs, desires, and emotions must surely be different from their genuine counterparts? How can imagination-generated states resemble genuine states so closely that similar decisions or new beliefs get outputted when the pretend states are inputted into similar cognitive mechanisms? Is there really evidence for a tight-enough similarity between pretend and genuine states to support high levels of mindreading accuracy? Yes. Cognitive science and neuroscience are replete with evidence that imagination is powerful enough to produce states that closely match their counterparts. This is most thoroughly researched in the domains of visual and motoric imagery. Neuroscientific studies confirm that visual imagery and motor imagery have substantial neurological correspondences with vision and motor execution respectively (Kosslyn, Thompson, & Alpert, 1997; Kosslyn, Pascual-Leone, Felician, & Camposano, 1999; Jeannerod, 2001). Chronometric studies of motor imagery are particularly striking (Decety, Jeannerod, & Preblanc, 1989; Currie & Ravenscroft, 2002: 75 ff). But just how powerful and accurate is imagination in non-perceptual and non-motoric cases? A recent study described in section 5 demonstrates the unexpected power of imagination, which should help deflate skepticism about simulation.

A major strand of the ST-TT debate has hinged on the plausibility of the thesis that a complex skill like mindreading is driven by a theory that unfolds during a child's early years. Early defenders of the child scientist version of TT claimed to find evidence that children revise their theory of belief between two and four years of age, yielding mature competence only around the age of four (Wellman, 1990; Perner, 1991; Gopnik & Meltzoff, 1997). The details of this claim, however, were blown out of the water when Onishi & Baillargeon (2005) found false-belief competence (in nonverbal tasks) at fifteen months of age. But now, in many parts of cognitive science, there is impressive evidence of statistical learning (specifically, Bayesian learning) in many areas of cognition. Does this support a new form of TT over the simulation hypothesis?

A study by Baker, Saxe, & Tenenbaum (2009) (also see Baker, Saxe, & Tenenbaum, forthcoming) is a good example of an empirical defense of theory-based mindreading supported by Bayesian inference. They propose "a computational framework based on Bayesian inverse planning for modeling human action understanding... which represents an intuitive theory of intentional agents' behavior.... The mental states that caused an agent's behavior are inferred by... Bayesian inference, integrating the likelihood of the observed actions with the prior over mental states" (2009: 329). If the cognitive reality of this framework is indeed empirically sustained as they claim, does it not decisively support TT over ST?

Interestingly, Baker and colleagues themselves concede that their findings do not favor TT over ST. The reason is that mindreaders who use Bayesian methods to ascribe mental states to others may simply be running their Bayesian reasoning capacity *as a simulation* of the target. Thus, as Baker et al. write:

> [T]he models we propose here could be sensibly interpreted under either account. . . . On a simulation account, goal inference is performed by inverting one's own planning process—the planning mechanism used in model-based reinforcement learning—to infer the goals most likely to have generated another agent's observed behavior. (2009: 347)

4. LOW-LEVEL SIMULATIONAL MINDREADING

The best evidence for low-level simulational mindreading, we submit, is found in research on *emotion* mirroring. First we examine evidence for the existence of mirroring (i.e., mental-state contagion). Then we present evidence that mirrored states are used as the causal basis for mindreading.

Both animal and human studies show that the anterior insula is the "gustatory cortex" and primary locus of the primitive distaste response, disgust (Rozin, Haidt, & McCauley, 2000; Phillips et al., 1997). Against this background, Wicker, Keysers, Plailly, Royet, Gallese, & Rizzolatti (2003) performed a functional imaging study in which participants first viewed movies of other people smelling the contents of a glass (disgusting, pleasant, or neutral) and displaying congruent facial expressions. After first serving in this observer capacity, the same participants then had their own brains scanned while inhaling disgusting or pleasant odorants through a mask over the nose and mouth. The core finding was that the left anterior insula and the right anterior cingulate cortex (ACC) were preferentially activated during both the inhaling of disgusting odorants and the observation of facial expressions of disgust. Thus, there is indeed mirroring (or contagion) for disgust.

This study presented no evidence concerning the mindreading of disgust (via observed facial expressions). For evidence that mindreading *is* based on mirrored disgust, however, we turn to neuropsychology. Patient NK, studied by Calder, Keane, Manes, Antoun, & Young (2000), suffered damage to the insula and basal ganglia. In questionnaire responses, NK showed himself to be selectively impaired in experiencing disgust. He was also significantly and selectively impaired in recognizing (i.e.,, attributing) disgust. It is hard to explain why NK would have this paired deficit unless experiencing disgust is (normally) causally involved in its attribution. A paired deficit in experience and attribution of disgust seems to be most readily explicable on the assumption that disgust attribution (in observational circumstances) is mediated by its experience. In other words, a normal person uses his intact disgust-experience system to attribute disgust to others. Note that NK was normal with respect to attributing other emotions via observation of facial expressions. Nor was a visual deficit a possible explanation, because NK had the same selective deficit in attributing disgust based on nonverbal sounds. Similar findings exist with respect to fear and the amygdala (Goldman & Sripada, 2005; Goldman, 2006: 115 ff.).

TT proponents have not offered any systematic account of these findings. One cannot appeal to damage to a hypothesized theorizing system to account for the disgust-attribution impairment, because the relevant patients performed normally when attributing other emotions based on facial or auditory stimuli. Is there a separate theorizing system for each distinct emotion, and was such a theorizing system coincidentally impaired when disgust experience was impaired? Recalling our criteria of adequacy proposed in section 1, the absence of any story of face-based emotion mindreading is a significant count against TT.

Sensations such as pain are another subcategory of low-level mindreading. The most relevant studies here are by Avenanti et al. (2005, 2006). When a participant experiences pain, motor evoked potentials (MEPs) elicited by transcranial magnetic stimulation (TMS) indicate a marked reduction of corticospinal excitability. Avenanti and colleagues found a similar reduction of excitability when participants merely observed someone else receiving a painful stimulus (e.g., a sharp needle being pushed into his hand). Thus, there appeared to be mirroring of pain in the observer. Moreover, when Avenanti and colleagues had participants judge the intensity of pain purportedly felt by a model, judgments of sensory pain seemed to be based on the mirroring of pain experienced by the participant.

The conclusion that mirrored pain is the causal basis of pain attribution is clouded somewhat, however, by Danziger et al.'s (2009) findings from twelve patients with congenital insensitivity to pain. Compared to normal controls in pain recognition tasks, these patients did not differ much in their estimates of the painfulness to other people of various verbally described events. Nor did they differ much from controls in their estimates of degree of pain judged on the basis of facial expression. However, as Carruthers (2011) points out, these individuals with congenital insensitivity to pain may have acquired a different route to pain mindreading than normal people. The findings do not disprove the hypothesis (which Danziger and colleagues embrace) that normal subjects use simulation in reaching their judgments of pain attribution.

Carruthers addresses problems for another putative example of low-level simulational mindreading, one concerning face-based mindreading of fear. Adolphs et al. (1994) studied a patient, SM, who had a paired deficit for fear perfectly analogous to the one for disgust displayed by NM. SM, who suffers from bilateral amygdala damage, lacks normal experience of fear and is also selectively impaired in fear attribution. This suggests that the mindreading of fear, like the mindreading of disgust, is ripe for interpretation in simulational terms. However, a later study of SM (Adolphs et al., 2005) showed that she was abnormal in scanning her target's eye areas. When she was directed to scan the eyes thoroughly, she improved on fear attribution. Thus, use of fear experience is apparently not strictly necessary for face-based fear attribution, which ostensibly runs counter to a low-level simulational story for fear attribution. However, we can make a similar hypothesis about this case as Carruthers did for those patients congenitally insensitive to pain. Perhaps SM simply developed (under instructions) a skill for face-based mindreading that differs from the simulation heuristic used by normal subjects.

Moreover, other patients with amygdala damage have been studied, with results that support the ST story. Sprengelmeyer et al. (1999) studied patient NM, who showed selective fear-recognition impairment not only using visual face observation but also using postural and vocal emotional stimuli. These recognition impairments of NM cannot be explained by appeal to inadequate facial scanning, because the target's eyes were not visible during the bodily posture task and played no role in the vocal expression task. So it seems that fear impairment due to amygdala damage does indeed provide a causal explanation of NM's recognition impairment, in conformity with the ST account.

Even if one grants that mental attribution in these cases is caused by the mirroring of others' emotions and sensations, one might balk at the idea that this qualifies as *simulation*-based attribution. Why does it so qualify? Here is our answer. Consider the diagram in figure 5.3, where an unshaded shape (on the left side of the diagram, depicting a mental state of the target) represents an actual occurrence of disgust, and a shaded shape (on the right side of the diagram, depicting a state of the observer) represents an observation-induced disgust experience. Just as the figure 5.2 mindreader imputes a specific decision to her target because she herself *makes* that very decision, so the figure 5.3 observer undergoes a mirrored experience of disgust, classifies it as an instance of disgust, and projects (i.e., imputes) it to the target. Such a projection of a self-experienced state is a signature of simulational mindreading. Thus, it seems reasonable to regard this as a species of simulation and simulation-based mindreading, even though it is distinguishable from high-level simulational mindreading in some respects (e.g., by not being a product of imagination). Both cases involve a process of (genuine or attempted) mental matching between attributor states and

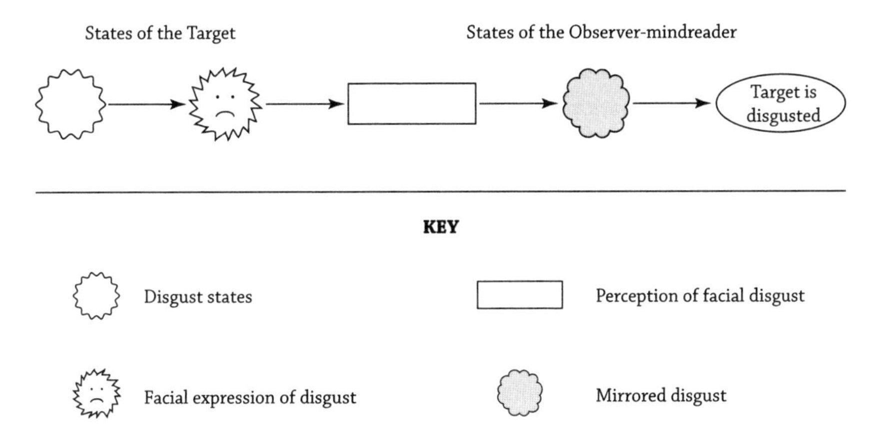

Figure 5.3: Low-level ST mindreading process.

target states. The mirror-produced mindreading may be called a *single-step* simulation process because simulation takes place more or less directly, whereas the decision case is a *multi-step* simulation process.

Our discussion thus far reviews older evidence pertaining to high- and low-level simulational mindreading. In the remainder of the chapter we adduce more recent lines of evidence and assess their bearing on ST and its competitors.

5. THE POWER OF IMAGINATION

As noted in section 3, ST implies that mindreaders' success in high-level mindreading depends on their ability to enact starting states that sufficiently match those of the target. It follows that if simulational mindreading is to succeed, imagination must be a highly precise mechanism, capable not only of generating suitable pretend states but also of firmly holding their progeny in mind while a multi-step simulational exercise unfolds. What is called for is no minor feat. Is the human imagination powerful enough to meet the challenge?

What exactly does it mean to say that we imagine things from another person's perspective? In what sense are mindreaders capable of imagining how a target thinks or feels? The sense of imagination we have in mind is a kind of *enactment imagination,* or *E-imagination* (Goldman, 2006: chap. 7). To E-imagine a state is to recreate the feeling of a state or conjure up what it is like to experience that state—in a sense, to enact that very state. To E-imagine feeling embarrassed involves using one's imagination to create inside oneself a pretend state that phenomenally feels somewhat like embarrassment. This enactment sense of imagination should be distinguished from another everyday notion of imagination that consists in imagining *that* such-and-such is the case (as if someone asked you to imagine *that* you are embarrassed). This ordinary sense of "imagine" means something like *suppose* or *assume* that you are embarrassed—which merely requires you to think about or consider a hypothetical situation of embarrassment. It does not require you to conjure up in yourself something resembling the feeling of embarrassment.[4]

Inspired by evidence of similarities between perception and mental imagery, researchers recently conducted a study to test the effect of imagined eating on actual subsequent eating (Morewedge et al., 2010). The results indicate a striking similarity between the results of actual eating and the results of merely imagined eating, namely both activities result in habituation to the presented stimulus. A series of experiments were conducted to ensure that it was in fact the act of imagining eating that led to a decrease in consumption, and that this decrease in consumption was indeed an effect of habituation. The important experiments and results for our purposes are summarized below.

The first experiment was designed to test whether repeatedly imagining consuming a particular food would influence subsequent consumption of that food. Participants were divided into three groups, each of which imagined performing thirty-three repetitive actions. The control group imagined putting thirty-three quarters into a laundry machine (an action similar to putting M&Ms in one's mouth); the second group imagined putting thirty quarters into a laundry machine and then eating three M&Ms; and the third group imagined putting three quarters into a laundry machine and then eating thirty M&Ms. All participants then ate freely from a bowl of M&Ms until they indicated that they were finished eating. How much each participant ate from the bowl was measured and compared. The results showed that participants who imagined eating thirty M&Ms subsequently ate significantly fewer M&Ms than participants in the other groups.

Another experiment tested whether the decrease in consumption was due to habituation or if it was a priming effect resulting from repeated exposure to the stimulus. This time participants imagined either eating three or thirty M&Ms, or putting three or thirty M&Ms into a bowl; then, as before, participants ate freely from a bowl of M&Ms. Again, results revealed that subjects who imagined eating thirty M&Ms ate significantly less than those who only imagined eating three; but results also showed that subjects who imagined putting thirty M&Ms in a bowl ended up eating significantly more than subjects who imagined putting only three in a bowl. This experiment strongly suggests that not only is priming not the cause of the decrease in consumption, but may even have the opposite effect of increasing subsequent food intake. A further experiment was designed to determine if imaginary eating was habituating people to particular food (causing them to eat less of it), or if it was an overall primed feeling of "fullness" responsible for the decrease in food intake. Here participants imagined eating either three or thirty M&Ms or cubes of cheese, and then ate freely from a bowl of cheese cubes. The participants who imagined eating thirty cubes ate significantly less than those who imagined eating three; but participants who imagined eating thirty M&Ms did not differ in subsequent cheese consumption from those who imagined eating three M&Ms. Thus, it seems that the effect of imaginary eating is *stimulus-specific*—providing additional evidence that the reduction in food intake is a result of habituation, and not of priming.

The results of this study are remarkable. Merely imagining eating can impact how much we actually eat. But how is a study concerning food consumption relevant to mindreading? We argue that this study is easily interpreted as a demonstration of the power of imagination, and to that extent, it supports our version of ST. In order for the case to be convincing, there are two things that need to be established: (1) The study's use of imaginary eating counts as an instance of imagination in our enactment sense of the word; and (2) the states generated by the imagination really do appropriately resemble their actual counterparts. Our first task is relatively straightforward given the study's experimental design. Participants in the first experiment were asked to repeatedly imagine themselves eating units of food one at a time, not merely to imagine that they had eaten a certain amount of food. Furthermore, results indicated that merely thinking about a particular food repeatedly was not enough. For the habituation effect to occur, a person had to actually imagine undergoing the experience of eating a particular food. But this is just how we have characterized an act of enactment imagination: as an attempt to reenact or re-experience a particular feeling or state.

What about our second task? Have Morewedge and colleagues shown that imagining eating is capable of producing accurate pretend states, similar to those that result from actually eating? When a person actually eats a particular food, they gradually habituate. Their desire to eat the food, along with the motivation to obtain it, gradually decreases. If presented with a different food, however, the person's desire and motivation recover. This implies that habituation effects are stimulus-specific (Epstein et al., 2003). If this is what happens when we actually eat, then something sufficiently similar to habituation should result when we repeatedly imagine ourselves eating. Additionally, it should be the case that if we imagine eating something only a few times, we do not habituate. The results of the Morewedge et al. study clearly demonstrate that imagined eating results in habituation, similar to when a person has actually eaten. Furthermore, given that habituation effects are stimulus-specific, the imagined consumption of a particular food should cause a person to habituate to that food only. As the study demonstrates, this is exactly what happens.

Imagined consumption is a clear instance of enactment imagination, as well as of the resemblance that can obtain between imagination-induced states and their genuine counterparts. But imaginary eating is not a case of mindreading. Can more be done to make the connection between this research and simulational mindreading clear? After all, if ST is right, mindreaders use their imagination in tasks involving a variety of mental states. So we need to establish that imagination can produce pretend states that closely resemble actual states across a respectable spectrum of cases. We maintain that research on imagined consumption gives us reason to think the imagination has this capacity.

According to Morewedge et al., "Habituation to food occurs too quickly for it to result from digestive feedback, so it is commonly thought to occur as a result of top-down cognitive processes (such as beliefs, memories, or expectations) or pre-ingestive sensory factors (such as texture or smell)" (2010: 1531). This study demonstrates that habituation can occur as a result of imagination alone, without any influence from sensory information. This is significant because habituation is a very general phenomenon, specific to neither food nor eating. Research indicates that we habituate to a wide range of complex emotions, attitudes, feelings, and moods: from states of happiness and love to states of fear and anxiety (Solomon, 1980). This study confirms that the power of imagination could be very general indeed.[5]

6. MINDREADING ACQUISITION

Recall that our second criterion of adequacy requires a comprehensive theory of mindreading to give a plausible, empirically sustainable story about how the mindreading capacity is acquired. Past research on the time course of childhood mentalizing had important implications for theories of how mindreading is acquired, such as the child scientist approach to TT. This approach claims that mental-state attributions are driven by naïve psychological theories that are initiated and gradually revised in the early years. This claim has been increasingly undercut, however, by recent research revealing sensitivity to false beliefs even in preverbal infants (e.g., Onishi & Baillargeon, 2005).

How does the simulation theory fit with such evidence? What is ST's position on the acquisition of the mentalizing capacity? ST does not take a firm stance vis-à-vis nativism. It is prepared to go with the flow of evidence. For example, it is prepared to say that the processes or methods of mindreading (or dispositions to use such processes) are part of our native endowment. It might be more skeptical about claims that particular mental-state concepts (belief, desire, pleasure, etc.) are all innate. It is prepared, however, to accommodate the former type of nativism if empirical studies provide warrant for this position. ST's theoretical apparatus does not preclude such strands of nativism. Indeed, one might say something stronger from the vantage point of ST. If imagination is an innate capacity, perhaps young infants automatically compute imaginary states for people around them. Then we would expect the practice of generating imaginary states to be no more cognitively demanding than one's own largely automatic production of mental states. This section will discuss how well such expectations comport with recent evidence in developmental psychology. Our primary focus is a compelling new study conducted by Kovacs et al. (2010) plus the simulational hypothesis we claim to be consistent with this study.[6]

Unlike standard false-belief tasks, this study was designed to investigate mindreading mechanisms *implicitly*—making no reference to others' beliefs and not requiring any behavioral predictions based on others' beliefs. The study had two components: one testing the reaction time (RT) of adult participants and the other measuring the looking time (LT) of seven-month-old infants. Participants watched a series of short movies involving an animated agent, a ball, and a table with an occluder. At the beginning of each movie, the animated agent entered the scene and placed the ball on the table in front of the occluder.[7] The ball then rolled behind the occluder. At this point, depending on the experiment, the ball either stayed in place or rolled off the screen. Then the agent left the scene. The ball's final location and the time the agent left the scene were varied, such that the agent would have a true belief about the ball's location if he left after the ball reached its final location and a false belief if he left before. The critical variables involved the participant's beliefs about the ball's presence or absence and the agent's "beliefs," such that the participant, the agent, both, or neither could believe the ball was behind the occluder (Kovacs, 2010: 1831). At the end of the movie, the agent reentered the scene and the occluder was lowered, revealing the ball to be either present or absent. Adult participants were instructed to press a button as soon as they detected the ball. Their RTs and the infants' LTs were measured in each of the four conditions.

The experimental conditions, for both adults and infants, were compared to a baseline condition where neither the participants nor the agent believed the ball to be behind the occluder.[8] The most important experiment with adult participants was one in which only the agent believed the ball to be behind the occluder. Results indicated that participants' RTs were faster in this case than in the baseline condition, despite no difference in the participants' beliefs in either condition. Kovacs et al. take this result to demonstrate that the participants not only automatically computed the agent's beliefs, but that these beliefs influenced the participants' behavior, despite the agent's beliefs being inconsistent with their own (1832). Additionally, the participants' RTs did not significantly differ when only the agent believed the ball to be behind the occluder and when they themselves believed it to be. Kovacs et al. further conclude: "Thus both types of belief representations speeded up the participants' RTs to similar extents, a result consistent with the view that the agent's beliefs are stored similarly to participants' own representations about the environment" (2010: 1832).

The crucial results with the infant participants similarly involved a comparison between the infants' LTs in two conditions: one in which only the agent believed the ball to be behind the occluder, and the other in which neither the infant nor the agent believed the ball to be there. When no ball appeared behind the occluder, the infants looked longer (indicating their "surprise" by the outcome) in the condition where *only* the agent believed, or expected, the ball to be there. Again, this suggests not only that the infants computed the agent's belief but also that this belief influenced the infants' behavior despite conflicting with their own (genuine) beliefs. It is similarly interesting that with both adults and infants, very similar results obtained even when the agent did not return to the scene and thus was not present when the occlusion was lowered. Infants and adults seemed to compute and maintain the agent's beliefs even when the agent was no longer present.

What do these results mean for the study of mindreading? More specifically, how do they fit with what ST says about mindreading? Concerning the question of acquisition, the infant results are of primary interest. How do they fit with the theory-driven versus simulation-driven mindreading processes depicted in figures 5.1 and 5.2 respectively? A figure 5.1-type story would say that at seven months of age, infants not only compute the beliefs of other agents but that these computations are based on the infants' beliefs about the beliefs of the agent. In other words, TT-type explanations rely on the infants' possession of relatively complex metarepresentational states, plus their possession of some body of psychological laws or generalizations.[9] Thus, TT's approach to mindreading is information-rich and requires a degree of cognitive or informational sophistication that one may be hesitant to attribute to seven-month-old infants.

In contrast, the figure 5.2-type story suggests that the same sort of tracking of the agent's thoughts has another, simpler interpretation. ST implies that infants track an agent's perspective in the same way they maintain their own perspective. Just as infants have their own current representations of the environment, they also track other possible states of the environment. This sort of explanation, in contrast to figure 5.1-type theories, is an information-poor approach, because it does not attribute to the infants any additional theoretical knowledge or metarepresentational states. To perform perspective computations, ST only requires that infants possess states with *object-level* representational content—information about the way the world seems from the shoes of the agent. This means that infants may track the content of an agent's belief (possible states of the environment) without encoding anything concerning the agent's beliefs or other mental states.

Given what we have said so far, ST is in as good a position as TT to account for the Kovacs study. Might there be reasons to think it may be in a *better* position to explain its findings? We argue that there are. First of all, TT has to say that preverbal infants compute metarepresentations. Is it psychologically plausible to impute such cognitively complex mental states to infants? Wouldn't it be preferable, if possible, to account for the infants' behavior without attributing to them such extra computational work or informational baggage? If so, then the ST explanation is clearly preferable, because it accounts for the evidence without positing the extra complexity of metarepresentational states or a body of psychological generalizations.

Concerning the question of acquisition, there are other reasons to think the Kovacs study supports a simulation story about mindreading. What this study shows, we have claimed, is that seven-month-old infants generate representations of the world that reflect another person's perspective—but to represent the world as it seems to another person *is* just to use one's imagination.[10] Although it seems unlikely that seven-month-old infants engage in explicit acts of mindreading (namely, attribution of mental states to others), they certainly appear to engage in mindreading-*like* activity. Furthermore, this mindreading-like activity involves use of their imagination. This means that before they ever engage in a single act of mindreading, infants are already experienced imaginers. By the time they get to the point of attributing mental states to other people, they have spent years spontaneously and automatically imagining the world from other people's perspectives.

7. THE NEURAL BASIS OF MINDREADING

Now we apply the third question of adequacy to our version of ST: Is this theory *neurally plausible*, given available empirical evidence? One issue is whether recent evidence from cognitive neuroscience supports (or is consistent with) the claim that simulation is a common method, if not the predominant method, of mindreading. A second issue is whether neuroscience supports our specific version of ST, namely a bi-level or duplex version of ST. Because neuroscientific evidence was already adduced in support of the existence of mirroring and the grounds for linking it to ST, we will not say more about the first issue. We shall concentrate on the second.

Waytz & Mitchell (2011) present the neuroscientific case for a duplex model of simulational mindreading as follows: First, they review the extensive evidence of multiple mirroring phenomena, sometimes referred to as "shared neural representations." These include regions in the inferior frontal cortex and superior parietal lobe (the parieto-frontal circuit) that are involved in the production and observation of goal-directed motor action.[11] They also include a wide range of regions for the mirroring of pain, touch, disgust, and fear (also see Rizzolatti & Sinigaglia, 2006). Networks in these areas are what we treated under the heading of low-level simulation.

Another set of brain regions has been identified, however, that serves as a substrate for what Buckner & Carroll (2007) call *self-projection*. These regions, known collectively as the "default network" (Raichle et al., 2001), consist of the medial prefrontal cortex, precuneus and posterior cingulate, and lateral parietal cortex. The default network has repeatedly been linked to tasks in which people imagine experiencing fictitious events, consider the possibility of experiencing specified events in the future, or recall their experiences from the past. It has also been reported by studies in which participants contemplated other people's mental states (Frith & Frith, 2003). Thus, the default network is an excellent candidate for the neural substrate of high-level simulational mindreading. Moreover, this network seems to be quite distinct from any of the circuits or processes involved in mirroring.

Finally, Waytz and Mitchell point out that there are dissociable functions of mirroring and self-projection. Perceivers mirror only when they see or hear another person's physical actions, observe an emotional expression, or witness a painful situation, such as a needle penetrating a hand. But mindreading also occurs when subjects represent targets who are not immediately present and hence are not observable. Such targets include fictitious individuals or individuals known only by description, where no observable cues are available. Waytz and Mitchell consider this a demonstration of dissociation between mirroring and self-projection. They cite a mentalizing study by Zaki et al. (2010) in which participants inferred a target's emotional state under three conditions: during perceptually cued trials, during context-only trials, and when participants had both perceptual and contextual information. Consistent with the proposed division of labor between two systems of mentalizing, they found that perceptual cues tended to elicit stronger activation in mirror-related brain regions (the fronto-parietal circuit), whereas contextual cues engaged the default network.

Lombardo et al. (2010) take a somewhat different perspective on the two-system approach based on their finding of functional connectivity patterns during mentalizing of both self and other. They do not deny the dissociability claim of Waytz and Mitchell, but they argue that the functional connectivity patterns revealed in their studies support a slightly different picture than the one offered here. Indeed, they advance the thesis that some aspects of both high-level and low-level social cognitive processes are "grounded" within a framework of embodied cognition. We do not believe that there are fundamental differences between their view and ours. At any rate, we find no reason to disagree with a very similar picture presented by Zaki & Ochsner (2012), who also stress functional connectivity between the two systems during experiences of empathy. As Zaki and Ochsner express it, "naturalistic" (i.e., ecologically valid) situations involve many dynamic social cues (featuring both sensorimotor and contextual information), and such cues unsurprisingly generate dynamical neural interactions among simpler processes (both low-level and high-level). These more complex processes could not be understood, Zaki and Ochsner acknowledge, without a prior understanding of the simpler processes in isolation, which are coupled during complex social tasks (2012: 678). By our lights, this is a reasonably clear recognition that there *are* simpler processes, which we take to be the low-level and high-level families of processes in our model.[12] It is the existence and distinctness of these simpler processes that comprise the core thesis defended in this section.

8. ST'S "MESH" WITH EVOLUTIONARY THEORY

We turn finally (and briefly) to the fourth question of section 1: Does our theory mesh with successful theories in other cognitive domains and with plausible accounts of the architecture and evolution of cognition?

Do successful theories of other parts of cognition invoke similar explanatory faculties or processes, and does a reasonable account of brain evolution find a natural home for simulationist stories of mindreading?

Begin with ST's account of high-level mindreading, in which imagination occupies a central role. It makes good sense, we submit, to assign a pivotal role to imagination because this faculty has demonstrated its power and versatility in many other domains of cognition. Its robust power and versatility are amply exhibited in such diverse phenomena as visual and motor imagery, the planning of action sequences, and the reduction of food consumption. Turning to low-level mindreading, the discovery of mirror neurons and mirror systems has revolutionized research about many aspects of low-level cognition (Rizzolatti & Craighero, 2004; Gallese et al., 2004). Contemporary social neuroscience is replete with new insights related to mirroring. A primitive kind of mindreading based on mirroring is a good fit with much of this literature.

ST also comports well with current understandings of brain evolution. As Anderson (2010) tells the story, it is very common for neural circuits originally established for one purpose to be exapted (i.e., exploited, recycled, redeployed) during evolution and put to different uses, without necessarily losing their original functions. Nature has a pervasive strategy of opportunistically exploiting existing neural hardware to solve new problems—or to create new solutions for old problems. Creating whole brain structures de novo in order to tackle problems would be expensive. Instead, nature prefers a redeployment strategy. This idea meshes well with ST's story of low-level mindreading. So, for example, suppose that nature had earlier hit upon the strategy of devising mechanisms by which shared representations are generated in the heads of two interacting individuals. A mental representation in one individual's brain leads to the generation of a matching representation in an observer. Once this kind of interpersonal transmission mechanism has evolved, members of the species can secure valuable information by piggybacking a mental *attribution* mechanism on top of the shared-representation, or mirroring, mechanism. This is a cheap way to create a reliable mindreading device. It would be unsurprising if something like this evolutionary story were true.

This is what we mean in saying that ST "meshes" well with what is known, or reasonably believed, about brain evolution. According to many philosophers of science, consilience with existing theory is one form of evidence for a new theory. Thus, another chunk of evidential support is added in favor of ST, on top of the more direct kinds of evidence presented in preceding sections.

NOTES

1. People read their own minds, as well as the minds of others. How first-person mindreading is executed is a question of equal importance and difficulty as the third-person mindreading question. For this reason it cannot be addressed within the confines of this chapter, so it is left for another day. (See Goldman, 2006: chaps. 9 and 10, for an earlier foray into this territory.)
2. A standard taxonomy of approaches to mindreading *other* than ST includes the rationality theory, the child scientist version of TT, and the modularity version of TT. For detailed expositions and critiques of these rivals, see Goldman, 2006: chaps. 3–5. Selected problems for some of these rivals are sprinkled throughout this chapter, but length limits preclude detailed treatments.
3. The reasoning relies on Bayesian conditionalization. When the likelihood of O given H_1 is greater than the likelihood of O given H_2, observation of O will increase the posterior probability of H_1 more than it increases the posterior probability of H_2. The present contrast between ST and TT is aimed mainly at the child scientist version of TT. As reported above, Leslie's form of modularity theory shares with ST a reliance on inhibitory mechanisms.
4. We intend our use of the word "imagination" to be understood in the E-imagination sense.
5. Additional confirmatory evidence comes from research on memory distortion involving imagination (Schacter, Guerin, & St. Jacques, 2011). A study by Mazzoni & Memon (2003) indicated that the strength of subjects' beliefs that events occurred increased more when they imagined events than when they simply read about them. Nash et al. (2009) showed that imagining that one has performed an act produces about as many false memories of actually having done it as viewing a doctored video that suggests that one did perform the act.
6. Although we focus on the results of the Kovacs et al. study, we find the wording of their study potentially misleading and worry that it may not convey the researchers' true intentions. Because of this, we will outline the results as reported in the study, but the conclusions drawn will be our own and are not intended to match those of the authors.
7. "Agent" always refers to the animated character in the film and "participant" refers to the adult or infant participant in the study.
8. However, recall that in none of the conditions were the agent's beliefs relevant to the task.
9. Depending on the particular TT-type approach we are discussing, such theories may also require that infants possess other complex theoretical beliefs about human psychology.
10. This is where our conclusions may come apart from those drawn by Kovacs et al. Although the conclusions drawn in the study seem to claim that seven-month-old infants are representing the beliefs of the agent alongside their own beliefs, we claim that the results of this study only demonstrate that infants are generating representations of the world that reflect what the target's or agent's beliefs *would be* (not that they represent them *as* beliefs). Furthermore, the study itself makes no mention of the imagination—rather, the results demonstrate that infants engage in an activity that epitomizes our conception of the use of imagination in simulation.
11. The mirroring theory has characteristically claimed that mirroring is used to understand the actions of others. There is continuing debate, however, over which specific brain networks comprise the action-observation system, and how exactly

they function. For example, Kilmer (2011) defends a two-pathway model of action understanding, featuring a dorsal pathway in addition to the initially discovered ventral pathway.

12. One nontrivial point of difference, however, is that Zaki and Ochsner identify the higher-level processes as "mentalizing" processes, implying that lower-level ("shared representation") processes are not involved in mentalizing. In contrast, we claim that the latter also serve as a causal basis of mentalizing.

REFERENCES

Adolphs, R., Gosselin, F., Buchanan, T. W., Tranel, D., Schyns, P., & Damasio, A. R. (2005). A mechanism for impaired fear recognition after amygdala damage. *Nature* 433:68–72.

Adolphs, R., Tranel, D., Damasio, H., & Damasio, A. (1994). Impaired recognition of emotion in facial expressions following bilateral damage to the amygdala. *Nature* 372:669–672.

Anderson, M. L. (2010). Neural reuse: A fundamental organizational principle of the brain. *Behavioral and Brain Sciences* 33:245–266.

Apperly, I. A. (2011). *Mindreaders: The Cognitive Basis of Theory of Mind*. Hove, East Sussex, UK: Psychology Press.

Apperly, I. A., & Butterfill, S. A. (2009): Do humans have two systems to track beliefs and belief-like states? *Psychological Review* 116(4):953–970.

Avenanti, A. Bueti, D., Galati, G., & Aglioti, S.M. (2005). Transcranial magnetic stimulation highlights the sensorimotor side of empathy for pain. *Nature Neuroscience* 8:955–960.

Avenanti, A., Paluello, I. M., Bufalari, J., & Aglioti, S.M. (2006). Stimulus-driven modulation of motor-evoked potentials during observation of others' pain. *NeuroImage* 32:316–324.

Baker, C. L., Saxe, R., & Tenenbaum, J. B. (2009). Action understanding as inverse planning. *Cognition* 113:329–349.

Baker, C.L., Saxe, R., & Tenenbaum, J. B. (forthcoming). Bayesian theory of mind: Modeling joint belief-desire attribution.

Baron-Cohen, S. (1995). *Mindblindness: An Essay on Autism and Theory of Mind*. Cambridge, MA: MIT Press.

Birch, S. A. J., & Bloom, P. (2003). Children are cursed: An asymmetric bias in mental-state attribution. *Psychological Science* 14:283–286.

Birch, S. A. J., & Bloom, P. (2004). Understanding children's and adults' limitations in mental state reasoning. *Trends in Cognitive Sciences* 8:255–260.

Buckner, R. L., & Carroll, D. C. (2007). Self-projection and the brain. *Trends in Cognitive Sciences* 11:49–57. .

Camerer, C., Loewenstein, G., & Weber, M. (1989). The curse of knowledge in economic settings: An experimental analysis. *Journal of Political Economy* 97:1232–1254.

Carruthers, P. (2011). *Opacity of Mind*. Oxford, UK: Oxford University Press.

Csibra, G., Biro, S., Koos, O., & Gergely, G. (2003). One-year-old infants use teleological representations of actions productively. *Cognitive Science* 27:111–133.

Currie, G., & Ravenscroft, I. (2002). *Recreative Minds*. Oxford, UK: Oxford University Press.

Danziger, N., Faillenot, I., & Peyron, R. (2009). Can we share a pain we never felt? Neural correlates of empathy in patients with congenital insensitivity to pain. *Neuron* 61(2):203–212.
Decety, J., Jeannerod, M., & Preblanc, C. (1989). The timing of mentally represented actions. *Behavioral and Brain Research* 34:35–42.
Decety, J. and Greze, J. (2006). The power of simulation: Imagining one's own and other's behavior. *Brain Research* 1079: 4–14.
Dennett, D. (1987). *The Intentional Stance*. Cambridge, MA: MIT Press.
Epstein, L. H., Saad, F. G., Handley, E. A., Roemmich, J. N., Hawk, L. W., & McSweeney, F. K. (2003). Habituation of salivation and motivated responding for food in children. *Appetite* 41(3):283–289.
Frith, U., & Frith, C. D. (2003). Development and neurophysiology of mentalizing. *Philosophical Transaction of the Royal Society of London. Series B: Biological Sciences*, 459:358.
Gallese, V. (2007). Before and below 'theory of mind': Embodied simulation and the neural correlates of social cognition. *Philosophical Transactions of the Royal Society of London. Series B: Biological Sciences* 362:659–669.
Gallese, V., & Goldman, A. (1998). Mirror neurons and the simulation theory of mindreading. *Trends in Cognitive Sciences* 2:493–501.
Gallese, V., Keysers, C., & Rizzolatti, G. (2004). A unifying view of the basis of social cognition. *Trends in Cognitive Sciences* 8:396–403.
Gergely, G., Nadasdy, Z., Csibra, G., & Biro, S. (1995). Taking the intentional stance at 12 months of age. *Cognition* 56:165–193.
Goldman, A. I. (1989). Interpretation psychologized. *Mind and Language* 4:161–185.
-----. (2006). *Simulating Minds: The Philosophy, Psychology, and Neuroscience of Mindreading*. New York: Oxford University Press.
Goldman, A. I., & Sripada, C. (2005). Simulationist models of face-based emotion recognition. *Cognition* 94:193–213.
Gordon, R. M. (1986). Folk psychology as simulation. *Mind and Language* 1:158–171.
Harris, P. L. (1992). From simulation to folk psychology: The case for development. *Mind and Language* 7:120–144.
Heal, J. (1986). Replication and functionalism. In J. Butterfield, ed., *Language, Mind, and Logic* (135–150). Cambridge, UK: Cambridge University Press.
Jeannerod, M. (2001). Neural simulation of action: A unifying mechanism for motor cognition. *NeuroImage* 14:S103–S109.
Kilmer, J. M. (2011). More than one pathway to action understanding. *Trends in Cognitive Sciences* 15(8):352–357.
Kosslyn, S. M., Pascual-Leone, A., Felician, O., & Camposano, S. (1999). The role of area 17 in visual imagery: Convergent evidence from PET from rTMS. *Science* 284:167–170.
Kosslyn, S. M., Thompson, W. L., & Alpert, N. M. (1997). Neural systems shared by visual imagery and visual perception: A positron emission tomography study. *Neuro-Image* 6: 320–334.
Kovacs, A. M., Teglas, E., & Endress, A. D. (2010). The social sense: Susceptibility to others' beliefs in human infants and adults. *Science* 330:1830–1834.
Leslie, A. (1994). Pretending and believing: Issues in the theory of ToMM. *Cognition* 50:211–238.
Leslie, A., German, T., & Polizzi, P. (2005). Belief-desire reasoning as a process of selection. *Cognitive Psychology* 50:45–85.

Lombardo, M. V., Chakraharti, B., Bullmore, E. T., Wheelwright, S. J., Sadek, S. A., Suckling, J., & Baron-Cohen, S. (2010). Shared neural circuits for mentalizing about the self and others. *Journal of Cognitive Neuroscience* 22(7):1623–1633.

Mazzoni, G., & Memon, A. (2003). Imagination can create false autobiographical memories. *Psychological Science* 14:186–188.

Morewedge, C. K., Huh, Y. E., & Vosgerau, J. (2010). Thought for food: Imagined consumption reduces actual consumption. *Science* 330:1530–1533.

Nash, R. A., Kimberly, A. W., & Lindsay, D. S (2009). Digitally manipulating memory: Effects of doctored videos and imagination in distorting beliefs and memories. *Memory and Cognition* 37(4):414–424.

Nickerson, R. S. (1999). How we know—and sometimes misjudge—what others know: Imputing one's own knowledge to others. *Psychological Bulletin* 125:737–759.

Perner, J. (1991). *Understanding the Representational Mind.* Cambridge, MA: MIT Press.

Phillips, M. L., Young, A. W., Senior, C., Brammer, M., Andrew, C., Calder, A. J., Bullmore, E. T., Perrett, D. I., Rowland, D., Williams, S. C. R., Gray, J. A., & David, S. (1997). A specific neural substrate for perceiving facial expressions of disgust. *Nature* 389:495–498.

Raichle, M. E., MacLeod, A. M., Snyder, A. Z., Powers, W. J., Gusnard, D. A., & Shulman, G. L. (2001). A default mode of brain function. *Proceedings of the National Academy of Sciences U.S.A.* 98:676–682.

Rizzolatti, G., & Craighero, L. (2004). The mirror-neuron system. *Annual Review of Neuroscience* 27:169–192.

Rizzolatti, G., & Sinigaglia, C. (2008). *Mirrors in the Brain: How Our Minds Share Actions and Emotions.* Oxford, UK: Oxford University Press.

Rozin P., Haidt J., & McCauley, C. (2000). Disgust. In M. Lewis & J. Haviland, eds., *Handbook of Emotions* (575–594). New York: Guilford Press.

Schacter, D. L., Guerin, S. A., & St. Jacques, P. L. (2011). Memory distortion: An adaptive perspective. *Trends in Cognitive Sciences* 15(10):467–506.

Solomon, R. L. (1980). The opponent-process theory of acquired motivation: The costs of pleasure and the benefits of pain. *American Psychologist* 35(8):691–712.

Sprengelmeyer R., Young, A. W., Schroeder, U., Grossenbacher, P. G., Federlein, J., Buttner, T., & Przuntek, H. (1999). Knowing no fear. *Proceedings of the Royal Society, B: Biology* 266:2451–2456.

Tager-Flusberg, H., & Sullivan, K. (2000). A componential view of theory of mind: Evidence from Williams syndrome. *Cognition* 76:59–89.

Waytz, A., & Mitchell, J. P. (2011). Two mechanisms for simulating other minds: Dissociations between mirroring and self-projection. *Current Directions in Psychological Science* 20(3):197–200.

Wellman, H. (1990). *The Child's Theory of Mind.* Cambridge, MA: MIT Press.

Wicker, B., Keysers, C., Plailly, J., Royet, J-P., Gallese, V., & Rizzolatti, G. (2003). Both of us disgusted in my insula: The common neural basis of seeing and feeling disgust. *Neuron* 40:655–664.

Zaki, J., Hennigan, K., Weber, J., & Ochsner, K. N. (2010). Social cognitive conflict resolution: Contributions of domain-general and domain-specific neural systems. *Journal of Neuroscience* 30:8481–8488.

Zaki, J., & Ochsner, K. N. (2012). The neuroscience of empathy: Progress, pitfalls and promise. *Nature Neuroscience* 15(5):675–680.

CHAPTER 6

The Psychology of Folk Psychology

1. INTRODUCTION

The central mission of cognitive science is to reveal the real nature of the mind, however familiar or foreign that nature may be to naïve preconceptions. The existence of naïve conceptions is also important, however. Prescientific thought and language contain concepts of the mental, and these concepts deserve attention from cognitive science. Just as scientific psychology studies folk physics (McCloskey, 1983; Hayes 1985), viz., the common understanding (or misunderstanding) of physical phenomena, so it must study folk psychology, the common understanding of mental states. This subfield of scientific psychology is what I mean by the phrase "the psychology of folk psychology."

The phrase "folk psychology" often bears a narrower sense than the one intended here. It usually designates a theory about mental phenomena that common folk allegedly hold, a theory in terms of which mental concepts are understood. In the present usage, "folk psychology" is not so restricted. It refers to the ordinary person's repertoire of mental concepts, whether or not this repertoire invokes a theory. Whether ordinary people have a theory of mind (ToM) (in a suitably strict sense of "theory") is controversial, but it is indisputable that they have a folk psychology in the sense of a collection of concepts of mental states. Yet people may not have, indeed, they probably do not have, direct introspective access to the contents (meanings) of their mental concepts, any more than they have direct access to the contents of their concepts of fruit or lying. Precisely for this reason we need cognitive science to discover what those contents are.

The study of folk psychology, then, is part of the psychology of concepts. We can divide the psychology of concepts into two parts: (1) the study of conceptualization and classification in general, and (2) the study of specific folk concepts or families of folk concepts, such as number concepts, material object concepts, and biological kind concepts. The study of folk psychology is a subdivision of (2), the one that concerns mental state concepts. It presupposes that mature speakers have a large number of mentalistic lexemes in their repertoire, such as happy, afraid, want, hope, pain (or hurt), doubt, intend, and so forth. These words are used in construction with other phrases and clauses to generate more complex mentalistic expressions. The question is: What is the meaning, or semantical content, of these mentalistic expressions? What is it that people understand or represent by these words (or phrases)?

This chapter advances two sorts of theses: methodological and substantive. The general methodological thesis is that the best way to study mental state concepts is through the theoretico-experimental methodology of cognitive science. We should consider the sorts of data structures and cognitive operations involved in mentalistic attributions (classifications), both attributions to oneself and attributions to others. Although this proposal is innocuous enough, it is not the methodology that has been followed, or even endorsed in principle, by philosophers, who have given these questions the fullest attention. And even the cognitive scientists who have addressed these questions empirically have not used the specific methodological framework I shall recommend below.

In addition to methodological theses, this chapter will advance some substantive theses, both negative and positive. On the negative side, some new and serious problems will be raised for the functionalist approach to mental-state concepts. Some doubts about pure computationalism will also be raised. On the positive side, the chapter will support a prominent role for phenomenology in our mental-state concepts. These substantive theses will be put forward tentatively because I have not done the kind of experimental work that my own methodological precepts would require for their corroboration; nor does existing empirical research address these issues in sufficient detail. Theoretical considerations, however, lend them preliminary support. I might state at the outset, however, that I am more confident of my negative thesis—about the problems facing the relevant form of functionalism—than my positive theses, especially the role of phenomenology in the propositional attitudes.

2. PROPOSED METHODOLOGY

Philosophical accounts of mental concepts have been strongly influenced by purely philosophical concerns, especially ontological and epistemological ones. Persuaded that materialism (or physicalism) is the only tenable ontology, philosophers have deliberately fashioned their accounts of the mental with an eye to safeguarding materialism. Several early versions of functionalism (Lewis, 1966; Armstrong, 1968, for example) were deliberately designed to accommodate type physicalism, and most forms of functionalism are construed as heavily physicalist in spirit. Similarly, many accounts of mental concepts have been crafted with epistemological goals in mind (e.g., to avoid skepticism about other minds).

According to my view, the chief constraint on an adequate theory of our commonsense understanding of mental predicates is not that it should have desirable ontological or epistemological consequences; rather, it should be psychologically realistic. Its depiction of how people represent and ascribe mental predicates must be psychologically plausible. An adequate theory need not be ontologically neutral, however. As we shall see in the last section, for example, an account of the ordinary understanding of mental terms can play a significant role in arguments about eliminativism. Whatever the ontological ramifications of the ordinary understanding of mental language, however, the nature of that understanding should be investigated purely empirically, without allowing prior ontological prejudices to sway the outcome.

In seeking a model of mental-state ascription (attribution), there are two types of ascriptions to consider: ascriptions to self and ascriptions to others. Here we focus primarily on self-ascriptions. This choice is made partly because I have discussed ascriptions to others elsewhere (Goldman, 1989, 1992, in press), and partly because ascriptions to others, in my view, are "parasitic" on self-ascriptions (although this is not presupposed in the present discussion).

Turning now to specifics, let us assume that a competent speaker/hearer associates a distinctive semantical representation with each mentalistic word, whatever form or structure this representation might take. This (possibly complex) representation, which is stored in long-term memory, somehow bears the "meaning" or other semantical properties associated with the word. Let us call this representation the category representation (CR), because it represents the entire category the word denotes. A CR might take any of several forms (see Smith & Medin, 1981), including the following: (1) a list of features treated as individually necessary and jointly sufficient for the predicate in question; (2) a list of characteristic features with weighted values, where classification proceeds by summing the weights of the instantiated features and determining whether the sum meets a specified criterion; (3) a representation of an ideal instance of the category, to which target instances are compared for similarity;

(4) a set of representations of previously encountered exemplars of the category, to which new instances are compared for similarity; or (5) a connectionist network with a certain vector of connection weights. The present discussion is intended to be neutral with respect to these theories. What interests us, primarily, is the semantical "contents" of the various mentalistic words, or families of words, not the particular "form" or "structure" that bears these contents.

Perhaps we should not say that a CR bears the "meaning" of a mental word. According to some views of meaning, after all, naïve users of a word commonly lack full mastery of its meaning; only experts have such mastery (see Putnam, 1975). But if we are interested in what guides or underpins an ordinary person's use of mental words, we want an account of what he understands or represents by that word. (What the expert knows cannot guide the ordinary person in deciding when to apply the word.) Whether or not this is the "meaning" of the word, it is what we should be after.

Whatever form a CR takes, let us assume that when a cognizer decides what mental word applies to a specified individual, active information about that individual's state is compared or "matched" to CRs in memory that are associated with candidate words. The exact nature of the matching process will be dictated by the hypothesized model of concept representation and categorization. As our present focus is self-ascription of mental terms, we are interested in the representations of one's own mental states that are matched to the CRs. Let us call such an active representation, whatever its form or content, an instance representation (IR). The content of such an IR will be something like "A current state (of mine) has features $f_1, \ldots, f_n$." Such an IR will match a CR having the content: "$f_1, \ldots, f_n$." Our aim is to discover, for each mental word M, its associated CR, or more generally, the sorts of CRs associated with families of mental words. We try to get evidence about CRs by considering what IRs are available to cognizers, IRs that might succeed in triggering a match.

To make this concrete, consider an analogous procedure in the study of visual object recognition; we will use the work of Biederman (1987) as an illustration. Visual object recognition occurs when an active representation of a stimulus that results from an image projected to the retina is matched to a stimulus category or concept category (e.g., chair, giraffe, or mushroom). The psychologist's problem is to answer three coordinated questions: (1) What (high-level) visual representations (corresponding to our IRs) are generated by the retinal image? (2) How are the stimulus categories represented in memory (do these representations correspond to our CRs)? (3) How is the first type of representation matched against the second in order to trigger the appropriate categories?

Biederman hypothesizes that stimulus categories are represented as arrangements of primitive components, viz., volumetric shapes such as cylinders or bricks, which he calls "geons" (for geometrical ions). Object recognition occurs by recovering arrangements of geons from the stimulus image and matching these to one of the distinguishable object models, which is paired with an "entry-level" term in the language (such as lamp, chair, giraffe, and so forth). The theory rests on a range of research supporting the notion that information from the image can be transformed (via edge extraction, etc.) into representations of geons and their relations. Thus, the hypothesis that, say, chair is represented in memory by an arrangement (or several arrangements) of geons is partly the result of constraints imposed by considering what information could be (A) extracted from the image (under a variety of viewing circumstances), and (B) matched to the memory representation. In similar fashion, I wish to examine hypotheses about the stored representations (CRs) of mental-state predicates by reflecting on the instance representations (IRs) of mental states that might actually be present and capable of producing appropriate matches.

Although we have restricted ourselves to self-ascriptions, there are still at least two types of cases to consider: ascriptions of current mental states ("I have a headache now") and ascriptions of past states ("I had a headache yesterday"). Instance representations in the two cases are likely to be quite different, obviously, so they need to be distinguished. Ascriptions of current mental states, however, have a kind of primacy, so these will occupy the center of our attention.

3. PROBLEMS FOR FUNCTIONALISM

In the cognitive scientific, as well as the philosophical, community, the most popular account of people's understanding of mental-state language is the theory of mind (ToM) theory, according to which naïve speakers, even children, have a theory of mental states and understand mental words solely in terms of that theory. The most precise statement of this position is the philosophical doctrine of functionalism. Functionalism says that the crucial or defining feature of any type of mental state is the set of causal relations it bears to (1) environmental or proximal inputs, (2) other types of mental states, and (3) behavioral outputs. (Detailed examples are presented below.) More precisely, because what is at stake is the ordinary understanding of the language of the mental, the doctrine in question is what is generally called analytic or commonsense functionalism. There are other doctrines that also go under the label "functionalism." In particular, there is scientific functionalism (roughly what Block (1978) calls psychofunctionalism), according to which it is a matter of scientific fact that mental states are functional states. That is, mental states have functional properties (i.e., causal relations to inputs, other mental states, and outputs) and should be studied in terms of their functional properties. I shall have nothing to say against scientific functionalism. I do not doubt that mental states have functional properties; nor do I challenge the proposal that mental states should be studied (at least in part) in terms of these properties. But this doctrine does not entail that ordinary people understand or represent mental words as designating functional properties and functional properties only. States designated by mental words might have functional properties without ordinary folk knowing this, or without their regarding it as crucial to the identity of the states. But since we are exclusively concerned here with the ordinary person's representations of mental states, only analytic functionalism is relevant to our present investigation.

Philosophers usually discuss analytic or commonsense functionalism quite abstractly, without serious attention to its psychological realization. I am asking us to consider it as a psychological hypothesis (i.e., a hypothesis about how the cognizer, or his cognitive system, represents mental words). It is preferable, then, to call the type of functionalism in question representational functionalism (RF). This form of functionalism is interpreted as hypothesizing that the CR associated with each mental predicate M represents a distinctive set of functional properties, or functional role, FM. Thus, RF implies that a person will ascribe a mental predicate M to himself when (and only when) an IR occurs in him bearing the message: "Role FM is now instantiated." That is, ascription occurs precisely when there is an IR that matches the functional-role content of the CR for M. (This may be subject to some qualification. Ascription may not require perfect or complete matching between IR and CR; partial matching may suffice.) Is RF an empirically plausible model of mental self-ascription? In particular, do subjects always get enough information about the functional properties of their current states to self-ascribe in this fashion (in real time)?

Before examining this question, let us sketch RF in more detail. The doctrine holds that folk wisdom embodies a theory, or a set of generalizations, that articulates an elaborate network of relations of three kinds: (A) relations between distal or proximal stimuli (inputs) and internal states, (B) relations between internal states and other internal states, and (C) relations between internal states and items of overt behavior (outputs). Here is a sample of such laws due to Churchland (1979). Under heading (A) (relations between inputs and internal states), we might have:

> When the body is damaged, a feeling of pain tends to occur at the point of damage. When no fluids are imbibed for some time, one tends to feel thirsty. When a red apple is present in daylight (and one is looking at it attentively), one will have a red visual experience.

Under heading (B) (relations between internal states and other internal states), we might have:

> Feelings of pain tend to be followed by desires to relieve that pain. Feelings of thirst tend to be followed by desires for potable fluids. If one believes that P, where P elementarily entails Q, one also tends to believe that Q.

Under heading (C) (relations between internal states and outputs), we might have:

> Sudden sharp pains tend to produce wincing. States of anger tend to produce frowning. An intention to curl one's finger tends to produce the curling of one's finger.

According to RF, each mental predicate picks out a state with a distinctive collection, or syndrome, of relations of types (A), (B), and/or (C). The term "pain," for example, picks out a state that tends to be caused by bodily damage, tends to produce a desire to get rid of that state, and tends to produce wincing, groaning, etc. The content of each mental predicate is given by its unique set of relations, or functional role, and nothing else. In other words, RF attributes to people a purely relational concept of mental states.

There are slight variations and important additional nuances in the formulations of functionalism. Some formulations, for example, talk about the causal relations among stimulus inputs, internal states, and behavioral outputs. Others merely talk about transitional relations (i.e., one state following another). Another important wrinkle in an adequate formulation is the subjunctive or counterfactual import of the relations in question. For example, part of the functional role associated with desiring water would be something like this: If a desire for water were accompanied by a belief that a glass of water is within arm's reach, then (other things being equal) it would be followed by extending one's arm. To qualify as a desire for water, an internal state need not actually be accompanied by a belief that water is within reach, nor need it be followed by extending one's arm. It must, however, possess the indicated subjunctive property: If it were accompanied by this belief, the indicated behavior would occur.

We are now in a position to assess the psychological plausibility of RF. The general sort of question I wish to raise is this: Does a subject who self-ascribes a mental predicate always (or even typically) have the sort of instance information required by RF? This is similar to an epistemological question sometimes posed by philosophers, viz., whether functionalism can give an adequate account of one's knowledge of one's own mental state. But the present discussion does not center on knowledge. It merely asks whether the RF model of the CRs and IRs in mental self-ascription is an adequate explanatory model of this behavior. Does the subject always have functional-role information about the target states—functional-role IRs—to secure a "match" with functional-role CRs?

There are three sorts of problems for the RF model. The first is ignorance of causes and effects (or predecessor and successor states). According to functionalism, what makes a mental state a state of a certain type (e.g., a pain, a feeling of thirst, a belief that 7 + 5 = 12, and so forth) is not any intrinsic property it possesses, but rather its relations to other states and events. What makes a state a headache, for example, includes the environmental conditions or other internal states that actually cause or precede it, and its actual effects or successors. There are situations, however, in which the self-ascription of headache occurs in the absence of any information (or beliefs) about relevant causes or effects, predecessors or successors. Surely there are cases in which a person wakes up with a headache and immediately ascribes this type of feeling to himself. Having just awakened, he has no information about the target state's immediate causes or predecessors. Nor need he have any information about its effects or successors. The classification of the state occurs "immediately," without waiting for any further effects, either internal or behavioral, to ensue. There are cases, then, in which self-ascription occurs in the absence of information (or belief) about critical causal relations.

It might be replied that a person need not appeal to actual causes or effects of a target mental state to type identify it. Perhaps he determines the state's identity by its subjunctive properties. This, however, brings us to the second problem confronting the RF model: ignorance of subjunctive properties. How is a person supposed to determine (form beliefs about) the subjunctive properties of a current state (instance or "token")? To use our earlier example, suppose the subject does not believe that a glass of water is within arm's reach. How is he supposed to tell whether his current state would produce an extending of his arm if this belief were present? Subjunctive properties are extremely difficult to get information about, unless the RF model is expanded in ways not yet intimated (a possible expansion is suggested in section 4). The subjunctive implications of

RF, then, are a liability rather than an asset. Each CR posited by RF would incorporate numerous subjunctive properties, each presumably serving as a necessary condition for applying a mental predicate. How is a cognizer supposed to form IRs containing properties that match those subjunctive properties in the CR? Determining that the current state has even one subjunctive property is difficult enough; determining many such properties is formidably difficult. Is it really plausible, then, that subjects make such determinations in type-identifying their inner states? Do they execute such routines in the brief timeframes in which self-ascriptions actually occur? This seems unlikely. I have no impossibility proof, of course, but the burden is on the RF theorist to show how the model can handle this problem.

The third difficulty arises from two central features of functionalism: (1) The type-identity of a token mental state depends exclusively on the type-identity of its relata (i.e., the events that are, or would be, its causes and effects, its predecessors and successors); and (2) the type-identity of an important subclass of a state's relata, viz., other internal states, depends in turn on their relata. To identify a state as an instance of thirst, for example, one might need to identify one of its effects as a desire to drink. Identifying a particular effect as a desire to drink, however, requires one to identify its relata, many of which would also be internal states whose identities are a matter of their relata, and so on. Complexity ramifies very quickly. There is no claim here of any vicious circularity or regress. If functionalism is correct, the system of internal state types is tacked down definitionally to independently specified external states (inputs and outputs) via a set of lawful relations. Noncircular definitions (so-called Ramsey definitions) can be given for each functional state-type in terms of these independently understood input and output predicates (see Lewis, 1970, 1972; Putnam, 1967; Block, 1978; Loar, 1981). The problem I am raising, however, concerns how a subject can determine which functional type a given state-token instantiates. There is a clear threat of combinatorial explosion: Too many other internal states will have to be type-identified in order to identify the target state.

This problem is not easily quantified with precision, because we lack an explicitly formulated and complete functional theory, so we don't know how many other internal states are directly or indirectly invoked by any single functional role. The problem is particularly acute, though, for beliefs, desires, and other propositional attitudes, which, under standard formulations of functionalism, have strongly "holistic" properties. A given belief may causally interact with quite a large number of other belief tokens and desire tokens. To type- identify that belief, it looks as if the subject must track its relations to each of these other internal states, their relations to further internal states, and so on, until each path terminates in an input or an output. When subjunctive properties are added to the picture, the task becomes unbounded, because there are infinitely many possible beliefs and desires. For each desire or goal state, there are indefinitely many beliefs with which it could combine to produce a further desire or subgoal. Similarly, for each belief, there are infinitely many possible desires with which it could combine to produce a further desire or subgoal, and infinitely many other beliefs with which it could combine to produce a further belief. If the type-identification of a target state depends on tracking all of these relations until inputs and outputs are reached, it is clearly unmanageably complex. At a minimum, we can see this as a challenge to an RF theorist, a challenge that no functionalist has tried to meet, and one that looks pretty forbidding.

Here the possibility of partial matching may assist the RF theorist. It is often suggested that visual object identification can occur without the IR completely matching the CR. This is how partially occluded stimuli can be categorized. Biederman (1987, 1990) argues that even complex objects, whose full representation contains six or more geons, are recognized accurately and fairly quickly with the recovery of only two or three of these geons from the image. Perhaps the RF theorist would have us appeal to a similar process of partial matching to account for mental-state classification.

Although this might help a little, it does not get around the fundamental difficulties raised by our three problems. Even if only a few paths are followed from the target state to other internal states and ultimately to inputs and/or outputs, the demands of the task are substantial. Nor does the hypothesis of partial matching address the problem of determining subjunctive properties of the target state. Finally, it does not help much when classification occurs with virtually no information about neighboring states, as in the morning headache example. Thus, the simple RF model of mental self-ascription seems distinctly unpromising.

4. A SECOND FUNCTIONALIST MODEL

A second model of self-ascription is available to the RF theorist, one that assumes, as before, that for each mental predicate there is a functional-role CR. (That's what makes an RF model functional.) The second model, however, tries to explain how the subject determines which functional role a given state token exemplifies without appealing to online knowledge of the state's current relations. How, after all, do people decide which objects exemplify other dispositional properties (e.g., being soluble in water)? They presumably do this by inference from the intrinsic and categorical (i.e., nondispositional) properties of those objects. When a person sees a cube of sugar, he may note that it is white, hard, and granular (all intrinsic properties of the cube), infer from background information that such an object must be made of sugar, and then infer from further background information that it must be soluble in water (because all sugar is soluble in water). Similarly, the RF theorist may suggest that a subject can detect certain intrinsic and categorical properties of a mental state, and from this he can infer that it has a certain functional property (i.e., a suitable relational and dispositional property).

Let us be a bit more concrete. Suppose that the CR for the word headache is the functional-role property F. Further suppose that there is an intrinsic (nonrelational) property E that mental states have, and the subject has learned that any state that has E also has the functional-role property F. Then the subject will be in a position to classify a particular headache as a headache without any excessively demanding inference or computation. He just detects that a particular state token (his morning headache, for example) has property E, and from this he infers that it has F. Finally, he infers from its having F that it can be labeled "headache."

Although this may appear to save the day for RF, it actually just pushes the problem back to what we may call the learning stage. A crucial part of the foregoing account is that the subject must know (or believe) that property E is correlated with property F—that whenever a state has E, it also has F. But how could the subject have learned this? At some earlier time, during the learning stage, the subject must have detected some number of mental states, each of which had both E and F. But during this learning period, he did not already know that E and F are systematically correlated. So he must have had some other way of determining that the E states in question had F. How did he determine that? The original difficulties we cited for identifying a state's functional properties would have been at work during the learning stage, and they would have been just as serious then as we saw them to be in the first model. So the second model of functionalist self-ascription is not much of an improvement (if any) over the first.

In addition, the second model raises a new problem (or question) for RF: What are the intrinsic properties of mental states that might play the role of property E? At this point, let us separate our discussion into two parts, one dealing with what philosophers call sensation predicates (roughly, names for bodily feelings and percepts), and the other dealing with propositional attitudes (believing that p, hoping that q, intending to r, etc.). In this section, we restrict attention to sensation predicates; in section 8, we turn to predicates for propositional attitudes.

What kinds of categorical, nonrelational properties might fill the role of E in the case of sensations? In addition to being categorical and nonrelational, such properties must be accessible to the system that performs the self-ascription. This places an important constraint on the range of possible properties.

There seem to be two candidates to fill the role of E: (1) neural properties, and (2) what philosophers call "qualitative" properties (the "subjective feel" of the sensation). Presumably any sensation state or event has some neural properties that are intrinsic and categorical, but do these properties satisfy the accessibility requirement? Presumably not. Certainly the naïve subject does not have "personal access" (in the sense of Dennett, 1969, 1978) to the neural properties of his sensations. That would occur only if the subject were, say, undergoing brain surgery and watching his brain in a mirror. Normally people don't see their brains; nor do they know much, if anything, about neural hardware. Yet they still identify their headaches without any trouble.

It may be replied that, although there is no personal access to neural information in the ordinary situation, the system performing the self-ascription may have subpersonal access to such information. To exclude neural properties (i.e., neural concepts) from playing the role of E, we need reasons to think that self-ascription does not use these properties of sensations. Now it goes without saying that neural events are involved in the relevant information processing; all information processing in the brain is, at the lowest level, neural processing. The question, however, is whether the contents (meanings) encoded by these neural events are contents about neural properties. This, to repeat, seems quite implausible. Neural events process visual information, but cognitive scientists do not impute neural contents to these neural events. Rather, they consider the contents encoded to be structural descriptions, things like edges and vertices (in low-level vision) or geons (in higher-level vision). When connectionists posit neurally inspired networks in the analysis of, say, language processing, they do not suppose that configurations of connection weights encode neural properties, but rather things like phonological properties.

There is more to be said against the suggestion that self-ascription is performed by purely subpersonal systems that have access to neural properties. Obviously a great deal of information processing does occur at subpersonal levels within the organism. But when the processing is purely subpersonal, no verbal labels seem to be generated that are recognizably "mental." There are all sorts of homeostatic activities in which information is transmitted about levels of certain fluids or chemicals; for example, the glucose level is monitored and then controlled by secretion of insulin. But we have no folk psychological labels for these events or activities. Similarly, there are information-processing activities in low-level vision and in the selection and execution of motor routines. None of these, however, are the subjects of primitive (pre-theoretic) verbal labeling, and certainly not "mentalistic" labeling. This strongly suggests that our spontaneous naming system does not have access to purely subpersonal information. Only when physiological or neurological events give rise to conscious sensations, such as thirst, perceived heat, or the like, does a primitive verbal label get introduced or applied. Thus, although there is subpersonal detection of properties such as excess glucose, these cannot be the sorts of properties to which the mentalistic verbal-labeling system has access.

We seem to be left, then, with what philosophers call "qualitative" properties. According to the standard philosophical view, these are indeed intrinsic, categorical properties that are detected "directly." Thus, the second model of functional self-ascription might hold that in learning to ascribe a sensation predicate like "itch," one first learns the functional role constitutive of that word's meaning (e.g., being a state that tends to produce scratching, and so forth). One then learns that this functional role is realized (at least in one's own case) by a certain qualitative property: itchiness. Finally, one decides that the word is self-ascribable whenever one detects in oneself the appropriate qualitative property, or quale, and infers the instantiation of its correlated functional role. This model still depicts the critical IR as a representation of a functional role, and similarly depicts the CR to which the IR is matched.

We have found a kind of property, then, that might plausibly fill the role of E in the second functionalist model. But is this a model that a true functionalist would welcome? Functionalists are commonly skeptical about qualia (e.g., Harman, 1990; Dennett, 1988, 1991). In particular, many of them wish to deny that there are any qualitative properties if these are construed as intrinsic, nonrelational properties. But this is precisely what the second model of RF requires: that qualitative properties be accepted as intrinsic (rather than functional) properties of mental states. It's not clear, therefore, how attractive the second model would be to many functionalists.

5. A QUALITATIVE MODEL OF SENSATION REPRESENTATION

Furthermore, although the second functionalist model may retain some appeal (despite its problems with the learning stage), it naturally invites a much simpler and more appealing model: one that is wholly nonfunctionalist! Once qualitative properties are introduced into the psychological story, what need is there for functional-role components in the first place? Why not drop the latter entirely? Instead, we hypothesize that both the CR and the IR for each sensation predicate are representations of a qualitative property, such as itchiness (or, as I suggest below, some microcomponents of the property of itchiness). This vastly simplifies our story for both the learning phase and the ascription phase itself. All one learns in the learning phase is the association between the term itch and the feeling of itchiness. At the time of self-ascription, all one detects (or represents) is itchiness, and then matches this IR to the corresponding CR. This is a very natural model for the psychology of sensation self-ascription, at least to anyone free of philosophical prejudice or preconception.

Of course, some philosophers claim that qualitative properties are "queer" and should not be countenanced by cognitive science. There is nothing objectionable about such properties, however, and they are already implicitly countenanced in scientific psychology. One major text, for example, talks of the senses producing sensations of different "quality" (Gleitman, 1981: 172). The sensations of pressure, A-flat, orange, or sour, for example, are sharply different in experienced quality (as Gleitman puts it). This use of the term "quality" refers to differences across the sensory domains, or sense modalities. It is also meaningful, however, to speak of qualitative differences within a modality (e.g., the difference between a sour and a sweet taste). It is wholly within the spirit of cognitive science, then, to acknowledge the existence of qualitative attributes and to view them as potential elements of systems of representation in the mind (see Churchland, 1985).

Although I think that this approach is basically on the right track, it requires considerable refinement. It would indeed be simplistic to suppose that for each word or predicate in the common language of sensation (e.g., itch) there is a simple, unanalyzable attribute (e.g., itchiness) that is the cognitive system's CR for that term. But no such simplistic model is required; most sensory or sensational experience is a mixture or compound of qualities, and this is presumably registered in the contents of CRs for sensations. Even if a person cannot dissect an experience introspectively into its several components or constituents, these components may well be detected and processed by the subsystem that classifies sensations.

Consider the example of pain. Pain appears to have at least three distinguishable dimensional components (see Rachlin, 1985; Campbell, 1985): intensity, aversiveness, and character (e.g., "stinging," "grinding," "shooting," or "throbbing"). Evidence for an intensity/aversiveness distinction is provided by Tursky et al. (1982), who found that morphine altered aversiveness reports from chronic pain sufferers without altering their intensity reports. In other words, although the pain still hurt as much, the subjects didn't mind it as much. Now it may well be that a subject would not, without instruction or training, dissect or analyze his pain into these microcomponents or dimensions. Nonetheless, representations of such components or dimensions could well figure in the CRs for pain and related sensation words; in particular, the subsystem that makes classification decisions could well be sensitive to these distinct components. The situation here is perfectly analogous to the phonological microfeatures of auditory experience that the phonologist postulates as the features used by the system to classify sequences of speech.

Granted that qualitative features (or their microcomponents) play some sort of role in sensation classification, it is (to repeat) quite parsimonious to hypothesize that such features constitute the contents of CRs for mental words. It is much less parsimonious to postulate functional-role contents for these CRs, with qualitative features playing a purely evidential or intermediate role. Admittedly, there are words in the language that do have a functional-style meaning, and their ascriptions must exemplify the sort of multistage process postulated by the complex version of functionalism. Consider the term "can-opener," for example. This probably means something like "a device capable of (or used for) opening cans." To identify something as a can-opener, however, one doesn't have to see it actually open a can. One can learn that objects having certain intrinsic and categorical properties (shape, sharpness, and so on) also thereby exemplify the requisite functional (relational, dispositional) property. So when one sees an object of the right shape (etc.), one classifies it as a can-opener.

Although this is presumably the right story for some words and expressions in the language, it isn't so plausible for sensation words. First, purely syntactic considerations suggest that can-opener is a functional expression, but there is no comparable suggestion of functionality for sensation words. Second, there are familiar difficulties from thought experiments, especially absent-qualia examples such as Block's Chinese nation (Block, 1978). For any functional description of a system that is in pain (or has an itch), it seems as if we can imagine another system with the same functional description but lacking the qualitative property of painfulness (or itchiness). When we do imagine this, we are intuitively inclined to say that the system is not in pain (has no itch). This supports the contention that no purely functional content exhausts the meaning of these sensation words; qualitative character is an essential part of that content.

On a methodological note, I should emphasize that the use of thought experiments, so routine in philosophy, may also be considered (with due caution) a species of psychological or cognitivist methodology, complementary to the methodology described earlier in this chapter. Not only do applications of a predicate to actual cases provide evidence about the correlated CR, but so do decisions to apply or withhold the predicate for imaginary cases. In the present context, reactions to hypothetical cases support our earlier conclusion that qualitative properties are the crucial components of CRs for sensation words.

Quite a different question about the qualitative approach to sensation concepts should now be addressed, viz., its compatibility with our basic framework for classification. This framework says that self-ascription occurs when a CR is matched by an IR, where an IR is a representation of a current mental state. Does it make sense, however, to regard an instance of a qualitative property as a representation of a mental state? Isn't it more accurate to say that it is a mental state, not a representation thereof? If we seek a representation of a mental state, shouldn't we look for something entirely distinct from the state itself (or any feature thereof)?

Certainly the distinction between representations and what they represent must be preserved. The problem can be avoided, however, by a minor revision in our framework. On reflection, self-ascription does not require the matching of an IR to a CR; it can involve the matching of an instance itself to a CR. The term "instance representation" was introduced because we wanted to allow approaches like functionalism, in which the state itself is not plausibly matched to a CR, only a representation of it. Furthermore, we had in mind the analogy of perceptual categorization, where the cognizer does not match an actual stimulus to a mental representation of the stimulus category, but an inner representation of the stimulus to a CR. In this respect, however, the analogy between perceptual recognition and sensation recognition breaks down. In the case of sensation, there can be a matching of the pain itself, or some features of the pain, to a stored structure containing representations of those features. Thus, we should revise our framework to say that categorization occurs when a match is effected between (1) a category representation, and (2) either (A) a suitable representation of a state, or (B) a state itself. Alternative (B) is especially plausible in the case of sensations because it is easy to suppose that CRs for sensations are simply memory "traces" of those sensations, which are easily activated by reoccurrences of those same (or similar) sensations.

This new picture might look suspicious because it seems to lead to the much-disparaged doctrines of infallibility and omniscience about one's own mental states. If a CR is directly matched to an instance of a sensation itself, isn't all possibility of error precluded? And won't it be impossible to be unaware of one's sensations, because correct matching is inevitable? Yet surely both error and ignorance are possible.

In fact, the proposed change implies neither infallibility nor omniscience. The possibility of error is readily guaranteed by introducing an assumption (mentioned earlier) of partial matching. If a partial match suffices for classification and self-ascription, there is room for inaccuracy. If we hypothesize that the threshold for matching can be appreciably lowered by various sorts of "response biases" (such as prior expectation of a certain sensation), this makes error particularly easy to accommodate. Ignorance can be accommodated in a different way, by supplementary assumptions about the role of attention. When attentional resources are devoted to other topics, there may be no attempt to match certain sensations to any CR. Even an itch or a pain can go unnoticed when attention is riveted on other matters. Mechanisms of selective attention are critical to a full story of classification, but this large topic cannot be adequately addressed here.

Even with these added points, some readers might think that our model makes insufficient room for incidental cognitive factors in the labeling of mental states. Doesn't the work on emotions by Schachter & Singer (1962), for example, show that such incidental factors are crucial? My first response is that I am not trying to address the complex topic of emotions, but restricting attention to sensations (in this section) and propositional attitudes (in section 8). Second, there are various ways of trying to accommodate the Schachter-Singer findings. One possibility, for example, is to say that cognitive factors influence which emotion is actually felt (e.g., euphoria or anger), rather than the process of labeling or classifying the felt emotion (see M. Wilson, 1991). So it isn't clear that the Schachter-Singer study would undermine the model proposed here, even if this model were applied to emotions (not my present intent).

6. THE CLASSICAL FUNCTIONALIST ACCOUNT OF SELF-ASCRIPTION

Classical functionalists, such as Putnam (1960); Sellars (1956); and Shoemaker (1975), have not been oblivious to the necessity of making room in their theory for self-ascriptions or self-reports of mental states. They thought this could be done without recourse to anything like qualitative properties. How, then, does their theory go, and why do I reject it?

According to the classical account, it is part of the specification of a mental state's functional role that having the state guarantees a self-report of it; or, slightly better, that it is part of the functional specification of a mental state (e.g., pain) that it gives rise to a belief that one is in that state (Shoemaker, 1975). If one adopts the general framework of RF that I have presented, however, it is impossible to include this specification. Let me explain why.

According to our framework, a belief that one is in state M occurs precisely when a match occurs between a CR for M and a suitable IR. (As we are now discussing functionalism again, we needn't worry about "direct" matching of state to CR.) But classical functionalism implies that part of the concept of being in state M (a necessary part) is having a belief that one is in M. Thus, no match can be achieved until the system has detected the presence of an M belief. However, to repeat, what an M belief is, according to our framework, is the occurrence of a match between the CR for M and an appropriate IR. Thus, the system can only form a belief that it is in M (achieve an IR-CR match) by first forming a belief that it is in M! Obviously this is impossible.

What this point shows is that there is an incompatibility between our general framework and classical functionalism. They cannot both be correct. But where does the fault lie? Which should be abandoned?

A crucial feature of classical functionalism is that it offers no story at all about how a person decides what mental state he is in. Being in a mental state just automatically entails, or gives rise to, the appropriate belief. Precisely this assumption of automaticity has until now allowed functionalism to ignore the sorts of questions raised in this chapter. In other words, functionalism has hitherto tended to assume some sort of "nonrecognitional" or "noncriterial" account of self reports. Allegedly, you don't use any criterion (e.g., the presence of a qualitative property) to decide what mental state you are in. Classification of a present state does not involve the comparison of present information with anything stored in long-term memory. Just being in a mental state automatically triggers a classification of yourself as being in that state.

It should be clear, however, that this automaticity assumption cannot and should not be accepted by cognitive science, for it would leave the process of mental state classification a complete mystery. It is true, of course, that we are not introspectively aware of the mechanism by which we classify our mental states. But we are likewise not introspectively aware of the classification processes associated with other verbal labeling, the labeling of things as birds or chairs, or as leapings or strollings. Lack of introspective access is obviously no reason for cognitive scientists to deny that there is a microstory of how we make—or how our systems make—mentalistic classifications. There must be some way a system decides to say (or believe) that it is now in a thirst state rather than a hunger state, that it is hoping for a rainy day rather than expecting a rainy day. That is what our general framework requires. In short, in a choice between our general framework and classical functionalism (with its assumption of automatic self-report), cognitive science must choose the former. Any tenable form of functionalism, at least any functionalism that purports to explain the content of naïve mental concepts, must be formulated within this general framework. That is just how RF has been formulated. It neither assumes automaticity of classification nor does it create a vicious circularity (by requiring the prior detection of a classification event—a belief—as a necessary condition for classification). So RF is superior to classical functionalism for the purposes at hand. Yet RF, we have seen, has serious problems of its own. Thus, the only relevant form of functionalism is distinctly unpromising.

7. DUAL REPRESENTATIONS

We have thus far assumed that there is a single CR (however complex) paired with each mental word. This assumption, however, is obviously too restrictive. Indeed, there are many lines of research that suggest the possibility of dual representations for a single word. For example, Biederman hypothesizes that a single word (e.g., piano) is associated with two or more visual object models (e.g., a model for a grand piano and a model for an upright). In addition, however, there may be a wholly nonvisual representation associated with "piano" (e.g., a keyboard instrument that has a certain characteristic sound). Some neuropsychological evidence for dual representations has been provided by Warrington and colleagues. Warrington & Taylor (1978) studied patients with right-hemisphere and left-hemisphere lesions. They were given two sorts of tasks: one involving perceptual categorization by physical identity, and the other involving semantic categorization by functional identity. The right-hemisphere group showed impairment on the perceptual categorization task. The left-hemisphere group showed no impairment on this task, but they did show impairment on the second task. For example, they could not pair (pictures of) two deckchairs that were physically dissimilar but had the same function. Warrington and Taylor accordingly postulate two anatomically separate, postsensory categorical stages in object recognition that occur in sequence. In a somewhat similar vein, Cooper, Schacter, and their colleagues have found evidence of distinct forms of representation of visual objects: a purely structural representation stored in implicit memory, and a semantical/functional representation available to explicit memory (Cooper, 1991; Schacter, Cooper, & Delaney, 1990; Schacter, Cooper, Delaney, et al., 1991). Other types of evidence or arguments for dual representations have been given by Landau, 1981; McNamara & Sternberg, 1983; Putnam, 1975; and Rey, 1983.

If the idea of dual representations is applied to sensation terms, or mental terms generally, it might support a hybrid theory that features both a qualitative representation and an independent, nonqualitative representation, which might be functional. The qualitative representation could accommodate self-ascriptions and thereby avert the problems we posed for pure functionalism. However, it is not clear how happy functionalists would be with this solution. As we have remarked, most functionalists are skeptics about qualia. Any dual- representation theory that invokes qualitative properties (not functionally reconstructed) is unlikely to find favor with them.

Furthermore, functionalist accounts of mental concepts face another difficulty, even if the functional representation is only one member of a subject's pair of representations. According to functional-style definitions of mental terms, the meaning of each term is fixed by the entire theory of functional relations in which it appears (see Lewis, 1972; Block, 1978; Loar, 1981). This implies that every little change anywhere in the total theory—every addition of a new law or revision of an old law—entails a new definition of each mentalistic expression. Even the acquisition of a single new mental predicate requires amending the definition of every other such predicate, because a new predicate introduces a new state type that expands the set of relations in the total theory. This holistic feature of functionalism entails all-pervasive changes in one's repertoire of mental concepts. Such global changes threaten to be as computationally intractable as the familiar "frame problem" (McCarthy & Hayes, 1969), especially because there is a potential infinity of mental concepts, owing to the productivity of "that clauses" and similar constructions. The problem of conceptual change for theory-embedded concepts is acknowledged and addressed by Smith, Carey, & Wiser (1985), but they are concerned with tracking descent for individual concepts, whereas the difficulty posed here is the computational burden of updating the vast array of mental concepts implicated by any theoretical revision.

8. A PHENOMENOLOGICAL MODEL FOR THE ATTITUDES?

Returning to my positive theory, I have thus far only proposed a qualitative approach to sensation concepts. Let us turn now from sensations to propositional attitudes (e.g., believing, wanting, intending, and so forth). This topic can be divided into two parts (Fodor, 1987): the representation of attitude types, and the representation of attitude contents. Wanting there to be peace and believing there will be peace are different attitudes because their types (wanting and believing) are different. Intending to go shopping and intending to go home are different attitudes because, although their type is the same, their contents differ. In this section, we consider how attitude types are represented; in the next, we consider attitude contents.

Philosophical orthodoxy favors a functionalist approach to attitude types. Even friends of qualia (e.g., Block, 1990a) feel committed to functionalism when it comes to desire, belief, and so forth. Our earlier critiques of functionalism, however, apply with equal force here. Virtually all of our antifunctionalist arguments (except the absent-qualia arguments) apply to all types of mental predicates, not just to sensation predicates. So there are powerful reasons to question the adequacy of functionalism for the attitude types. How, then, do people decide whether a current state is a desire rather than a belief, a hope rather than a fear?

In recent literature, some philosophers use the metaphor of "boxes" in the brain (Schiffer, 1981). To believe something is to store a sentence of mentalese in one's "belief box"; to desire something is to store a sentence of mentalese in one's "desire box"; and so on. Should this metaphor be taken seriously? I doubt it. It is unlikely that there are enough "boxes" to have a distinct one for each attitude predicate. Even if there are enough boxes in the brain, does the ordinary person know enough about these neural boxes to associate each attitude predicate with one of them (the correct one)? Fodor (1987) indicates that box-talk is just shorthand for treating the attitude types in functional terms. If so, this just reintroduces the forbidding problems already facing functionalism.

Could a qualitative or phenomenological approach work for the attitude types? The vast majority of philosophers reject this approach out of hand, but this rejection is premature. I shall adduce several tentative arguments in support of this approach.

First, a definitional point: The terms qualia and qualitative are sometimes restricted to sensations (percepts and somatic feelings), but we shouldn't allow this to preclude the possibility of other mental events (beliefs, thoughts, etc.) having a phenomenological or experiential dimension. Indeed, at least two cognitive scientists (Jackendoff, 1987; Baars, 1988) have recently defended the notion that "abstract" or "conceptual" thought often occupies awareness or consciousness, even if it is phenomenologically "thinner" than modality-specific experience. Jackendoff appeals to the tip-of-the-tongue phenomenon to argue that phenomenology is not confined to sensations. When one tries to say something but can't think of the word, one is phenomenologically aware of having requisite conceptual structure, that is, of having a determinate thought-content one seeks to articulate. What is missing is the phonological form: the sound of the sought-for word. The absence of this sensory quality, however, does not imply that nothing (relevant) is in awareness. Entertaining the conceptual unit has a phenomenology, just not a sensory phenomenology.

Second, in defense of phenomenal "parity" for the attitudes, I present a permutation of Jackson's (1982, 1986) argument for qualia (cf. Nagel, 1974). Jackson argues that qualitative information is a kind that cannot be captured in physicalist (including functionalist) terms. Imagine, he says, that a brilliant scientist named Mary has lived from birth in a cell where everything is black, white, or gray. (Even she herself is painted all over.) She watches a black-and-white television, reads books, engages in discussions, and observes experiments. Suppose that by this means, Mary learns all physical and functional facts concerning color, color vision, and the brain states produced by exposure to colors. Does she therefore know all facts about color? There is one kind of fact about color perception, says Jackson, of which she is ignorant: what it is like (i.e., what it feels like) to experience red, green, etc. These qualitative sorts of facts she will come to know only if she actually undergoes spectral experiences.

Jackson's example is intended to dramatize the claim that there are subjective aspects of sensations that resist capture in functionalist terms. I suggest a parallel style of argument for attitude types. Just as someone deprived of any experience of colors would learn new things upon being exposed to them, viz., what it feels like to see red, green, and so forth, so (I submit) someone who had never experienced certain propositional attitudes (e.g., doubt or disappointment) would learn new things on first undergoing these experiences. There is "something it is like" to have these attitudes, just as much as there is "something it is like" to see red. In the case of the attitudes, just as in the case of sensations, the features to which the system is sensitive may be microfeatures of the experience. This still preserves parity with the model for sensations.

My third argument is from the introspective discriminability of attitude strengths. Subjects' classificational abilities are not confined to broad categories such as belief, desire, and intention; they also include intensities thereof. People report how firm is their intention or conviction, how much they desire an object, and how satisfied or dissatisfied they are with a state of affairs. Whatever the behavioral predictive power of these self-reports, their very occurrence needs explaining. Again, the functionalist approach seems fruitless. The other familiar device for conceptualizing the attitudes—viz., the "boxes" in which sentences of mentalese are stored—would also be unhelpful even if it were separated from functionalism, because box storage is not a matter of degree. The most natural hypothesis is that there are dimensions of awareness over which scales of attitude intensity are represented.

The importance of attitude strength is heightened by the fact that many words in the mentalistic lexicon ostensibly pick out such strengths. Certain, confident, and doubtful represent positions on a credence scale; delighted, pleased, and satisfied represent positions on a liking scale. Because we apparently have introspective access to such positions, self-ascription of these terms invites an introspectivist account (or a quasi-introspectivist account that makes room for microfeatures of awareness).

One obstacle to a phenomenological account of the attitudes is that stored (or dispositional) beliefs, desires, and so on are outside awareness. However, there is no strain in the suggestion that the primary understanding of these terms stems from their activated ("occurrent") incarnations; the stored attitudes are just dispositions to have the activated ones.

A final argument for the role of phenomenology takes its starting point from still another trouble with functionalism, a trouble not previously mentioned here. In addition to specific mental words like hope and imagine, we have the generic word "mental." Ordinary people can classify internal states as mental or non-mental. Notice, however, that many non-mental internal states can be given a functional-style description. For example, having measles might be described as a state that tends to be produced by being exposed to the measles virus, which produces an outbreak of red spots on the skin. So having measles is a functional state; clearly, though, it isn't a mental state. Thus, functionalism cannot fully discharge its mission simply by saying that mental states are functional states; it also needs to say which functional states are mental. Does functionalism have any resources for marking the mental/non-mental distinction? The prospects are bleak. By contrast, a plausible-looking hypothesis is that mental states are states having a phenomenology, or an intimate connection with phenomenological events. This points us again in the direction of identifying the attitudes in phenomenological terms.

Skepticism about this approach has been heavily influenced by Wittgenstein (1953, 1967), who questioned whether there is any single feeling or phenomenal characteristic common to all instances of an attitude like intending or expecting. (A similar worry about sensations is registered by Churchland & Churchland, 1981.) Notice, however, that our general approach to concepts does not require there to be a single defining characteristic for each mentalistic word. A CR might be, for example, a list of exemplars (represented phenomenologically) associated with the word, to which new candidate instances are compared for similarity. Thus, even if Wittgenstein's (and the Churchlands') worries about the phenomenological unity of mental concepts are valid, this does not exclude a central role for phenomenological features in CRs for attitude words.

9. CONTENT AND COMPUTATIONALISM

The commonsense understanding of the contents of the propositional attitudes is an enormous topic; we shall touch on it here only lightly. The central question concerns the "source" of contentfulness for mental states. Recent theories have tended to be externalist, claiming that content arises from causal or causal-historical interactions between inner states (or symbols in the language of thought) and external objects. It is highly doubtful, however, whether any of the most developed externalist theories gives an adequate account of the naïve cognizer's understanding or representation of content.

Fodor currently advocates a complex causal-counterfactual account (Fodor, 1987, 1990). Roughly, a mental symbol C means cow if and only if (1) C tokens are reliably caused by cows, and (2) although non-cows (e.g., horses) also sometimes cause C tokens, non-cows wouldn't cause C tokens unless cows did, whereas it is false that cows wouldn't cause C tokens unless non-cows did. Clause (2) of this account is a condition of "asymmetric dependence," according to which there being non-cow-caused C tokens depends on there being cow-caused C tokens, but not conversely. It seems most implausible, however, that this sort of criterion for the content of a mental symbol is what ordinary cognizers have in mind. Similarly implausible for this purpose are Millikan's evolutionary account of mental content (Millikan, 1984, 1986) and Dretske's learning-theoretic (i.e., operant conditioning) account of mental content (1988). Most naïve cognizers have never heard of operant conditioning, and many do not believe in evolution. Nevertheless, these same subjects readily ascribe belief contents to themselves. So did our sixteenth-century ancestors, who never dreamt of the theory of evolution or operant conditioning. (For further critical discussion, see Cummins, 1989.) Millikan and Dretske probably do not intend their theories as accounts of the ordinary understanding of mental contents. Millikan (1989), for one, expressly disavows any such intent. But then we are left with very few detailed theories that do address our question. Despite the popularity of externalist theories of content, they clearly pose difficulties for self-ascription. Cognizers seem able to discern their mental contents—what they believe, desire, or plan to do—without consulting their environment.

What might a more internalist approach to contents look like? Representationalism, or computationalism, maintains that content is borne by the formal symbols of the language of thought (Fodor, 1975, 1981, 1987; Newell & Simon, 1976). But even if the symbolic approach gives a correct de facto account of the working of the mind, it does not follow that the ordinary concept of mental content associates it with formal symbols per se. I would again suggest that phenomenological dimensions play a crucial role in our naïve view. Only what we are aware or conscious of provides the primary locus of mental content.

For example, psycholinguists maintain that in sentence processing, there are commonly many interpretations of a sentence that are momentarily presented as viable, but we are normally aware of only one: the one that gets selected (Garrett, 1990). The alternatives are "filtered" by the processing system outside of awareness. Only in exceptional cases, such as "garden path" sentences (e.g., "Fatty weighed 350 pounds of grapes"), do we become aware of more than one considered interpretation. Our view of mental content is, I suggest, driven by the cases of which we are aware, although they may be only a minority of the data structures or symbolic structures that occupy the mind.

Elaboration of this theme is not possible in the present chapter, but brief comment about the relevant conception of "awareness" is in order. Awareness, for these purposes, should not be identified with accessibility to verbal report. We are often aware of contents that we cannot (adequately) verbalize, either because the type of content is not easily encoded in linguistic form or because its mode of cognitive representation does not allow full verbalization. The relevant notion of awareness, or consciousness, then, may be that of qualitative or phenomenological character (there being "something it is like") rather than verbal reportability (see Block, 1990b, 1991).

The role I am assigning to consciousness in our naïve conception of the mental bears some similarity to that assigned by Searle (1990). Unlike Searle, however, I see no reason to decree that cognitive science cannot legitimately apply the notion of content to states that are inaccessible (even in principle) to consciousness. First, it is not clear that the ordinary concept of a mental state makes consciousness a "logical necessity" (as Searle says). Second, even if mental content requires consciousness, it is inessential to cognitive science that the non-conscious states to which contents are ascribed should be considered mental. Let them be "psychological" or "cognitive" rather than "mental"; this doesn't matter to the substance of cognitive science. Notice that the notion of content in general is not restricted to mental content; linguistic utterances and inscriptions are also bearers of content. So even if mental content is understood to involve awareness, this places no constraints of the sort Searle proposes on cognitive science.

10. EMPIRICAL RESEARCH ON THE THEORY-THEORY

The idea underlying functionalism, that the naïve cognizer has a theory of mind (ToM), goes increasingly by the label "the theory-theory" (TT) (Morton, 1980). Although originally proposed by philosophers, this idea is now endorsed by a preponderance of empirical researchers, especially developmental psychologists and cognitive anthropologists. Does their research lend empirical support to the functionalist account of mental concepts?

Let us be clear about exactly what we mean by functionalism, especially the doctrine of RF that concerns us here. There are two crucial features of this sort of view. The first feature is pure relationalism. RF claims that the way subjects represent mental predicates is by relations to inputs, outputs, and other internal states. The other internal-state concepts are similarly represented. Thus, every internal-state concept is ultimately tied to external inputs and outputs. What is deliberately excluded from our understanding of mental predicates, according to RF, is any reference to the phenomenology or experiential aspects of mental events (unless these can be spelled out in relationalist terms). No "intrinsic" character of mental states is appealed to by RF in explaining the subject's basic conception or understanding of mental predicates. The second crucial feature of RF is the appeal to nomological (law-like) generalizations in providing the links between each mental-state concept and suitably chosen inputs, outputs, and other mental states. Thus, if subjects are to exemplify RF, they must mentally represent laws of the appropriate sort. Does empirical research on ToM support either of these two crucial features? Let us review what several leading workers in this tradition say on these topics. We shall find that very few of them, if any, construe ToM in quite the sense specified here. They usually endorse vaguer and weaker views.

Premack & Woodruff (1978), for example, say that an individual has a ToM if he simply imputes mental states to himself and others. Ascriptions of mental states are regarded as "theoretical" merely because such states are not directly observable (in others) and because such imputations can be used to make predictions about the behavior of others. This characterization falls short of RF because it does not assert that the imputations are based on law-like generalizations, and it does not assert that mental-state concepts are understood solely in terms of relations to external events.

Wellman (1988, 1990) also conceives of TT quite weakly. A body of knowledge is theory-like, he says, if it has (1) an interconnected ("coherent") set of concepts, (2) a distinctive set of ontological commitments, and (3) a causal-explanatory network. Wellman grants that some characterizations of theories specify commitments to nomological statements, but his own conception explicitly omits that provision (Wellman. 1990: chap. 5). This is one reason why his version of TT falls short of RF. A second reason is that Wellman explicitly allows that the child's understanding of mind is partly founded on firsthand experience: "The meanings of such terms/constructs as belief, desire and dream may be anchored in certain firsthand experiences, but by age three children have not only the experiences but the theoretical constructs" (Wellman, 1990: 195). Clearly, then, Wellman's view is not equivalent to RF, and the evidence he adduces for his own version of TT is not sufficient to support RF.

Similarly, Rips & Conrad (1989) present evidence that a central aspect of people's beliefs about the mind is that mental activities are interrelated, with some activities being kinds or parts of others. For example, reasoning is a kind of thinking and a part of problem- solving. The mere existence of taxonomies and partonomies (part-whole hierarchies), however, does not support RF, because mental terms could still be represented in introspective terms, and such taxonomies may not invoke laws.

D'Andrade (1987) also describes the "folk model of the mind" as an elaborate taxonomy of mental states, organized into a complex causal system. This is no defense of functionalism, however, as D'Andrade expressly indicates that concepts like emotion, desire, and intention are "primarily defined by the conscious experience of the person" (1987: 139). The fact that laymen recognize causal relations among mental events does not prove that they have a set of laws. Whether or not belief in causal relations requires belief in laws is a controversial philosophical question. Nor does the fact that people use mental concepts to explain and predict the behavior of others imply the possession of laws, as we shall see below.

The TT approach to mental concepts is, of course, part of a general movement toward understanding concepts as theory-embedded (Carey, 1985; Gopnik, 1993; Karmiloff-Smith & Inhelder, 1975; Keil, 1989; Murphy & Medin, 1985). Many proponents of this approach acknowledge, however, that their construal of "theory" is quite vague, or remains to be worked out. For example, Murphy & Medin (1985: 290) simply characterize a theory as "a complex set of relations between concepts, usually with a causal basis"; and Keil (1989: 279–280) says: "So far we have not made much progress on specifying what naïve theories must look like or even what the best theoretical vocabulary is for describing them." Thus, commitment to a TT approach does not necessarily imply commitment to RF in the mental domain; nor would evidential corroboration of a TT approach necessarily corroborate RF.

A detailed defense of TT is given by Gopnik, who specifically rejects the classical view of direct or introspective access to one's own psychological states. However, even Gopnik's view is significantly qualified, and her evidential support is far from compelling. First, although her main message is the rejection of an introspective or "privileged access" approach to self-knowledge of mental states, she acknowledges that we use some mental vocabulary "to talk about our phenomenologically internal experiences, the Joycean or Woolfean stream of consciousness, if you will." This does not sound like RF. Second, Gopnik seems to concede privileged access, or at least errorless performance, for subjects' self-attributions of current mental states. At any rate, all of her experimental data concern self-attributions of past mental states: She nowhere hints that subjects make mistakes about their current states as well. But how can errorless performance be explained based on her favored inferential model of self-attribution? If faulty theoretical inference is rampant in children's self-attribution of past states, why don't they make equally faulty inferences about their current states? Third, there is some evidence that children's problems with reporting their previous thoughts are just a problem of memory failure. Mitchell and Lacohee (1991) found that such memory failure could be largely alleviated with a little help. Fourth, Gopnik's TT does not explain very satisfactorily why children perform well on self-attributions of past pretense and imaging. Why are their inferences so much more successful for those mental states than for beliefs? Finally, how satisfactory is Gopnik's explanation of the "illusion" of first-person privileged access? If Gopnik were right that this illusion stems from expertise, why shouldn't we have the same illusion in connection with attribution of mental states to others? If people were similarly positioned vis-à-vis their own mental states and those of others, they would be just as expert for others as for themselves, and should develop analogous illusions. But there is no feeling of privileged access to others' mental states.

At this point, the tables might be turned on us. How are we to account for attributions to others if subjects don't have a theory (i.e., a set of causal laws) to guide their attributions? An alternative account of how such attributions might be made is the "simulation" or role-taking theory (Goldman, 1989, 1992a, 1992b ; Gordon, 1986, 1992; Harris 1989, 1991, 1992; Johnson, 1988), according to which a person can predict another person's choices or mental states by first imagining himself in the other person's situation and then determining what he himself would do or how he would feel. For example, to estimate how disappointed someone will feel if he loses a certain tennis match or does poorly on a certain exam you might project yourself into the relevant situation and see how you would feel. You don't need to know any psychological laws about disappointment to make this assessment. You just need to be able to feed an imagined situation as input into some internal psychological mechanism that then generate a relevant output state. Your mechanism can "model" or mimic the target agent's mechanism even if you don't know any laws describing these mechanisms.

To compete with TT, the simulation theory (ST) must do as well in accounting for the developmental data, such as three-year-olds' difficulties with false-belief ascriptions (Wimmer & Perner, 1983; Astington, Harris, & Olson, 1988). Defenders of TT usually postulate a major change in children's theory of mind: from a primitive theory—variously called a "copy theory" (Wellman, 1990), a "Gibsonian theory" (Astington & Gopnik, 1991), a "situation theory" (Perner, 1991), or a "cognitive connection theory" (Flavell, 1988)—to a full representational theory. Defenders of ST might explain these developmental data in a different fashion, by positing not fundamental changes of theory but increases in flexibility of simulation (Harris, 1992). Three- year-olds have difficulty in imagining states that run directly counter to their own current states; but by age four, children's imaginative powers overcome this difficulty. ST also comports well with early propensities to mimic or imitate the attitudes or actions of others, such as joint visual attention and facial imitation (Butterworth, 1991; Meltzoff & Moore, 1977, 1983; Harris, 1992; Goldman, 1992a). Thus, ST provides an alternative to TT in accounting for attributions of mental states to others.

11. PSYCHOLOGICAL EVIDENCE ABOUT INTROSPECTION AND THE ROLE OF CONSCIOUSNESS

The positive approach to mental concepts I have tentatively endorsed has much in common with the classical doctrine of introspectionism. Doesn't this ignore empirical evidence against introspective access to mental states? The best-known psychological critique of introspective access is Nisbett & Wilson's (1977); let us briefly review the question of how damaging that critique is and where the discussion now stands.

The first sentence of Nisbett and Wilson's abstract reads: "Evidence is reviewed which suggests that there may be little or no direct introspective access to higher order cognitive processes" (1977: 231). At first glance, this suggests a sweeping negative thesis. What they mean by "process," however, is causal process; and what their evidence really addresses is people's putative access to the causes of their behavior. This awareness-of-causes thesis, however, is one that no classical introspectionist, to my knowledge, has ever asserted. Moreover, Nisbett and Wilson explicitly concede direct access to many or most of the private states that concern us here and that concern philosophy of mind in general:

> We do indeed have direct access to a great storehouse of private knowledge.... The individual knows a host of personal historical facts; he knows the focus of his attention at any given point of time; he knows what his current sensations are and has what almost all psychologists and philosophers would assert to be "knowledge" at least quantitatively superior to that of observers concerning his emotions, evaluations, and plans. (Nisbett & Wilson, 1977: 255)

Their critique of introspectionism, then, is hardly as encompassing as it first appears (or as citations often suggest). As White (1988) remarks, "causal reports could turn out to be a small island of inaccuracy in a sea of insight" (37).

Nisbett and Wilson's paper reviewed findings from several research areas, including attribution, cognitive dissonance, subliminal perception, problem-solving, and bystander apathy. Characteristically, the reported findings were of manipulations that produced significant differences on behavioral measures, but not on verbal self-report measures. In Nisbett and Wilson's position effect study, for example, passersby appraised four identical pairs of stockings in a linear display and chose the pair they judged to be of best quality. The results showed a strong preference for the rightmost pair. Subjects did not report that position had influenced their choice and vehemently denied any such effect when the possibility was mentioned.

However, as Bowers (1984) points out, this sort of finding is not very damaging to any sensible form of introspectionism. As we have known since Hume (1748), causal connections between events cannot be directly observed, nor can they be introspected. A sensible form of introspectionism, therefore, would not claim that people have introspective access to causal connections. But this leaves open the idea that they do have introspective access to the mere occurrence of certain types of mental events.

Other critics, such as Ericsson & Simon (1980), complain that Nisbett and Wilson fail to investigate or specify the conditions under which subjects are unable to make accurate reports. Ericsson & Simon (1980, 1984) themselves develop a detailed model of the circumstances in which verbal reports of internal events are likely to be accurate. In particular, concurrent reports about information that is still in short-term memory (STM) and fully attended are more likely to be reliable than retrospective reports. In most of the studies reviewed by Nisbett and Wilson, however, the time lag between task and probe was sufficiently great to make it unlikely that relevant information remained in STM. A sensible form of introspectionism would restrict the thesis of privileged access to current states and not extend it to past mental events. Of course, people often do have long-term memories of their past mental events. But their direct access is then to these memories, not to the original mental events themselves.

In more recent work, one of the two authors, T. D. Wilson, has been very explicit in accepting direct access. He writes: "[P]eople often have direct access to their mental states, and in these cases the verbal system can make direct and accurate reports. When there is limited access, however, the verbal system makes inferences about what these processes and states might be" (Wilson, 1985: 16). He then explores four conditions that foster imperfect access, with the evident implication that good access is the default situation. This sort of position is obviously quite compatible with the one advocated in the present chapter.

With its emphasis on conscious or phenomenological characteristics, the present chapter appears to be challenged by Velmans (1991), who raises doubts about the role of consciousness in focal-attentive processing, choice, learning and memory, and the organization of complex, novel responses. His target article seems to conjecture that consciousness does not enter causally into human information-processing at all.

However, as many of his commentators point out, Velmans's evidence does not support this conclusion. Block (1991) makes the point particularly clearly. Even if Velmans is right that consciousness is not required for any particular sort of information-processing, it does not follow that consciousness does not in fact figure causally. Block also sketches a plausible model, borrowed from Schacter (1989), in which consciousness does figure causally. In the end, this debate may be misplaced, because Velmans, in his response to commentators, says that he didn't mean to deny that consciousness has causal efficacy.

Velmans's views aside, the position sketched in the present chapter invites questions about how, exactly, qualitative or phenomenological properties can figure in the causal network of the mind. This raises large and technical issues pertaining not only to cognitive science but also to philosophical questions about causation, reduction and identity, supervenience, and the like. Such issues require an entirely separate address (or more than one) and cannot be discussed here.

12. ONTOLOGICAL IMPLICATIONS OF FOLK PSYCHOLOGY

These technical issues aside, the content of folk psychology may have significant ontological implications about, for example, the very existence of the mental states for which common speech provides labels. Eliminativists maintain that there are no such things as beliefs, desires, thoughts, hopes, and other such intentional states (Churchland, 1981; Stich, 1983). They are all, like phlogiston, caloric, and witches, the mistaken posits of a radically false theory, in this case, a commonsense theory. The argument for eliminativism proceeds from the assumption that there is a folk theory of mind. The present chapter counters this form of argument by denying (quite tentatively, to be sure) that our mental concepts rest on a folk theory.

It should be stressed that the study of folk psychology does not by itself yield ontological consequences. It just yields theses of the form: "Mental (or intentional) states are ordinarily conceptualized as states of kind K." This sort of thesis, however, may appear as a premise in arguments with eliminativist conclusions, as we have just seen. If kind K features nomological relations R, for example, one can defend eliminativism by holding that no states actually instantiate relations R. On the other hand, if K just includes qualitative or phenomenological properties, it is harder to see how a successful eliminativist argument could be mounted. One would have to hold that no such properties are instantiated or even exist. Although qualitative properties are indeed denied by some philosophers of mind (e.g., Dennett, 1988, 1991; Harman, 1990), the going is pretty tough (for a reply to Harman, see Block, 1990a). The general point, however, should be clear. Although the study of folk psychology does not directly address ontological issues, it is indirectly quite relevant to such issues.

Apart from ontological implications, the study of folk psychology also has intrinsic interest, as an important subfield of cognitive science. This, of course, is the vantage point from which the present discussion has proceeded.

ACKNOWLEDGMENTS

I am grateful to the following people for very helpful comments and discussion: Kent Bach, Paul Bloom, Paul Boghossian, Robert Cummins, Carl Ginet, Christopher Hill, John Kihlstrom, Mary Peterson, Sydney Shoemaker, and Robert Van Gulick, as well as the BBS referees. I also profited from discussions of earlier versions of the chapter at the Society for Philosophy and Psychology, the Creighton Club, CUNY Graduate Center, Cornell University, Yale University, University of Connecticut, and Brown University.

REFERENCES

Armstrong, D. M. (1968). *A Materialist Theory of the Mind*. London: Routledge and Kegan Paul.

Astington, J. W., & Gopnik, A. (1991). Theoretical explanations of children's understanding of the mind. *British Journal of Developmental Psychology* 9:7–31.

Astington, J. W., Harris, P. L., & Olson, D. R. (eds.) (1988). *Developing Theories of Mind*. Cambridge, UK: Cambridge University Press.

Baars, B. (1988). *A Cognitive Theory of Consciousness*. Cambridge University Press.

Biederman, I. (1987). Recognition by components: A theory of human image understanding. *Psychological Review* 94:115–147.

———. (1990). Higher-level vision. In *Visual Cognition and Action*, eds. D. Osherson, S. Kosslyn, & J. Hollerbach (41–72). Cambridge, MA: MIT Press.

Block, N. (1978). Troubles with functionalism. In *Minnesota Studies in Philosophy of Science* IX. ed. C. Savage (261–325). Minneapolis: University of Minnesota Press.

———. (1990a). Inverted earth. In *Philosophical Perspectives* IV, ed. J. Tomberlin (51–79) Atascadero, CA: Ridgeview.

———. (1990b). Consciousness and accessibility. *Behavioral and Brain Sciences* 13:596–598.

———. (1991). Evidence against epiphenomenalism. *Behavioral and Brain Sciences* 14(4): 670–672.

Bowers, K. (1984). On being unconsciously influenced and informed. In *The Unconscious Reconsidered*, eds. K. Bowers & D. Meichenbaum. New York: Wiley.

Butterworth, G. (1991). The ontogeny and phylogeny of joint visual attention. In *Natural Theories of Mind*, ed. A. Whiten. Oxford, UK: Basil Blackwell.

Campbell, K. (1985). Pain is three-dimensional, inner, and occurrent. *Behavioral and Brain Sciences* 8:56–57.

Carey, S. (1985). *Conceptual Change in Childhood*. Cambridge, MA: MIT Press.

Churchland, P. M. (1979). *Scientific Realism and the Plasticity of Mind*. Cambridge, UK: Cambridge University Press.

Churchland, P. M. (1981). Eliminative materialism and the propositional attitudes. *Journal of Philosophy* 78:67–90.

———. (1985). Reduction, qualia, and the direct introspection of brain states. *Journal of Philosophy* 82(1):8–28.

Churchland, P. M., & Churchland, P. S. (1981). Functionalism, qualia and intentionality. *Philosophical Topics* 12:121–145.

Cooper, L. (1991). Dissociable aspects of the mental representation of visual objects. In *Mental Images in Human Cognition*, eds. R. Logie & M. Denis. New York: Elsevier.

Cummins, R. (1989). *Meaning and Mental Representation*. Cambridge, MA: MIT Press.

D'Andrade, R. (1987). A folk model of the mind. In *Cultural Models of Language and Thought*, eds. D. Holland & N. Quinn. Cambridge, UK: Cambridge University Press.

Dennett, D.C. (1969). *Content and Consciousness*. Atlantic Highlands, NJ: Humanities Press.

———. (1978). *Brainstorms*. Cambridge, MA: MIT Press.

———.(1988). Quining qualia. In *Consciousness in Contemporary Science*, eds. A. Marcel & E. Bisiach (42–77). Oxford, UK: Oxford University Press.

———. (1991). *Consciousness Explained*. New York: Little, Brown.

Dretske, F. (1988). *Explaining Behavior*. Cambridge, MA: MIT Press.

Ericsson, K., & Simon, H. (1980). Verbal reports as data. *Psychological Review* 87:215–251.

———. (1984). *Protocol Analysis: Verbal Reports as Data*. Cambridge, MA: MIT Press.

Flavell, J. (1988). The development of children's knowledge about the mind: From cognitive connections to mental representations. In *Developing Theories of Mind*, eds. J. W. Astington, P. L. Harris, & D. R. Olson (244–267). Cambridge, UK: Cambridge University Press.

Fodor, J. (1975). *The Language of Thought*. New York: Crowell.

———. (1981). *Representations*. Cambridge, MA: MIT Press.

———. (1987). *The Modularity of Mind: An Essay on Faculty Psychology*. Cambridge, MA: MIT Press.

Garrett, M. (1990). Sentence processing. In *Language*, eds. D. Osherson & H. Lasnik (133–175). Cambridge, MA: MIT Press.

Gleitman, H. (1981). *Psychology*. New York: W. W. Norton.

Goldman, A. (1989). Interpretation psychologized. *Mind and Language* 4:161–185.

———. (1992a). In defense of the simulation theory. *Mind and Language* 7(1–2):104–119.

———. (1992b). Empathy, mind and morals. *Proceedings and Addresses of the American Philosophical Association*, vol. 66, no. 3.

Gopnik, A. (1993). How we know our minds: The illusion of first-person knowledge of intentionality. *Behavioral and Brain Sciences* 16: 1–14.

Gordon, R. M. (1986). Folk psychology as simulation. *Mind and Language* 1:158–171.

———. (1992). Reply to Stich and Nichols. *Mind and Language* 7:81–91.

Harman, G. (1990). The intrinsic quality of experience. In *Philosophical Perspectives* 4, ed. J. Tomberlin. Atascadero, CA: Ridgeview.

Harris, P. L. (1989). *Children and Emotion: The Development of Psychological Understanding*. Oxford, UK: Basil Blackwell.

———. (1991). The work of the imagination. In *Natural Theories of Mind*, ed. A. Whiten. Oxford, UK: Basil Blackwell.

———. (1992). From simulation to folk psychology: The case for development. *Mind and Language* 7:120–144.

Hayes, P. (1985). The second naïve physics manifesto. In *Formal Theories of the Commonsense World*, eds., J. Hobbs & R. Moore. New York: Ablex.

Hume, D. (1748). *An Enquiry Concerning Human Understanding*. Oxford, UK: Clarendon Press.

Jackendoff, R. (1987). *Consciousness and the Computational Mind*. Cambridge, MA: MIT Press.

Jackson, F. (1982). Epiphenomenal qualia. *Philosophical Quarterly* 32:127–136.

———. (1986). What Mary didn't know. *Journal of Philosophy* 83:291–95.

Johnson, C. (1988). Theory of mind and the structure of conscious experience. In *Developing Theories of Mind*, eds. J. Astington, P. Harris, & D. Olson (47–63). Cambridge, UK: Cambridge University Press.

Landau, B. (1981). Will the real grandmother please stand up? The psychological reality of dual meaning representations. *Journal of Psycholinguistic Research* 11:47–62.

Lewis, D. (1966). An argument for the identity theory. *Journal of Philosophy* 63:17–25.

———. (1970). How to define theoretical terms. *Journal of Philosophy* 67:427–446.

———. (1972). Psychophysical and theoretical identifications. *Australian Journal of Philosophy* 50:249–258.

Loar, B. (1981). *Mind and Meaning*. Cambridge, UK: Cambridge University Press.
McCarthy, J., & Hayes, P. (1969). Some philosophical problems from the standpoint of artificial intelligence. In *Machine Intelligence*, eds. B. Meltzer & D. Michie. New York: Elsevier.
McCloskey, M. (1983). Naïve theories of motion. In *Mental Models*, eds. D. Gentner & A. Stevens. Mahwah, NJ: Erlbaum.
McNamara, T., & Sternberg, R. (1983). Mental models of word meaning. *Journal of Verbal Learning and Verbal Behavior* 22:449–474.
Meltzoff, A., & Moore, M. (1977). Imitation of facial and manual gestures by human neonates. *Science* 198:75–78.
———. (1983). Newborn infants imitate adult facial gestures. *Child Development* 54:702–709.
Millikan, R. (1984). *Language, Thought, and Other Biological Categories*. Cambridge, MA: MIT Press.
———. (1986). Thoughts without laws, cognitive science with content. *Philosophical Review* 95:47–80.
———. (1989). In defense of proper functions. *Philosophy of Science* 56:288–302.
Mitchell, P., & Lacohee, H. (1991). Children's early understanding of false belief. *Cognition* 39:107–127.
Morton, A. (1980). *Frames of mind*. Oxford, UK: Oxford University Press.
Nagel, T. (1974). What is it like to be a bat? *Philosophical Review* 83:435–450,
Newell, A., & Simon, H. (1976). Computer science as empirical inquiry: Symbols and search. *Communications of the Association of Computing Machinery* 19:113–126,
Nisbett, R. E., & Wilson, T. D. (1977). Telling more than we can know: Verbal reports on mental processes. *Psychological Review* 84:231–259.
Perner, J. (1991). *Understanding the Representational Mind*. Cambridge, MA: MIT Press.
Premack, D., & Woodruff, G. (1978). Does the chimpanzee have a theory of mind? *Behavioral and Brain Sciences* 1:515–526.
Putnam, H. (1960). Minds and machines. In *Dimensions of Mind*, ed. S. Hook. New York: New York University Press.
———. (1967). The mental life of some machines. In *Intentionally, Minds and Perception*, ed. H. Castaneda. Detroit: Wayne State University Press.
———. (1975). The meaning of 'Meaning.' In *Language, Mind, and Knowledge*, ed. K. Gunderson. Minneapolis: University of Minnesota Press.
Rachlin, H. (1985). Pain and behavior. *Behavioral and Brain Sciences* 8:43–53.
Rey, G. (1983). Concepts and stereotypes. *Cognition* 15:237–262.
Rips, L., & Conrad, F. (1989). Folk psychology of mental activities. *Psychological Review* 96:187–207.
Schachter, S., & Singer, J. (1962). Cognitive, social and physiological determinants of emotional state. *Psychological Review* 69:379–399.
Schacter, D. (1989). On the relation between memory and consciousness. In *Varieties of Memory and Consciousness: Essays in Honor of Endel Tulving*, eds. H. Roediger III & F. Craik. Mahwah, NJ: Erlbaum.
Schacter, D., Cooper, L., & Delaney, S. (1990). Implicit memory for visual objects and the structural description system. *Bulletin of the Psychonomic Society* 28:367–372.
Schacter, D., Cooper, L., Delaney, S., Peterson, M., & Tharan, M. (1991). Implicit memory for possible and impossible objects. Constraints on the construction of structural descriptions. *Journal of Experimental Psychology: Learning, Memory and Cognition* 17:3–19.

Schiffer, S. (1981). Truth and the theory of content. In *Meaning and Understanding*, eds. H. Parrett & J. Bouveresse. Berlin: de Gruyter.

Searle, J. (1990). Consciousness, explanatory inversion, and cognitive science. *Behavioral and Brain Sciences* 13:585–642.

Sellars, W. (1956). Empiricism and the philosophy of mind. In *Minnesota Studies in Philosophy of Science 1*, eds. H. Feigl & M. Scriven (253–329). Minneapolis: University of Minnesota Press.

Shoemaker, S. (1975). Functionalism and qualia. *Philosophical Studies* 27:291–315.

Smith, C., Carey, S., & Wiser, M. (1985). On differentiation: A case study of the development of the concepts of size, weight, and density. *Cognition* 21:177–237.

Smith, E., & Medin, D. (1981). *Categories and Concepts*. Cambridge, MA: Harvard University Press.

Stich, S. (1983). *From Folk Psychology to Cognitive Science: The case against belief*. Cambridge, MA: MIT Press.

Tursky, B., Jamner, L., & Friedman, R. (1982). The pain perception profile: A psychophysical approach to the assessment of pain report. *Behavior Therapy* 13:376–394.

Velmans, M. (1991). Consciousness from a first-person perspective. *Behavioral and Brain Sciences* 14(4):702–719.

Warrington, E., & Taylor, A. (1978). Two categorical stages of object recognition. *Perception* 7:695–705.

Wellman, H. (1988). First steps in the child's theorizing about the mind. In *Developing Theories of Mind*, eds. J. Astington, P. Harris, & D. Olson (64–92). Cambridge, UK: Cambridge University Press.

———. (1990). *The Child's Theory of Mind*. Cambridge, MA: MIT Press.

White, A. (1988). Knowing more than what we can tell: "Introspective access" and causal report accuracy 10 years later. *British Journal of Psychology* 79:13–45.

Wilson, M. (1991). Privileged access. Paper, department of psychology, University of California, Berkeley.

Wilson, T. D. (1985). Strangers to ourselves: The origins and accuracy of beliefs about one's own mental states. In *Attribution: Basic Issues and Applications*, eds. J. Harvey & G. Weary. New York: Academic Press.

Wimmer, H., & Perner, J. (1983). Beliefs about beliefs: Representation and constraining function of wrong beliefs in young children's understanding of deception. *Cognition* 13:103–128.

Wittgenstein, L. (1953). *Philosophical Investigations*. London: Macmillan.

———. (1967). *Zettel*, eds. G. Anscombe & G. von Wright. Oxford, UK: Basil Blackwell.

PART TWO

Empathy and Embodied Cognition

CHAPTER 7

Empathy, Mind, and Morals

1. PHILOSOPHY AND COGNITIVE SCIENCE

Early Greek philosophers doubled as natural scientists; that is a commonplace. It is equally true, though less often noted, that numerous historical philosophers doubled as cognitive scientists. They constructed models of mental faculties in much the spirit of modern cognitive science, for which they are widely cited as precursors in the cognitive science literature. Today, of course, there is more emphasis on experiment and greater division of labor. Philosophers focus on theory, foundations, and methodology, while cognitive scientists are absorbed by experimental techniques and findings. Nonetheless, there are sound reasons for massive communication between philosophy and cognitive science, which happily proceeds apace.

On this occasion I shall not try to enumerate or delineate these lines of communication in any comprehensive fashion. I just wish to *illustrate* the benefits to philosophy in two domains: the theory of mind (ToM) and moral theory. Though this may sound like an ambitious agenda, I in fact examine just a single phenomenon: empathy. Using that term first broadly and later narrowly, I argue that empathy may be the key to one sector of the philosophy of mind and to several sectors of moral theory. But whether empathy can in fact unlock any doors depends heavily on the outcome of empirical research in cognitive science.

2. RATIONALITY AND THE THEORY-THEORY

Only one topic in the philosophy of mind is addressed here: the ordinary or naïve understanding of mental states, commonly called "folk psychology."

The general problem is how ordinary people conceive of mental states and deploy mental-state language. The problem will not be addressed in full generality. I won't try to say what people mean or understand by mental words, nor how they apply these words to themselves. Attention will be focused exclusively on third-person ascriptions. Assuming a grasp of mental-state predicates and an ability to self-ascribe them, how do people go about ascribing these predicates to others? Further, how are mental ascriptions to others used to predict their behavior? People are strikingly successful in predicting the conduct of their peers, and ostensibly good at explaining them as well. Such predictive and explanatory success seems impossible unless attributors possess an impressive storehouse of knowledge about others, or some remarkably apt heuristic. What, then, is the content of this knowledge, or what is the heuristic?

My framing of the problem deliberately construes it as an empirical question about the psychology of attributors: What actually goes on in the heads of attributors that accounts for their attributions? This psychological construal is proper, I submit, though it hasn't been standard within the philosophy of mind. The task of attributing mental states to others is just one cognitive task among others. We should seek to identify the cognitive processes involved in this task just as we seek to identify the processes involved in perception, language acquisition, and probabilistic reasoning. Moreover, people have rather limited introspective access to *how* they execute the task of mental attribution (though they do have access to its outputs). Thus, we need scientific methods to get at the processes in question.

Until recently, two approaches to the nature of mental attribution have dominated contemporary theory. The first, focusing on intentional states, says that the language of intentionality embodies a "stance" adopted toward others, one that presupposes or imputes to them a high degree of rationality (Dennett, 1971, 1987; Davidson, 1984; Taylor, 1985). The second approach says that the language of the mental presupposes a commonsense theory of mind, a set of causal laws that interrelates stimulus inputs, internal states, and behavioral outputs. Framing these two approaches as empirical hypotheses about attributors, we obtain the following claims. The "rationality" or "charity" approach claims that the naïve attributor uses some sort of "rationality heuristic" in making third-person attributions. It begins with the assumption that people's beliefs, desires, and choices conform to certain canons of rationality. The theory-theory (TT) claims that the naïve attributor has a mentally stored set of functional laws, and makes inferences from these laws plus observed properties of the agent to the agent's internal states and future behavior.

Viewing these approaches as empirical hypotheses, how well do they fare? They should be judged by at least two criteria of empirical adequacy. First, do they make correct predictions about the verbal behavior or beliefs of attributors? Do they tell us correctly when an attributor A will ascribe particular mental states to an agent or believe that the agent is in those states? Second, each approach owes us an account of how the hypothesized knowledge or heuristic is acquired, and this acquisition hypothesis must accord with the observed development of "mentalizing" skills in human beings. On both counts, I believe, the two approaches have problems. I have discussed these before (Goldman, 1989, 1995), so here I shall be very brief.[1]

At least in its usual forms, the rationality approach predicts an unwillingness or reluctance on the part of attributors to ascribe logical deficiencies to agents (e.g., inconsistency in beliefs). In fact, however, attributors are often ready to accuse others of contradicting themselves and to hold these self-contradictions to be expressions of genuine beliefs (rather than mere verbal slips or changes in view). So it doesn't look as if attributors in fact employ the rationality heuristic. Of course, there are many variants of the rationality approach, but I don't know any variant that is sufficiently specific and yet survives this type of criticism.

As a test of TT, consider the results of an experiment by Kahneman & Tversky (1982). The subjects in this experiment were given the following example:

> Mr. Crane and Mr. Tees were scheduled to leave the airport on different flights, at the same time. They traveled from town in the same limousine, were caught in a traffic jam, and arrived at the airport thirty minutes after the scheduled departure time of their flights. Mr. Crane is told that his flight left on time. Mr. Tees is told that his was delayed and just left five minutes ago. Who is more upset?

Ninety-six percent of the subjects in the experiment said that Mr. Tees would be more upset. How did they arrive at their answers? TT would say that each subject must have possessed some lawful generalization from which relative degrees of upsetness were inferred. Did the subjects really possess some relevant generalization? Certainly none of the platitudes standardly listed by proponents of TT would do the job. Did they have other generalizations about upsetness or disappointment that dictated their conclusions? Perhaps, but there is considerable room for doubt.

Turning to the matter of acquisition, our two approaches must each defend the claim that its favored heuristic or knowledge structure is acquired by the age at which children in fact become fully competent wielders of mentalistic language (i.e., age four or five). This requirement appears troublesome for these approaches. Do children of this age really understand the putative canons of rationality? Have they acquired the relevant set of causal laws? How would these acquisitions have been accomplished?

Some defenders of TT suggest that we learn the folk-psychological framework "at our mother's knee," that it embodies the accumulated wisdom of thousands of generations' attempts to understand how humans work.[2] This implies acquisition by cultural transmission. That model, however, is extremely dubious. Very few children have mothers, or any other adult, who utter folk-psychological platitudes. Philosophers are virtually the only people who have attempted to give the platitudes explicit formulation, and few children have contact with philosophers.

More defensible is the view that acquisition of such laws mirrors the tacit acquisition of grammar by young children. However, although this is congenial from a cognitive science point of view, the details of how children might learn the folk-psychological laws are very problematic. Struck by problems of acquisition, a leading proponent of TT, Jerry Fodor, writes: "[T]here are, thus far, precisely no suggestions about how a child might acquire the apparatus of intentional explanation 'from experience' " (Fodor, 1987: 133). Fodor therefore postulates innate possession of the folk-psychological laws. In the mode of creation myth, Fodor says that he would have made these laws innate if he had been designing homo sapiens for social coordination. I do not doubt that evolution used some stratagem to help people interpret and predict others. But was it the hardwiring of state-specific causal generalizations?

3. THE SIMULATION HEURISTIC

If the two traditional approaches seem empirically shaky, as I believe they do, it is worth considering a third approach. Recall the airport example. An attractive explanation of why 96 percent of the subjects agreed in thinking that Tees would be more upset than Crane is that they imaginatively projected themselves into the shoes of the two protagonists and all decided that *they themselves* would be more upset in Tees's situation than in Crane's.[3] They therefore concluded that Tees would be more upset than Crane. The deployment of imaginative projection, or "empathy" (in a broad sense of the term), is the third approach I commend to your attention.

This approach to third-person interpretation has, of course, a significant philosophical history, often under the label of "Verstehen." Early exponents included Vico, Dilthey, Weber, Schutz, and Collingwood. These figures promoted Verstehen as the proper methodology for history or social science. Collingwood (1946), for example, viewed "dramatic re-enactment" as the distinctive method of historical understanding. In the present discussion, empathy is not being floated as a preferred method for the human *sciences*, but as a hypothesis about people's *naïve* heuristic for interpreting, explaining, and predicting others. (Its potential value in the human sciences is not excluded, but neither is it being endorsed as a primary scientific method.)

Much disparaged by positivistic philosophers, the Verstehen idea has recently enjoyed a quiet resurgence in a few circles. In *Word and Object* (1960), Quine explained indirect quotation in terms of an "essentially dramatic act" in which we project ourselves into the speaker's mind. He continued this in *Pursuit of Truth* (Quine 1990: 42), where he wrote: "Empathy dominates the learning of language, both by child and by field linguist. In the child's case it is the parent's empathy. The parent assesses the appropriateness of the child's observation sentence by noting the child's orientation and how the scene would look from there. In the field linguist's case it is empathy on his own part when he makes his first conjecture about 'Gavagai' on the strength of the native's utterance and orientation, and again when he queries 'Gavagai' for the native's assent in a promising subsequent situation. We all have an uncanny knack for empathizing another's perceptual situation, however ignorant of the physiological or optical mechanism of his perception." Two other Harvard philosophers, Putnam (1978) and Nozick (1981), have also given at least partial endorsements to the Verstehen idea, though they do not develop the theme very far.

My own treatment of this topic (Goldman, 1989, 1995) has been most directly influenced by Robert Gordon's (1986) essay, which formulated the Verstehen heuristic as an alternative to TT, and endorsed it partly on the basis of recent psychological literature. Gordon also dubbed it "the simulation heuristic," which terminology I have followed.[4] Psychologists use the label "perspective-taking" (Flavell, 1977) for this type of heuristic. However, the dominant approach within current cognitive psychology, especially developmental psychology, is still TT.

Let us now describe the simulation heuristic in more detail. The initial step, of course, is to imagine being "in the shoes" of the agent (e.g., in the situation of Tees or Crane). This means pretending to have the same initial desires, beliefs, or other mental states that the attributor's background information suggests the agent has. The next step is to feed these pretend states into some inferential mechanism, or other cognitive mechanism, and allow that mechanism to generate further mental states as outputs by its normal operating procedure. For example, the initial states might be fed into the practical reasoning mechanism that generates as output a choice or decision. In the case of simulating Tees and Crane, the states are fed into a mechanism that generates an affective state, a state of annoyance or "upsetness." More precisely, the output state should be viewed as a pretend or surrogate state, because presumably a simulator doesn't feel the *very same* affect or emotion as a real agent would. Finally, upon noting this output, one ascribes to the agent an occurrence of this output state. Predictions of behavior would proceed similarly. In trying to anticipate your chess opponent's next move, you pretend you are on his side of the board with his strategy preferences. You then feed these beliefs, goals, and preferences into your practical reasoning mechanism and allow it to select a move. Finally, you predict that *he* will make this move. In short, you let your own psychological mechanism serve as a "model" of his.

In assigning mental states to others, we often have their observed behavior as evidence. We test the plausibility of candidate state assignments by asking whether those states would have been compatible with the observed behavior. Here again we can operate by simulation: "If I had been in M, would I have performed that action?"

The example of Anita Hill is a good case in point. We know that she moved with Clarence Thomas to the EEOC. Is this behavior compatible with her claim that he had sexually harassed her and that she therefore had mental states consequent on such harassment? A natural heuristic to try to settle this question is to imagine oneself suffering and recalling the alleged harassment and seeing whether this would produce Hill's actual behavior. Many Americans reported using this heuristic as they reflected on the charges and denials in question.

The Hill-Thomas case highlights another important aspect of the simulation heuristic: There is no guarantee of *faithful* imagination of the agent's initial states. It was widely alleged in the Hill case that men in particular could not adequately project themselves into Anita Hill's shoes. They would not appreciate just how constrained her choice situation was, or how constrained she herself would perceive it to be. In other words, they would be liable to misrepresent the other components of her mental state that might produce the decision in question. I suspect that this is right. In any case, it seems to be a general fact, congruent with the simulation theory, that when an agent's life experience is very different from our own, we have a harder time predicting or explaining their behavior.

Because competing theories of mentalistic ascription, in my view, are competing cognitive hypotheses, it helps to have a cognitive model of how the simulation heuristic would work. Stich & Nichols (1995), though

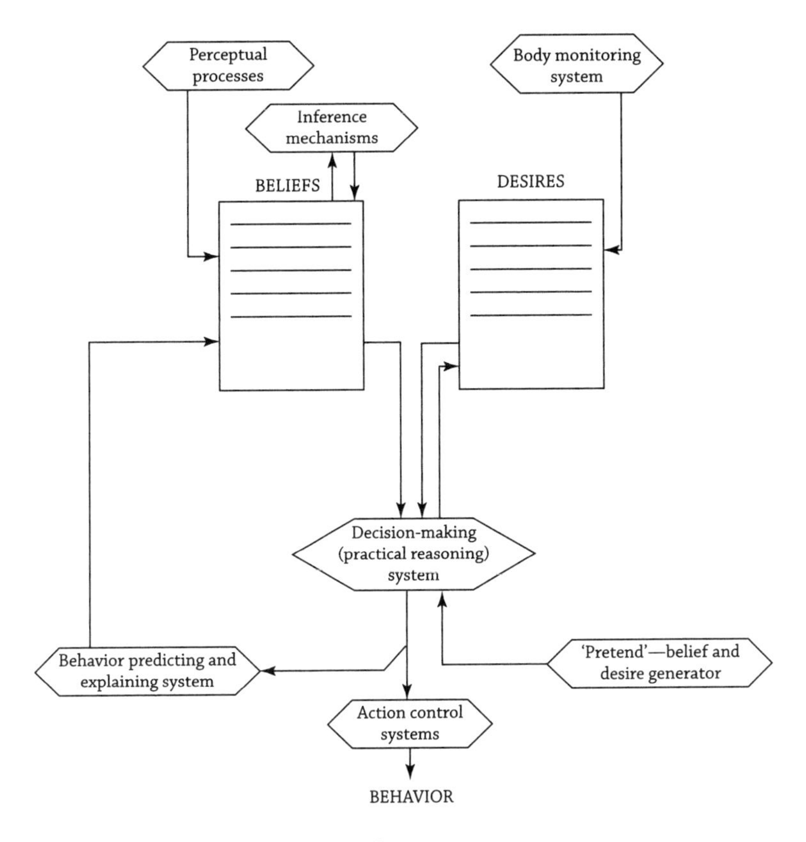

Figure 7.1: Flowchart of the simulation heuristic.
Source: Stich & Nichols, 1995.

critical of the simulation heuristic, construct a flowchart of its operation (figure 7.1), based on the Gordon-Goldman analyses.

The bulk of figure 7.1 depicts the ordinary mechanisms of belief and desire acquisition, plus the mechanism for combining desires and beliefs to produce decisions and ultimately behavior. At the bottom of the diagram, however, simulation-relevant mechanisms appear. In the lower right corner is a generator of "pretend" beliefs and desires. The outputs of this device can be fed into the decision-making (DM) system. When the DM system gets inputs from this device, however, it does not feed its outputs into the action-control systems. In the case of simulation, the DM outputs are taken "offline" and simply sent back to the belief box as beliefs about what the targeted agent will do. Notice that there is no assumption that these processes are always conscious or phenomenologically salient.

One thing neglected by figure 7.1 that is important for successful simulation is the need for the simulator to *isolate* or *quarantine* any beliefs or desires of his own that are not shared by the target agent. If such beliefs or desires are mistakenly fed into the DM system, the output of the simulation is unlikely to match the agent's decision. This need for isolating or quarantining unshared attitudes will play an important role in my later interpretation of certain empirical findings.

Implicit in our model (but only implicit) is what crucially distinguishes the simulation theory (ST) from TT. ST does not require a predictor to know the nomological properties that govern his DM system (or his other psychological mechanisms). It only requires the predictor to *possess* a DM system and to be capable of feeding it appropriate inputs and interpreting its outputs. This is a much weaker commitment than that of TT.

Nothing in our discussion, of course, excludes the mixed hypothesis that people use *both* the simulation heuristic *and* some forms of nomological information. Some kind of mixed theory, I suspect, is unavoidable. Surely there is knowledge of some behavior patterns, for example, that enable predictions to be made completely without simulation. When Jones approaches a drinking fountain and bends over it, I can safely predict that he will drink without performing a simulation, as this stereotypic behavior pattern is widely observed. Similarly, I might use past simulations to construct generalizations about mental-state sequences. Third, I can use inductive information to modify my simulations. If I wish to simulate Tom's current deliberation, I might recall his past impulsiveness and adjust the "impulsiveness parameter" of my DM system.

Granted, then, that pure simulation is unlikely to be the *whole* story of mental and behavioral predictions of others, but what is its extent and significance? There is obviously a spectrum of possibilities. A weak hypothesis would say that it is just a rare or infrequent heuristic. A strong hypothesis would say that it is the primary heuristic at work in interpreting others, a heuristic that comes into play early in life and forms the basis for mentalizing about others. I am attracted by the strong form of the simulation theory; but no view should be firmly adopted until more evidence is examined. In pursuit of this goal, let me sketch some of the relevant evidence, attending to theoretical, as well as experimental, considerations.

To assuage total skepticism about simulation, we first need to be satisfied that simulation can account for *some* cases of successful prediction. One thing essential for successful simulation is that the relevant cognitive mechanisms be capable of operating on pretend states just as they operate on genuine, unfeigned states. Second, the simulator must be able to take the output offline and use it, not to guide his behavior, but to predict others' behavior. Are such capacities actually in place?

The existence of such capacities is confirmed by the phenomena of hypothetical planning and counterfactual reasoning. A planning mechanism can take not only actual goals or intentions as inputs, but also feigned ones. If your family is debating whether to spend the day at the lake, you can hypothetically take this destination as a goal and reflect on the optimal driving route. Your planning mechanism operates on the feigned goal just as it does on a genuine goal. The capacity to take outputs offline is equally unassailable. When I select a route for a hypothetical destination, I don't begin to follow that route until I genuinely adopt the destination as my goal. Counterfactual reasoning also displays these features. When considering the truth value of "If X were the case, then Y would obtain," a reasoner feigns a belief in X and reasons about Y under that pretence.

Some of the required apparatus for successful simulation, then, is in place. This supports a weak simulation hypothesis, at a minimum. What about a strong simulation hypothesis? Could simulation be the fundamental heuristic (or *one* fundamental heuristic) that guides interpretation? This would require simulation to be developmentally *early*. Does that hypothesis comport with the evidence?

4. DEVELOPMENTAL FINDINGS AND THEIR EXPLANATION

At present, few developmental psychologists favor ST, but that is not surprising because this option is *very* new and only beginning to receive attention in the literature.[5] For developmentalists, the main question is whether ST can accommodate the interesting empirical findings about children's acquisition of the mentalizing repertoire, and how well it accommodates these findings as compared with the rival TT. Let me therefore present a simulation-based developmental scenario sketched by Harris (1995). He first presents a theoretically neutral description of the principal developmental stages in the normal child's mentalizing ability, based on a wide range of studies.

Toward the middle or end of the first year of life, children begin to adjust their intentional attitudes to those displayed by adults. This is the phenomenon of "joint visual attention." When an adult turns to look at an object, the infant will look toward that same object. Further, when the adult looks at an object and expresses an emotion toward it facially and/or vocally, the infant adopts a similar emotional stance, avoiding it if the adult looks apprehensive, approaching it if the adult looks positive.

Toward the end of the first year and increasingly during the second year, the child begins to *act* on another's current attitude. The child seeks to redirect another's gaze by pointing at, showing, or giving an object of interest. The child also seeks to alter another's current emotional state by deliberate acts of teasing or comforting.

Three-year-olds can anticipate or enact the reactions of people or toy characters whose current mental stance differs from their own. The target agent can diverge from the child with respect to the actual or potential objects or situations that are seen, wanted, liked, expected, or known.

The three-year-old stage, however, is characterized by a notable deficiency, the discovery of which has fueled much of the research in this area. Three-year-olds are weak at ascribing false beliefs to others. In a seminal experiment by Wimmer & Perner (1983), children were shown a puppet character, Maxi, who put his chocolate in a box and went out to play. Next they saw Maxi's mother transfer the chocolate to the cupboard while Maxi was still out. Where will Maxi look for the chocolate when he returns? Five-year-olds responded, "in the box," a response typical of adults. Children of three, however, indicated the cupboard, where they themselves believed it to be. So three-year-olds have not fully mastered the art of belief ascription.

This three-year-old phenomenon is quite robust and applies even to their own previous false beliefs. Shown a closed candy box, children express a belief that candies are inside. The box is opened and seen to contain only pencils. The children are then asked what they had originally thought was inside. Three-year-olds typically insist that they thought there were pencils, whereas four- and five-year-olds admit that they thought there was candy (Astington & Gopnik, 1988; Wimmer & Hartl, 1991).

How are these phenomena, especially the three-year-olds' performance on false-belief tasks, to be explained? Most theory-theorists explain the phenomena by saying that two- and three-year-olds do not yet have a *full* theory of mind, or, more precisely, they have a *different* theory of mind. Perner (1991) articulates this view most explicitly. He argues that two- and three-year-olds have something like a theory of "direct reference," in which propositional attitudes connect people directly to actual or possible situations without the mediation of "representations."[6] They conceptualize knowledge or belief as connecting a person with an external or upcoming situation, but not as having a descriptive content about that situation. At about four years of age, children achieve a crucial insight. They realize that mental states have both a referent and a content that may not match one another. This enables the four-year-old to understand how someone might hold a false belief, and to realize that the content of the belief may depict reality in a misleading or outdated fashion. Thus, the three-year-old to four-year-old transition is allegedly marked by a *radical shift in theory*, from a nonrepresentational to a representational conception of beliefs and other propositional attitudes.

Let us now look at Harris's alternative theoretical account of the empirical findings, an account that is consonant with the simulation approach. In the first year of life, he says, the child echoes or simulates another person's attentional or emotional stance. At the next stage, the child attributes the simulated stance to the other person, and uses this attribution to guide action vis-à-vis a targeted object.

By age three, the child can set aside her own intentional stance and simply imagine another person's stance. For example, she can imagine another person seeing an object that is invisible to her, or wanting an object that she doesn't want. As yet, however, she is not good at imagining someone believing or seeing a state of affairs that runs directly *counter* to what she herself takes to be the case, as in the false-belief task.

The transition to the four-year-old level, which is critical in the present controversy, primarily features an increase in *imaginative power*. The four-year-old can use hypothetical situations as pretend inputs although they run counter to the existing situation as she understands it. Thus, the transition from the three-year-old to the four-year-old stage is not a radical transformation of the child's conceptualization of mental states, as theory-theorists contend, but just an increase in imaginative flexibility.

Let me suggest a slightly different explanation of the three-year-olds' deficit, which I think provides an even stronger explanation of their poor performance on false-belief tasks from the vantage point of ST. Recall that effective simulation requires isolating or quarantining the simulator's own attitudes, which are unshared by the agent. It is quite plausible, however, that three-year-olds are weak at isolating or quarantining their own actual beliefs. They tend to allow their actual beliefs to "slip in" to the simulation procedure. Precisely what the literature reports is that three-year-olds standardly give answers in false- belief tasks that reflect their *own* current beliefs. This is just what ST would predict when it is conjoined with the hypothesis that younger children are weak at the "isolation" or "suspension" component of the simulation procedure.

We must next consider how satisfactory is TT's explanation of three-year-olds' observed behavior. Does the postulation of a nonrepresentational conception of belief at age three really comport with empirical findings? I shall now show that new findings conflict with this picture. At least two studies indicate that three-year-olds *are* capable of correct performance on false-belief tasks, and hence they must have a conceptualization of belief that allows the possibility of false belief. Wellman (1990: chap. 6) asked children to explain why characters engaged in certain actions, where some of the actions were inappropriate given the facts. For example, children were first told the following: "Jane is looking for her kitten. The kitten is hiding under a chair, but Jane is looking for her kitten under the piano." They were then asked, "Why do you think Jane is doing that?" Sixty-five percent of the three-year-olds gave at least one false-belief explanation in response to such a question (e.g., "She thinks the kitten's under the piano"). Apparently, then, three-year-olds are capable of conceptualizing false belief, contrary to the dominant TT approach!

Second, Mitchell & Lacohée (1991) found that, with a little mnemonic help, three-year-olds can correctly report their own previous false beliefs. After children expressed their false expectation of what was in a box (or tube), they selected a picture of what they thought was inside and "mailed" it. This overt action apparently strengthened their memory of what they had expected, because most of these three-year-olds later acknowledged that they had mistakenly thought that candy was in the box, although they now knew that it was pencils. Evidently, these young children do not have the sort of deep-seated deficit that theory-theorists commonly postulate. They must not have a nonrepresentational conception of mental states that doesn't admit the possibility of false belief. Thus, the transition from the three-year-old to the four-year-old stage cannot be a radical shift in theory, as theory-theorists propose.

Harris further elaborates and supports the simulative approach to young children's understanding of mind by reference to studies of autistic children. Recall that simulation processes appear to be set in motion by special-purpose inbuilt mechanisms for establishing joint attention and joint emotional stance. Impairment of those mechanisms, then, can be expected to disrupt the developmental process that builds on them, and such impairment is what seems to occur in autism. Autistic children are poor at following another person's direction of gaze even when it is emphasized by a pointing gesture. They are also poor at taking measures to influence another's visual attention. Strikingly, older autistic children show the same deficit at false-belief tasks as normal three-year-olds (Baron-Cohen, Leslie, & Frith, 1985). Finally, autistic children show a variety of difficulties in talking about and diagnosing the mental states of others. Elsewhere (Goldman, 1995), I argue that several of the reported difficulties of autistic children in diagnosing mental states can best be explained by reference to deficits in simulation.

The simulation approach, then, has excellent prospects for explaining many of the empirical data about mentalizing in the developmental and clinical literature. Indeed, it can explain them better than the competing ("radical shift in theory") account that theory-theorists typically offer. It therefore stands as a viable account of the *basic* processes of interpersonal interpretation.[7]

Additional evidence for the prevalence of simulation comes from the phenomenon of motor mimicry. Adam Smith, who along with Hume was one of the most astute observers and theorists of empathy (sympathy), made early observations of motor mimicry:

> When we see a stroke aimed, and just ready to fall upon the leg or arm of another person, we naturally shrink and draw back on our leg or our own arm.... The mob, when they are gazing at a dancer on the slack rope, naturally writhe and twist and balance their own bodies, as they see him do. (1759/1976: 10)

Numerous other writers commented on the same phenomenon, including Spencer (1870), Darwin (1872/1965), McDougall (1908), and Mead (1934). Hull (1933) and Allport (1961) brought the phenomenon into the experimental laboratory and recorded it. Plausibly, motor mimicry is a manifestation of mental mimicry, or simulation, in which simulated plans are *not* taken offline, but manage to seize motor control.

Although adult motor mimicry probably reflects mental mimicry, there is a primitive phenomenon of motor mimicry even in infants, which may suggest that mimicry is an evolutionarily favored trait. Striking feats of infant mimicry have been identified by Meltzoff and colleagues. Meltzoff & Moore (1977) found that neonates between twelve and twenty-one days old could imitate certain facial actions such as lip protrusion, mouth opening, tongue protrusion, and sequential finger movement, as shown in figure 7.2. A few years later, Meltzoff & Moore (1983) reported similar facial imitation by newborns with a mean age of thirty-two hours, of which the youngest subject was only forty-two *minutes* old! (See Meltzoff, 1990, for a review.) Thus, facial imitation appears to be an innate facility.[8]

What are the consequences of ST for the philosophy of mind? If simulation is what primarily accounts for third-person mental ascription, there is no need to posit an extensive knowledge of mental laws on the part of ordinary people. Nor is it necessary to invoke a knowledge of any such laws or theory in explaining how people conceptualize mental-state concepts.[9] But if our ordinary concepts of the mental are not theoretical concepts, their soundness cannot so readily be challenged by the possible falsity or inadequacy of the putative theory in question. These morals are obviously significant for the philosophy of mind.

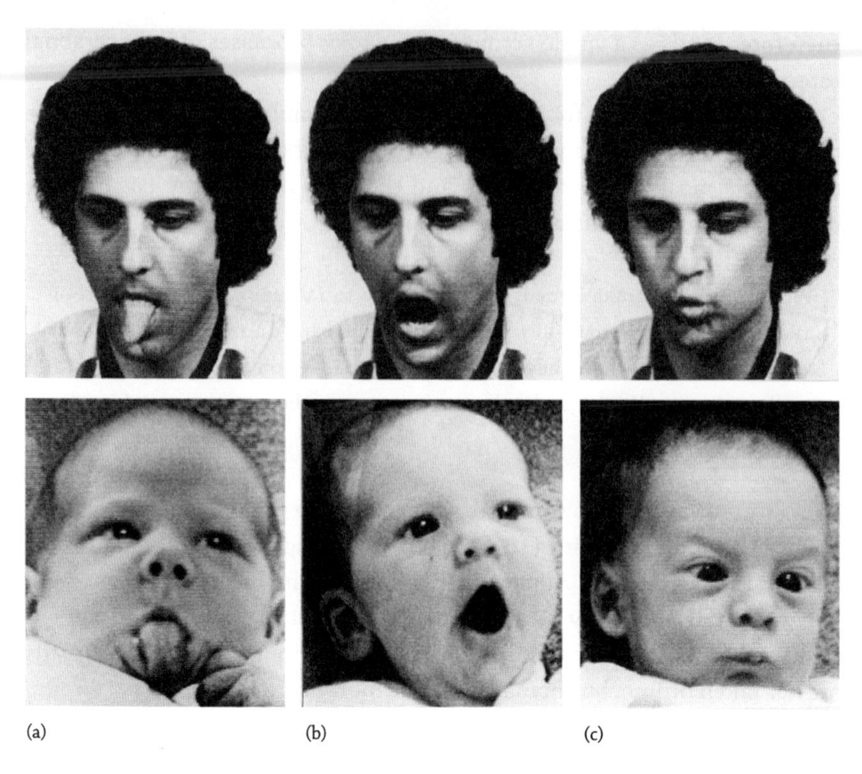

Figure 7.2: Photographs of two-to-three-week-old infants imitating facial gestures presented to them by an adult experimenter.
Source: Meltzoff & Moore, 1977.

5. AFFECTIVE EMPATHY AND COGNITIVE SCIENCE

Until now I have used the term "empathy" to denote the process of simulation. But empathy typically has a narrower meaning, one specifically concerned with *affective* or *emotional* states. To empathize with someone, in its most frequent sense, is to sympathize or commiserate, which involves shared attitudes, sentiments, or emotions. In the remainder of the discussion, I shall focus on this narrower sense of empathy. Central cases of empathy, so construed, may arise from simulation (i.e., from imaginatively adopting the perspective of another). Such initial "pretend" states are then operated upon by psychological processes that generate feelings, attitudes, or affects that are similar to, or homologous to, the target individual's states. Furthermore, just as the simulator is generally aware of his states *as* simulations of the target, so the empathizer is presumed to be aware of his vicarious affects and emotions *as* representatives of the affects and emotions of the target. Thus, empathy consists of a sort of "mimicking" of one person's affective state by that of another.

Although perspective-taking is the standard way by which vicarious or resonant emotions are generated, there may be resonance phenomena not preceded by perspective-taking. These cases can also be considered cases of empathy in a *wide* sense. This would comport with certain definitions in the psychological literature. Barnett (1987), for example, defines empathy as "the vicarious experiencing of an emotion that is congruent with, but not necessarily identical to, the emotion of another individual."

Although almost everyone experiences empathy from time to time, it remains to be seen how fundamental and robust a phenomenon it is and whether the foregoing description is psychologically sustainable. Is it true, for example, that empathic states are "similar," "congruent," or "homologous" to genuine affective states of a target agent? And what exactly are the respects of similarity or congruence?

At present I don't think we can properly specify the respects of similarity between original and vicarious affective states. This is insufficient reason, however, to cede this territory to the skeptic. Cognitive science is also currently unable to specify all points of resemblance between vision and visual imagery. Nonetheless, in the last fifteen years, subtle experiments have demonstrated very considerable respects of similarity between the two domains (see Finke & Shepard, 1986, and Kosslyn, 1980, 1990). If comparable experimental creativity is invested in the field of vicarious affect, I would expect analogous points of similarity to be found. At any rate, that is the assumption on which I shall proceed.

Let us return to emotional contagion. Although this phenomenon is familiar enough—consider, for example, the infectiousness of smiles and laughter—its primitive basis has received experimental support in the reactive crying of newborns (Simner, 1971). Another example of resonant emotion, briefly alluded to earlier, occurs in the context of *social referencing*. Klinnert (1981) presented twelve- and eighteen-month-old children with novel and somewhat forbidding toys in their mothers' presence. Mothers were instructed to pose facial expressions conveying either fear, joy, or neutral emotion. For those children who referenced their mother (i.e., looked at her to "check" on her attitude), maternal facial expressions had a profound effect. Infants were significantly more likely to approach the toy when the mother was smiling, but to retreat to the mother when she was displaying fear. There is additional evidence that such behavior was mediated through the arousal of a resonant emotion in the children, who themselves showed negative affect.

Although the foregoing cases may not involve perspective-taking, there is other experimental work in which congruent emotion is produced by perspective-taking. Stotland (1969) had subjects watch someone else whose hand was strapped in a machine that they were told generated painful heat. Some were told just to watch the man carefully; some were told to imagine the way he was feeling; and some were told to imagine themselves in his place. Using both physiological and verbal measures of empathy on the part of the subjects, the experimental results showed that deliberate acts of imagination produced a greater response than just watching.

These results are not restricted to painful or distressing experiences. Krebs (1975), for example, found that participants reported feeling bad when watching someone whom they thought was about to receive an electric shock, but they reported feeling good when watching someone about to receive a reward.

An insightful observation of "positive" empathy was presented by Adam Smith:

> When we have read a book or poem so often that we can no longer find any amusement in reading it by ourselves, we can still take pleasure in reading it to a companion. To him it has all the graces of novelty; we enter into the surprise and admiration which it naturally excites in him, but which it is no longer capable of exciting in us; we consider all the ideas which it presents rather in the light in which they appear to him, than in that in which they appear to ourselves, and we are amused by sympathy with his amusement which thus enlivens our own. (1759/1976: 14)

6. EMPATHY IN DESCRIPTIVE AND PRESCRIPTIVE ETHICS

Let us assume, then, that empathy is a genuine and fairly pervasive facet of human life. What are the consequences for moral theory? And what relevance can further empirical investigation of empathy have to moral theory? Let me first divide moral or ethical theory into two components: *descriptive* and *prescriptive* ethics. Descriptive ethics in turn has two branches. Branch 1 would seek to describe and explain the acceptance of the various moral codes in different cultures and subcultures. Branch 2 of descriptive ethics would seek to describe and explain the extent of *conformity* with each code by people who subscribe to it. This second branch would focus heavily on *motivational* factors. What enables or inhibits an agent from acting on her moral creed? Prescriptive ethics would, of course, be concerned with the formulation and justification of a "proper" or "correct" moral system.

The empirical study of empathy is relevant to all of these branches and sub-branches of ethics. Historically, a key role for empathy in descriptive ethics was championed by Schopenhauer (1841/1965). The primary ethical phenomenon, according to Schopenhauer, is compassion, which he characterized as the (vicarious) "participation" in the suffering of another (1841/1965: 143–144). He divided ethical duties (as formulated in many codes) into duties of justice and duties of philanthropy. Duties of justice are "negative" duties to refrain from injuring others, and are ultimately based on feelings of compassion. Duties of philanthropy are "positive" duties to help that are also based on compassion (1841/1965: 148–149). Finally, Schopenhauer assigns compassion a critical place in explaining the cross-cultural display of moral behavior in human life:

> [T]he foundation of morals or the incentive to morality as laid down by me is the only one that can boast of a real, and extensive, effectiveness.... [D]espite the great variety of religions in the world, the degree of morality, or rather immorality, shows absolutely no corresponding variety, but is essentially pretty much the same everywhere.... [Unlike the ineffectiveness of religion] the moral incentive that is put forward by me [viz., compassion]... displays a decided and truly wonderful effectiveness at all times, among all nations, in all the situations of life, even in a state of anarchy and amid the horrors of revolutions and wars. (1841/1965: 170, 172)

Schopenhauer is far from saying, of course, that compassion is the predominant motivation in our lives. Nonetheless, he views it as the source of moral principles and the ultimate root of compliance with such principles. A similar line is taken by Rousseau:

> Mandeville has rightly recognized that, with all their morality, men would never have been anything but hideous monsters, had not nature given them compassion as a support for their faculty of reason. But he did not see that from this one quality spring all the social virtues that he wishes to deny men. In fact, what are generosity, clemency, and humanity if not compassion that is applied to the weak, the guilty, or even the entire human race? Properly understood, benevolence and even friendship are the result of a constant pity that is fixed on a particular object.... The commiseration will be the more energetic, the more intimately the spectator identifies himself with the sufferer. (1755: 94)

It is not implied, of course, that compassion or empathy plays an exclusively direct or simple role in influencing moral codes or eliciting compliant behavior. Even Schopenhauer grants that compassion commonly operates indirectly, by means of principles (see Schopenhauer, 1841/1965: 151).

The task of descriptive ethics, we have said, is to identify and explain the moral systems that are found in various cultures. In recent years, however, many writers point to differences in moral systems or orientations even within a single culture. In particular, they claim to find *gender* differences in moral orientation. Gilligan's *In a Different Voice* (1982) is probably the most influential statement of such a hypothesis. The question of whether there are such differences, and if so what is their source, is a good example of a subject ripe for empirical inquiry. Gilligan claims that women have a moral orientation that focuses on "caring" and "connecting," rather than abstract rights or justice. This thesis, however, has been criticized on empirical grounds. In a series of studies, Walker and his colleagues found no statistically significant gender differences as measured within Kohlberg's widely used moral stage framework (e.g., Walker, 1984). In a more recent study (Walker, DeVries, & Trevarthen, 1987), Walker did find that females were more likely to choose personal over impersonal dilemmas as problems to talk about, and problems they claimed to confront. Moreover, personal dilemmas were more likely to elicit a "care" response, rather than a "justice" or "rights" response. Controlling for dilemma *content*, however, sex differences were still not found to be significant.[10]

Gilligan's thesis, and similar theses advanced by other feminist writers, is particularly relevant to us because a focus on "caring" and "connecting" might stem from more frequent or more salient empathy. Indeed, Gilligan quotes with approval Haan's (1975) and Holstein's (1976) research that indicates "that the moral judgments of women differ from those of men in the greater extent to which women's judgments are tied to feelings of empathy and compassion" (Gilligan, 1982: 69). This naturally raises the question of whether there is a psychological difference between the genders in the incidence or strength of empathy, which common sex stereotypes, of course, suggest. This is a heavily researched topic, but the results are complex and inconclusive.

Eisenberg and Lennon (1983, 1987) survey the field as follows. A principal complication in empathy research is the variety of measures used in its detection. The most popular method of assessing empathy in young children uses picture/story stimuli and operationalizes empathy as the degree of match between the self-report of emotion and the emotion appropriate to the protagonist in the vignette. In twenty-eight studies using this measure, most found no significant gender differences. In studies of school-age children and adults, the most widely used index is a self-report questionnaire. In sixteen studies of this sort, females scored significantly higher in all. However, these differences may be due to biases in self-reports. Females are *expected* to be more concerned with others, as well as more emotional, than males, so both females and males may respond in ways consistent with sex-role stereotypes. Other measures of empathy include facial/

gestural and vocal measures of empathy, as well as physiological measures of empathy. Eisenberg and Lennon conclude that no significant gender differences are found on these measures.

Let me turn now to prescriptive ethics. Although the meta-constraints on prescriptive ethics are of course highly controversial, most writers would agree that a satisfactory prescriptive theory should be firmly rooted in human nature. It would be hard to defend any moral system as prescriptively valid that did not make important contact with human moral psychology. Much of ethical theory has focused on the human capacity for reason, a tradition most vividly exemplified by Kant. In recent literature, there is also a tendency to associate moral rationalism with highly *universalistic* moral norms and to associate emotionalism (as the contrasting approach might be dubbed) with a *particularist* point of view. Universalism requires the moral agent to consider everyone's pleasure or pain equally and impartially. By contrast, particularism allows the agent to display some degree of partiality toward individuals with whom one has a personal affinity, such as family members, friends, students, or comrades. If we now consider the prospects of an empathy-based view of morality, it might seem natural for it to tilt toward particularism, as Blum (1980, 1987), for example, has suggested This is because empathy inclines an agent toward actions that are responsive to those with whom he empathizes, and these are most likely to be people with whom personal contact is established.

However, it is not clear that an emphasis on empathy or sympathy necessarily dictates a particularist or "agent-centered" morality. A universalist may point out that empathy can be extended beyond personal contacts, for example, to characters in fiction or history. In fact, we can readily think of sympathy-based theories that are quite universalistic. Hume's theory of justice, at least as reconstructed by Rawls (1971: 185–186), is both sympathy-based and highly universalistic. Again, Hare's (1963) highly universalistic theory acknowledges the instrumental value of sympathetic imagination in people's readiness to universalize.

Let me turn now to a more concrete way in which psychological facts may impinge on prescriptive ethics, namely by setting constraints of realism or feasibility. A moral code that is psychologically unrealizable by human beings, or just too demanding or difficult for people to follow, might be rejected on meta-ethical grounds. Not all moral theorists would accept these constraints, as Scheffler (1986) points out. Nonetheless, it is plausible to impose a constraint like Flanagan's principle of minimal psychological realism: "[M]ake sure when constructing a moral theory or projecting a moral ideal, the character, decision processing, and behavior prescribed are possible for creatures like us" (1991: 32). Moral theories like utilitarianism may fail to satisfy this principle because they require more altruistic behavior, or more universalism, than is feasible for human beings.

This raises the question of people's capacities for altruism, and their capacities for serving everyone's welfare equally, as opposed to their own welfare and that of specially related others. Here, empathy again becomes particularly relevant, because it seems to be a prime mechanism that disposes us toward altruistic behavior. The question then arises: What exactly is the potential scope, extent, or power of empathy? Can we empathize with everyone equally? As psychologist Hoffman (1987) points out, the research evidence shows that empathy tends to be biased. Observers are more empathic toward victims who are familiar and similar to themselves than to those who are different. Second, people are more apt to be empathically aroused by someone's distress in the immediate situation than by distress they know of being experienced by someone elsewhere.[11] But these issues need more empirical investigation. More generally, cognitive science needs to give us a systematic account of the properties of the empathizing process. What targets and circumstances encourage the initiation of empathy? What variables affect the vividness or strength of empathic feelings? How do empathic feelings combine with other cognitions to influence an agent's conduct? These and other parameters concerning empathy need to be better understood.

I have said that psychological realism might be relevant to prescriptive ethics by excluding moral systems that are too demanding. But equally, psychological realism may help exclude moral or social systems that aren't *sufficiently* constraining. Hume remarks that institutions of justice would not be necessary at all if the human mind were "so replete with friendship and generosity, that every man has the utmost tenderness for every man, and feels no more concern for his own interest than for that of his fellows" (1777/1972: section 146). In our actual state, there do seem to be limits to our benevolence and empathic powers. These limits create the need for legal and political institutions that will be sufficiently constraining to shape the conduct of creatures like us.

Finally, to the extent that both descriptive and prescriptive ethics should seek to place human moralizing within the context of human biology, it is instructive to inquire into the evolutionary origins of moral psychology (a topic recently broached by Gibbard, 1990, and others). Empathy might well figure centrally in an evolutionary account. Empathizing with one's kin or one's neighbor would promote mutual aid and inhibit injurious behavior, thereby contributing to biological fitness. Of course, empathy also advantageously promotes mutual interpretation and prediction, but this *might* just be a serendipitous byproduct. More plausibly, the interpretive property of empathy may itself be fitness-enhancing, because it encourages beneficial social coordination. There are numerous ways, then, in which a mechanism for mimicking, simulating, or resonating to the states of others could be part of an evolutionarily stable strategy. Thus, the mechanism of empathy, in both its effective and interpretive manifestations, may be a critical aspect of human nature.

What I have been arguing is that moral theory needs to be sensitive to the phenomenon of empathy. The precise impact of this sensitivity on moral theory will depend on specific properties of empathy, properties that can only be firmly identified and established through empirical research. Thus, moral theory stands to benefit from the work of cognitive science.[12] At the same time, cognitive science can profit from the observations and suggestions that philosophers have been making about empathy for more than two hundred years.[13]

NOTES

1. Also see Stich (1981) and Thagard & Nisbett (1983).
2. See Churchland, 1988: 59.
3. Although I have borrowed the Crane-Tees example from Kahneman & Tversky (1982) and the title of their article is "The simulation heuristic," their discussion of this heuristic is quite different from mine. They do not much address *interpersonal* simulation, which is my topic. Instead, they are interested in how people simulate counterfactual scenarios and compare them to actual events. In particular, they are interested in how such counterfactual scenarios influence people's judgments and feelings (e.g., how a counterfactual scenario generated by someone who has missed his flight would affect his degree of disappointment).
4. Other recent writers who endorse the projection or Verstehen idea include Heal (1986), Ripstein (1987), and Loar (1987). Although Stich (1983) has been interpreted by some as a proponent of this idea, he disavows any such intention and has more recently become an emphatic critic of the idea (Stich & Nichols, 1995).
5. One proponent of ST among the developmentalists is Harris (1989, 1995). Furthermore, although few "cognitive" developmentalists have examined the simulation approach until recently, developmentalists in the clinical or social branches of the field have studied simulation or perspective-taking for a number of years (e.g., Selman,1980). Extended discussion of ST in the current literature is found in Davies & Stone (1995).
6. Perner doesn't use the phrase "direct reference"; he calls children of this age "situation theorists." For similar views, see Flavell (1988), Wellman (1990), Astington & Gopnik (1991), and Gopnik (1993).
7. There is no room here to survey criticisms of the simulation approach from the vantage point of TT. For such criticisms, see the papers by Stich & Nichols; Perner, Gopnik & Wellman; and others in Davies &Stone (1995).
8. One thing particularly striking about Meltzoff's findings from a psychological point of view is that, while infants see the facial expressions of the adult, they

don't see their own mouths. Indeed, the youngest infants will never have seen their own mouths, even in a mirror. Thus, to select movements that correspond to what they see in the adult's face, they must have an *innate* "cross-modal" (visual-motor) mapping. It should also be emphasized that infant mimicry is not a mere reflex. On the contrary, infants can remember the display and imitate it even after delays or intervening motor tasks. To illustrate this, pacifiers were put in the infants' mouths while they watched the adult gestures; the infants could observe the gestures but could not duplicate them online. At the end of the stimulus-presentation period, the experimenter assumed a passive-face pose and only then removed the pacifiers. Even with this pacifier technique, the infants were found to imitate the display.

9. For further problems about TT as an account of the ordinary understanding of mental concepts, see Goldman (1993).
10. For an excellent review of this literature, see Flanagan (1991).
11. Similar points are made by Hume in the *Treatise*: "We sympathize more with persons contiguous to us, than with persons remote from us: With our acquaintances, than with strangers: With our countrymen, than with foreigners" (Hume, 1739/1888: 581).
12. The best collection of articles on the psychology of empathy is Eisenberg & Strayer (1987). Also see Wispe (1991). Brandt (1976) has useful discussion of the relation between benevolence, empathy, and moral theory, but with too much emphasis, in my opinion, on the role of conditioning. Empathy and morality are also discussed by Boer & Lycan (1986: chap. 7).
13. Thanks to Holly Smith and Ronald Milo for helpful comments on an earlier draft of this chapter, and to Lynn Nadel for reference to Meltzoff's research.

REFERENCES

Allport, G. W. 1961. *Pattern and Growth in Personality*. New York: Holt, Rinehart, and Winston.

Astington, J. W., & Gopnik, A. 1988. Knowing you've changed your mind: Children's understanding of representational change. In J. W. Astington, P. L. Harris, & D. R. Olson (eds.), *Developing Theories of Mind* (193–206) Cambridge, UK: Cambridge University Press.

——. 1991. Theoretical explanations of children's understanding of the mind. *British Journal of Developmental Psychology* 9:7–31.

Barnett, M. A. 1987. Empathy and related responses in children. In N. Eisenberg & J. Strayer. (eds.), *Empathy and Its Development*. Cambridge, UK: Cambridge University Press.

Baron-Cohen, S., Leslie, A., & Frith, U. 1985. Does the autistic child have a theory of mind? *Cognition* 21:37–46.

Blum, L. A. 1980. *Friendship, Altruism, and Morality*. London: Routledge and Kegan Paul.

——. 1987. Particularity and responsiveness. In J. Kagan & S. Lamb (eds.), *The Emergence of Morality in Young Children*. Chicago: University of Chicago Press.

Boer, S. E., & Lycan, W. G. 1986. *Knowing Who*. Cambridge, MA: MIT Press.

Brandt, R. B. 1976. The psychology of benevolence and its implications for philosophy."*Journal of Philosophy* 73:429–453.

Churchland, P. M. 1988. *Matter and Consciousness*, 2nd ed. Cambridge, MA: MIT Press.
Collingwood, R. G. 1946. *The Idea of History*. Oxford, UK: Clarendon Press.
Darwin, C. 1872/1965. *The Expression of the Emotions in Man and Animals*. Chicago: University of Chicago Press.
Davidson, D. 1984. *Inquiries into Truth and Interpretation*. Oxford, UK: Oxford University Press.
Davies, M., & Stone, T. (eds.) 1995. *Folk Psychology: The Theory of Mind Debate*. Oxford, UK: Blackwell.
Dennett, D. C. 1971. Intentional systems. *Journal of Philosophy* 68:87–106.
——. 1987. *The Intentional Stance*. Cambridge, MA: MIT Press.
DeVries, B., & Trevethan, S. D. 1987. Moral stages and moral orientations in real-life and hypothetical dilemmas. *Child Development* 58:842–858.
Eisenberg, N., & Lennon, R. 1983. Sex differences in empathy and related capacities. *Psychological Bulletin* 94:100–131.
Eisenberg, N., & Strayer, J. (eds.) 1987. *Empathy and Its Development*. Cambridge, UK: Cambridge University Press.
Finke, R. A., & Shepard, R. N. 1986. Visual functions of mental imagery. In K. R. Boff, L. Kaufman, & J. P. Thomas (eds.), *Handbook of Perception and Human Performance*. New York: Wiley.
Flanagan, O. 1991. *Varieties of Moral Personality: Ethics and Psychological Realism*. Cambridge, MA: Harvard University Press.
Flavell, J. H. 1977. *Cognitive Development*. Engelwood Cliffs, NJ: Prentice-Hall.
——. 1988. The development of children's knowledge about the mind. In J. W. Astington, P. L. Harris, & D. R. Olson (eds.), *Developing Theories of Mind* (244–267). Cambridge, UK: Cambridge University Press.
Fodor, J. A. 1987. *Psychosemantics*. Cambridge, MA: MIT Press.
Gibbard, A. 1990. *Wise Choices, Apt Feelings*. Cambridge, MA: Harvard University Press.
Gilligan, C. 1982. *In a Different Voice*. Cambridge, MA: Harvard University Press.
Goldman, A. I. 1989. Interpretation psychologized. *Mind and Language* 4:161–185.
——. 1992. *Liaisons: Philosophy Meets the Cognitive and Social Sciences*. Cambridge, MA: MIT Press.
——. 1993. The psychology of folk psychology. *Behavioral and Brain Sciences* 16:15–28.
——. 1995. In defense of the simulation theory. *Mind and Language* 7(1–2):104–119.
Gopnik, A. 1993. How we know our minds: The illusion of first-person knowledge of intentionality. *Behavioral and Brain Sciences* 16:1–14.
Gordon, R. 1986. Folk psychology as simulation. *Mind and Language* 1:158–171.
Haan, N. 1975. Hypothetical and actual moral reasoning in a situation of civil disobedience. *Journal of Personality and Social Psychology* 32:255–270.
Hare, R. M. 1963. *Freedom and Reason*. Oxford, UK: Clarendon Press.
Harris, P. L. 1989. *Children and Emotion: The Development of Psychological Understanding*. Oxford, UK: Blackwell.
——. 1995. From simulation to folk psychology: The case for development. *Mind and Language* 7:120–144.
Heal, J. 1986. Replication and functionalism. In J. Butterfield (ed.), *Language, Mind, and Logic* (135–150). Cambridge, UK: Cambridge University Press.
Hoffman, M. L. 1987. The contribution of empathy to justice and moral judgment. In Eisenberg & Strayer.(eds.), *Empathy and Its Development*. Cambridge, UK: Cambridge University Press.
Holstein, C. 1976. Development of moral judgment: A longitudinal study. *Child Development* 47:51–61.

Hull, C. L. 1933. *Hypnosis and Suggestibility*. New York: Appleton-Century.

Hume, D. 1739/1888. *A Treatise of Human Nature*, ed. L. A. Selby-Bigge. Oxford, UK: Clarendon Press.

——. 1777/1972. *Enquiry Concerning the Principles of Morals*, 2nd ed., ed. L. A. Selby-Bigge. Oxford, UK: Clarendon Press.

Kahneman, D., & Tversky, A. 1982. The simulation heuristic. In D. Kahneman, P. Slovic & A. Tversky (eds.), *Judgement under Uncertainty: Heuristics and Biases* (201–208). Cambridge, UK: Cambridge University Press.

Klinnert, M. 1981. Infants' use of mothers' facial expressions for regulating their own behavior. Paper presented to the meeting of the Society for Research in Child Development, Boston.

Kosslyn, S. M. 1980. *Image and Mind*. Cambridge, MA: Harvard University Press.

——. 1990: Mental imagery. In D. N. Osherson, S. M. Kosslyn, & J. M. Hollerbach (eds.), *Visual Cognition and Action* (73–97). Cambridge, MA: MIT Press.

Krebs, D. 1975. Empathy and Altrusm. *Journal of Personality and Social Psychology* 32: 1134–1146.

Lennon, R., & Eisenberg, N. 1987. Gender and age differences in empathy and sympathy. In N. Eisenberg & J. Strayer.(eds.), *Empathy and Its Development*. Cambridge, UK: Cambridge University Press.

Loar, B. 1987. Subjective intentionality. *Philosophical Topics* 15:89–124.

McDougall, W. 1908. *Introduction to Social Psychology*. London: Methuen.

Mead, G. H. 1934. *Mind, Self, and Society*. C. M. Morris (ed.), Chicago: University of Chicago Press.

Meltzoff, A. N. 1990. Towards a developmental cognitive science: The implications of cross-modal matching and imitation for the development of representation and memory in infancy. In A. Diamon (ed.), *The Development and Neural Bases of Higher Cognitive Functions*. Annals of the New York Academy of Sciences, vol. 608. New York: New York Academy of Sciences.

Meltzoff, A. N., & Moore, M. K. 1977. Imitation of facial and manual gestures by human neonates. *Science* 198:75–78.

Meltzoff, A. N., & Moore, M. K. 1983. Newborn infants imitate adult facial gestures. *Child Development* 54:702–709.

Mitchell, P., & Lacohée, H. 1991. Children's early understanding of false belief. *Cognition* 39:107–127.

Nozick, R. 1981. *Philosophical Explanations*. Cambridge, MA: Harvard University Press.

Perner, J. 1991. *Understanding the Representational Mind*. Cambridge, MA: MIT Press.

Putnam, H. 1978. *Meaning and the Moral Sciences*. London: Routledge and Kegan Paul.

Quine, W. V. 1960. *Word and Object*. Cambridge, MA: MIT Press.

——. 1990. *Pursuit of Truth*. Cambridge, MA: Harvard University Press.

Rawls, J. 1971. *A Theory of Justice*. Cambridge, MA: Harvard University Press.

Ripstein, A. 1987. Explanation and empathy. *Review of Metaphysics* 40:465–482.

Rousseau, J. J. 1755/1950. *Discourse on the Origin of Inequality*. In *The Social Contract and Discourses*. G. D. H. Cole, trans. (175–282). New York: E. P. Dutton.

Scheffler, S. 1986. Morality's demands and their limits. *Journal of Philosophy* 83:531–537.

Schopenhauer, A. 1841/1965. *On the Basis of Morality*. A. F. J. Payne, trans. Indianapolis: Bobbs-Merrill.

Selman, R. L. 1980. *The Growth of Interpersonal Understanding*. New York: Academic Press.

Simner, M. L. 1971. Newborn's response to the cry of another infant. *Developmental Psychology* 5:136–150.
Smith, A. 1759/1976. *A Theory of Moral Sentiments*. D. D. Raphael & A. L. Macfie eds. Oxford, UK: Clarendon Press.
Spencer, H. 1870. *The Principles of Psychology*, 2nd ed., vol. 1. London: Williams and Norgate.
Stich, S. P. 1981. Dennett on intentional systems. *Philosophical Topics* 12:39–62.
——. 1983: *From Folk Psychology to Cognitive Science: The Case Against Belief*. Cambridge, MA: MIT Press.
Stich, S. P., & Nichols, S. 1995. Folk psychology: Simulation or tacit theory?
Stotland, E. 1969. Exploratory investigations of empathy. In L. Berkowitz (ed.), *Advances in Experimental Social Psychology*, vol. 4. New York: Academic Press.
Taylor, C. 1985. *Philosophy and the Human Sciences*. Cambridge, UK: Cambridge University Press.
Thagard, P., & Nisbett, R. 1983. Rationality and charity. *Philosophy Science* 50:250–267.
Walker, L. J. 1984. Sex differences in the development of moral reasoning critical review. *Child Development* 55:677–691.
Wellman, H. 1990. *The Child's Theory of Mind*. Cambridge, MA: MIT Press.
Wimmer, H., & Hartl, M. 1991. The child's understanding of own false beliefs. *British Journal of Developmental Psychology* 9:125–138.
Wimmer, H., & Perner, J. 1983. Beliefs about beliefs: Representation and constraining function of wrong beliefs in young children's understanding of deception. *Cognition* 13:103–128.
Wispe, L. 1991. *The Psychology of Sympathy*. New York: Plenum Press.

CHAPTER 8

Two Routes to Empathy

Insights from Cognitive Neuroscience

1. DEFINITIONAL OVERVIEW

The concept of empathy has a considerable history in both philosophy and psychology, and may currently be enjoying an apex of attention in both. It is certainly receiving close attention in cognitive neuroscience, which brings fresh discoveries and perspectives to the subject. The term "empathy" however, does not mean the same thing from every mouth. Nor is there a single, unified phenomenon that uniquely deserves the label. Instead, numerous empathy notions or phenomena prance about in the same corral, and part of the present task is to tease some of these notions apart. More important, there are fascinating new findings that should be reported, analyzed, and mutually integrated, whether one's interest in empathy is primarily driven by pure science, philosophy of mind, moral philosophy, or aesthetic theory.

As a first step in distinguishing multiple senses, grades, or varieties of empathy, consider a definition offered by Vignemont & Singer (2006):

> There is empathy if: (i) one is in an affective state; (ii) this state is isomorphic to another person's affective state; (iii) this state is elicited by the observation or imagination of another person's affective state; (iv) one knows that the other person is the source of one's own affective state. (2006: 435)

Issues can be raised about this definition that might motivate alternative definitions. For example, clause (i) restricts empathic states to affective or emotional states, but this is too narrow for some purposes. Cognitive neuroscientists talk of empathy for touch (Keysers et al., 2004) and empathy for pain (Singer et al., 2004; Jackson et al., 2004; Morrison et al., 2004), but neither touch nor pain is usually considered an emotion (although pain has an affective dimension, as well as a sensory one). Concerning clause (ii), it should be asked exactly what is meant by "isomorphic." If it means a state of one person that matches a state of the target, then that requirement is more restrictive than definitions offered by others. Hoffman (2000), for example, defines empathy as "an affective response more appropriate to another's situation than one's own." This doesn't imply that the receiver's affective state matches (or is isomorphic to) that of the target.

Clause (iii) might be questioned on a rather different ground. It seems right to restrict empathic states to ones acquired by observation or imagination of the target individual. But shouldn't the elicitation process be constrained even further? For example, Hume wrote:

> 'Tis indeed evident, that when we sympathize with the passions and sentiments of others, these movements appear at first in *our* mind as mere ideas, and are conceiv'd to belong to another person, as we conceive any other matter of fact.... No passion of another discovers itself immediately to the mind. We are only sensible to its causes or effects. From *these* we infer the passion: and consequently *these* give rise to our sympathy. (1739–1740 (1978): 319, 576)

Hume (using the term "sympathy" rather than "empathy") apparently endorses a three-stage hypothesis: One observes another person's movements, one infers from those movements a certain passion in the person, and the inferred belief causes a matching passion in oneself. If this is right, the process satisfies the Vignemont-Singer definition because the affect is elicited—albeit indirectly—by observation. But many people conceptualize empathy as a spontaneous, non-inferential process. If they wish to define empathy in that fashion, the previous definition would have to be amended to exclude inferential steps.

Another dimension of empathy important to many theorists is "care" or "concern" for the target. This dimension is omitted in the Vignemont-Singer definition. Social psychologists are traditionally interested in empathy as the basis of altruistic behavior, and many would want to highlight that component of it. Other investigators are interested in empathy as a key to mindreading and might even use the term "empathy" to describe (what they take to be) the most common form of mindreading. In other words, they use the term "empathize" as roughly equivalent to "simulate" (in an intersubjective fashion). I myself am a partisan of this position (Goldman,

2006), but this will play only a secondary role in the present chapter. The proffered definition is neutral on the question of mindreading, and that's fine for present purposes.

It is easy to conflate different features of empathy, so readers can sometimes be mystified as to exactly how a given writer uses the term. For example, in Baron-Cohen's (2003) account of autism, or Asperger's syndrome, the linchpin of the account is a deficiency in "empathizing." But in reading Baron-Cohen, it is often difficult to tell which of three possible senses of "empathizing" he primarily has in mind: (1) using simulation when engaging in mentalizing, (2) being curious about others' mental states, or (3) feeling concern about other people's feelings. Correspondingly, a deficiency in empathizing might consist of a sparse use of simulation, a dearth of curiosity about others' mental states, or a low level of concern about other people's feelings.

These preliminary comments should alert the reader to the fact that different writers and researchers exhibit different approaches to empathy. In addition, however, research findings can contribute to an understanding of how empathy is produced. Is there exactly one route to empathy, that is, one cognitive system—or one *type* of cognitive system—that produces empathy, or is there more than one? How exactly does this system (or these systems) work? What different consequences might ensue as upshots of different modes of empathizing? These are the primary questions that this chapter addresses.

2. THE MIRRORING ROUTE TO EMPATHY

In an earlier era, one might have been skeptical about the isomorphism or matching condition we provisionally accepted in the definition of empathy. Do empathizers really undergo states that match those of their targets? Are the feeling states of receivers exactly the same as those of their targets? Since the discovery of mirror neurons and mirroring processes, however, there is much less room for skepticism. There is little doubt about the existence of processes through which patterns of neural activation in one individual lead, via their observed manifestations (e.g., behavior or facial expressions), to matching patterns of activation in another individual. If the corresponding patterns of activation are not perfect duplicates, at least they resemble their corresponding states in the target in terms of the kinds or types of mental or brain activity involved. Some might balk at calling the resonant states "mental" states, because the mirroring episodes commonly occur below the threshold of consciousness even when the episodes being mirrored are fully conscious. If the term "mental" is used broadly, however, they are processes of "mental mimicry."

Mirror neurons and mirroring processes were first discovered in monkeys, and subsequently in humans, in connection with preparation for motor action (Rizzolatti et al., 1996; Gallese et al., 1996). When a monkey plans a certain type of goal-related hand action (e.g., tearing, holding, or grasping) neural cells in its premotor cortex dedicated to the chosen type of action are activated. Surprisingly, when a monkey merely observes another monkey or human perform a similar hand action, the same cells coded for that type of action are also selectively activated. Thus, for certain neurons, there is a sort of neural mirroring; one thing that occurs in the actor's brain is (more or less) replicated in the brain of the observer. These kinds of cells were therefore dubbed "mirror neurons." There are many details concerning the precise activation properties of mirror neurons in an observer versus an actor (Rizzolatti et al., 2001). But the basic finding is that there is robust, selective activation of the same cells in both execution and observation modes.

Using different techniques, an action-related mirror system has been found in humans, centered on the inferior parietal lobule and the premotor cortex, including Brodmann's area 44 (see Rizzolatti & Sinigaglia, 2008). Cochin et al., (1998) showed that the same μ rhythm that is blocked or desynchronized when a human performs a leg or finger movement is also blocked when he merely observes a similar movement by another person. Similar results were obtained from research studies using magnetoencephalography (MEG) and transcranial magnetic stimulation (TMS). Fadiga et al. (1995) recorded the motor evoked potentials (MEPs) induced by magnetic stimulation of the left motor cortex in various muscles of the contralateral (right) hands and arms of subjects who were watching the experimenter either grasp objects with his hand or make movements unrelated to any object. In both cases, a selective increase in MEPs was found in the recorded muscles. Thus, mirroring properties were detected for both the observation of goal-related actions, found in monkeys, and for non-object-related arm movements, not found in monkeys.

A study by Buccino et al. (2001) showed that mirroring for action isn't restricted to actions of the hand or arm. Subjects were shown action stimuli of the following sorts: biting an apple, grasping a cup, kicking a ball, and non-object-related actions involving the mouth, hand, or foot. The results showed that observing both object-related and non-object-related actions led to the somatotopic activation of the premotor cortex, with the mouth represented laterally and the foot medially.

Which mental states are activated in the case of motor mirroring? As I have said, it is presumably plans or intentions to do specific actions. Matching motor plans are activated in the observer, but they don't normally lead to imitation. Their outputs are usually inhibited downstream. There is mental mimicry, one might say, but not behavioral mimicry.

Mimicry of action-planning states doesn't naturally invite the label of empathy. But many other mental states that partake of mirroring more naturally invite talk of empathy. Some writers might prefer other labels. One might speak of "resonance," for example, or "contagion." But I think that "empathy" is a reasonable choice. It must be stressed, however, that in many mirroring activities, the receiving end of the mirroring relationship may not be conscious. The receiver may not be aware, or not fully aware, of the mental event she is undergoing that happens to be congruent with an event in the sender. This may raise issues concerning condition (iv) of the definition discussed earlier. I think it is fair to require a receiver to have some sort of intentional attitude directed toward the target by which the resonating state is linked to him. Otherwise, it doesn't seem like a case of empathy. I suspect that condition (iv) is too strong an intentional condition of this kind, but I don't have a wholly suitable replacement for it.

Even if a suitable replacement for condition (iv) is found, "empathy" might not be a tempting term for mental mimicry of action planning. Let us therefore examine other categories, starting with the sensation of touch. Keysers et al. (2004) found that when a person watches another person being touched, the same brain areas are activated as those in the person being touched. More specifically, they found that touching a subject's own legs activated the primary and secondary somatosensory cortex of the subject. Large extents of the secondary somatosensory cortex also responded to the sight of someone else's legs being touched. Films used with control subjects in which the same legs were approached by an object, but never touched, produced much smaller activations. This phenomenon is naturally described as empathy for touch.

Another mirroring domain involves the sensation of pain. Pain is a complex sensory and emotional mental state associated with actual or potential body damage. Sensory components of pain evaluate the locus, duration, and intensity of a pain stimulus, and affective components evaluate the unpleasantness of the noxious stimulus. These are mapped in different nodes of the so-called pain matrix. Sensorimotor cortices process sensory features of pain and display somatotopical organization (mapping locations of the stimuli in brain tissue). Affective and motivational components of pain are coded in the affective node of the pain matrix, which includes anterior cingulate cortex (ACC) and anterior insula (AI). The subjective feeling of unpleasantness is strictly associated with neural activity in these structures.

In 2004, mirroring for pain was established in three articles: Singer et al. (2004), Jackson et al. (2004), and Morrison et al. (2004). In each of these studies, empathy for pain elicited neural activity mainly in the affective division of the pain matrix, suggesting that only emotional components of pain are shared between self and other. However, using TMS, Avenanti et al. (2005, 2006) found that the direct observation of painful stimulations on a model elicits inhibitory responses in the observer's corticospinal motor system similar to responses found in subjects who actually experience painful stimulations. When participants watched a video showing a sharp needle being pushed into someone's hand, there was a reduction in corticospinal excitability in related muscles. No change in excitability occurred when they saw a Q-tip pressing the hand or a needle being pushed into a tomato. These mirror responses were specific to the body part that the subjects observed being stimulated and correlated with the intensity of the pain ascribed to the model, thus hinting at the sensorimotor side of empathy for pain.

The best example of mirroring in the sphere of emotions features the emotion of disgust, and the clearest evidence comes from an fMRI study by Wicker et al. (2003). Participants were scanned while passively inhaling disgusting or pleasant odorants through a mask and, separately, while observing movies of individuals who smelled the contents of a glass (disgusting, pleasant, or neutral) and spontaneously manifested appropriate facial expressions. The core finding was that the left AI—previously known to be implicated in disgust experience—and the right ACC were preferentially activated both during the inhaling of disgusting odorants (compared with pleasant and neutral odors) and during the observation of disgust facial expressions (compared with pleasure-expressive and neutral faces). This shows that observing a disgust-expressive face produces mental mimicry, or empathy, in an observer of the model. To use another expression very common in the literature, part of the observer's brain *simulates* the activity of a corresponding part of the model's brain.

In addition to the fMRI demonstration of matching experiences in observers and models in the gustatory cortex, researchers have used another measure of empathy to test whether observers experienced empathy. Jabbi et al. (2007) examined whether the IFO (anterior insula and adjacent frontal operculum) was associated with observers' self-reported empathy, measured by the Interpersonal Reactivity Index (IRI). They found that participant observers' empathy scores were predictive of their gustatory IFO activation while witnessing both the pleased and the disgusted facial expressions of others.

As is evident from the foregoing, a variety of systems in the human brain have mirror properties. They do not all use the same neural network or hardware. In particular, the mirror systems associated with sensations and emotions do not use the same neural hardware as the motor mirror system, nor as one another. Nonetheless, I shall treat them all as similar in having significant mirror properties, for present purposes. Ascending to an appropriate level of abstraction, we can consider them all to instantiate a single *type* of route to empathy, namely a mirroring route. This does not imply that they all use the very same cytoarchitectural pathway.[1]

3. A RECONSTRUCTIVE ROUTE TO EMPATHY

Granted that mirroring constitutes *one* (type of) route to empathy, is it the only type? This section presents two reasons to suspect otherwise. Mirroring seems to be, at least in one respect, automatic. The nature and content of mirroring events seem to be "pre-packaged"; they are not constructed on the fly. The disgust system, for example, is ready to respond to appropriate facial stimuli in a disgust-production mode. It doesn't have to manufacture a novel response to simulate the corresponding disgust experience in a model. Similarly, the action repertoire susceptible to motor mirroring is presumably pretty well fixed early in life. Although there is no consensus about the origins of mirror neurons, one promising hypothesis posits the work of the associative mechanism of "Hebbian learning" (Heyes, 2005; Keysers & Perrett, 2004; Keysers & Gazzola, 2006). According to this hypothesis, mirror properties of visuomotor neurons are shaped in infancy, as a result of synchronous firing, and their subsequent activation should not require substantial online construction. In contrast with this automaticity of mirror-based empathy, a large chunk of empathy seems to involve a more effortful or constructive process. When empathizing with another, you often reflect on that person's situation, construct in imagination how things are (were, or will be) playing out for him, and imagine how you would feel if you were in his shoes. This process of perspective-taking is the stuff of which most conscious empathizing, at any rate, is made. It doesn't have the effortless, automatic quality of mirroring (if mirroring is describable as having any "quality" at all, i.e., any phenomenological "feel"). This suggests that there is, indeed, a different kind of empathy in addition to mirroring. In fact, this other kind of empathy is more detectable in daily life than the mirroring kind, as mirroring is largely inaccessible to introspective awareness. Such a distinction between two types of empathy is embraced by Stueber (2006), who calls them "basic" and "reenactive" empathy respectively.

One must be careful in presenting the foregoing argument because some findings indicate that mirroring is not automatic in all respects. Singer et al. (2006) and Vignemont & Singer (2006) found that empathic responses to pain are modulated by learned preferences, and hence are not purely automatic. In the Singer et al. experiment, participants played a prisoner's dilemma game in which confederates of the experimenters played either fairly or unfairly. Participants then underwent functional imaging while observing the confederates receive pain stimuli. The mirroring responses of the male participants were of special interest. Their level of pain mirroring was significantly reduced when observing painful stimuli being applied to individuals who had played unfairly. Thus, their level of pain activation was not automatic in the sense of being purely stimulus-driven. Rather, it was modulated by internal preferences acquired from information about the targets. A similar result was obtained by Lamm et al. (2007), who found that subjects have a weaker empathetic response in pain-related areas when they know that the pain inflicted on another is useful as a cure.

However, the fact that mirroring can be modulated does not imply that it is a constructive activity comparable to creating an imagined scenario or adopting another person's perspective. Modulation of pain responses is inhibitory activity, something much less complex than the construction of an imagined scenario. It is the constructional aspect of many instances of empathizing that I mean to highlight here. Mirroring is subject to modulation, but this doesn't make it a constructive or effortful activity, like some form of empathizing appears to be. In view of these features of the second type of empathizing, I shall call it *reconstructive* empathy (cf. Vignemont, 2008).

Another argument for this second route to empathy proceeds as follows: Assume that this kind of empathizing involves adopting the perspective of the empathic target. It is widely thought that such perspective-taking (arguably a form of simulation) is a crucial part of mindreading, or theory of mind (ToM). We can then argue from functional neuroimaging data about ToM that this kind of empathizing is probably not the same as mirroring, because the brain regions subserving ToM have minimal overlap with either motoric mirror areas or areas involved in the mirroring of sensations or emotions. Of course, we have previously argued that mirroring is not subserved by a unique set of brain regions. In principle, then, areas involved in ToM might be mirror areas. This is possible in principle, but there is no evidence to support it. Thus, if empathizing is involved in these other types of mindreading (e.g., attribution of beliefs and thoughts), it is likely to be a different type of empathizing process than mirroring.

Which brain regions are implicated in the mindreading of beliefs and other propositional attitudes? According to a number of researchers, they include the medial prefrontal cortex (MPFC), the temporoparietal junction (right and left), and the temporal poles. Some authors contend that one area in particular, the right temporoparietal junction (RTPJ), is specifically involved in tasks concerning belief attribution (Saxe & Kanwisher, 2003; Saxe & Powell, 2006; Saxe, 2006). Assuming that these brain regions are indeed involved in mindreading, what reason is there to suspect a connection between them and empathizing? What is the connection, after all, between mindreading and empathizing—especially reconstructive empathizing?

According to the simulation approach to mindreading, especially as developed in my *Simulating Minds* (Goldman, 2006), there is a very tight connection. In what I call high-level simulation (Goldman, 2006: chap. 7), mindreading another person's mental state involves an attempt to replicate or re-experience the target's state via a constructive process. Exploiting prior information about the target, the mindreader uses "enactment imagination" to reproduce in his own mind what might have transpired, or may be transpiring, in the target. This coincides with the reconstructive type of empathizing proposed here. The only difference is that mindreading involves an additional final step in which one or more of the constructed mental states are categorized (commonly, in terms of both mental type and propositional content) and assigned to the target. This final stage—especially the categorization element—may be absent in empathizing.

4. A POSSIBLE NEURAL SYSTEM SUBSERVING RECONSTRUCTIVE EMPATHIZING

Is there a neural system that subserves a process of reconstructive empathizing? Let us reconnoiter the subject by starting at what seems like a great distance: episodic memory. Episodic memory allows individuals to project themselves backward in time and recollect aspects of their previous experience (Tulving, 1983; Addis et al., 2007). A growing number of investigators, however, have begun to approach episodic memory in a broader context, one that emphasizes people's ability both to re-experience episodes from the past and also imagine or "pre-experience" episodes that may occur in the future (Atance & O'Neill, 2005; D'Argembeau & Van der

Linden, 2004; Gilbert, 2006; Klein & Loftus, 2002; Schacter & Addis, 2007; Schacter, Addis, & Buckner, 2007; Buckner & Carroll, 2007). Evidence for a linkage between representations of past events and future events initially comes from studies of patients with episodic memory deficits. Tulving's (1985) patient KC suffered from total loss of episodic memory due to damage to the medial temporal and frontal lobes. KC was also unable to imagine specific events in his personal future, despite no loss in general imagery abilities. A second amnesic patient, DB, also exhibited deficits in both retrieving past events and imagining future events (Klein & Loftus, 2002). DB's deficit in imagining the future was also specific to his personal future; he could still imagine possible future events in the public domain (e.g., political events and issues). In general, projecting one's thoughts backward or forward in time is referred to as "mental time travel."

Hassabis et al. (2007) examined the ability of five amnesic patients with bilateral hippocampal damage to imagine novel experiences (see the summary by Schacter et al., 2007). The imaginary constructions by four of the five patients were greatly reduced in richness and content compared with those of control subjects. Because this study did not specifically require patients to construct scenes pertaining to future events, they seem to suffer from a more general deficit to construct novel scenes. Recent neuroimaging studies provide insight into whether common brain systems are used while remembering the past and imagining the future. In a PET study by Okuda et al. (2003), participants talked freely about either the near or distant past or future. The scans showed evidence of shared activity during descriptions of past and future events in a set of regions that included the prefrontal cortex and parts of the medial temporal lobe—namely the hippocampus and the parahippocampal gyrus.

Drawing on these and related studies (see also Schacter et al., 2007, 2008; Schacter & Addis, 2009), Buckner & Carroll (2007) have proposed a core brain system that subserves as many as four forms of self-projection. These include remembering the past, thinking about the future (prospection), conceiving the viewpoint of others (ToM), and navigation. What these mental activities all share is a shift of perspective from the immediate environment to an alternative situation. All four forms rely on autobiographical information and are constructed as a "perception" of an alternative perspective. (This brain system also goes under the label of "the default network.")

The hypothesized core brain system involves frontal lobe systems traditionally associated with planning and medial temporal-parietal lobe systems associated with memory. How does ToM fit into this picture neuroanatomically? Buckner and Carroll suggest that Saxe & Kanwisher's (2003) findings on the role of right TPJ in ToM provide further evidence that the core system extends to ToM. In the Saxe & Kanwisher (2003) study, individuals answered questions about stories that required participants to conceive a reality that was different from the current state of the world. In one condition, the conceived state was a belief; in the other, it was an image held by an inanimate object (e.g., a camera). Conceiving of the beliefs of another person strongly activated the network shared by prospection and remembering, whereas the control condition did not. Buckner and Carroll also cite Gallagher & Frith's (2003) proposal that the frontopolar cortex contributes to ToM. In particular, the paracingulate cortex, the anterior-most portion of the frontal midline, is recruited in executive components of simulating others' perspectives. Thus, the Buckner-Carroll suggestion is that the core brain system is used by many diverse types of tasks that require mental simulation of alternative perspectives, and this includes thinking about the perspectives of other people.

Shanton (unpublished) follows up the hypothesis of Schacter, Addis, Buckner, and Carroll by identifying an assortment of experimentally confirmed parallels between episodic memory and ToM. She begins by explaining how each can be understood as a form of "enactment imagination," in the sense of Goldman (2006: chaps. 2, 7). Enactment imagination is a species of imagination in which one tries to match a mental state or sequence of mental states in another by recreating or pre-creating this state or states in oneself. Shanton (unpublished; Shanton & Goldman, 2010) argues that if the same type of simulation strategy is used for both episodic memory and mindreading tasks, there should be parallels in terms of various cognitive parameters. She reviews evidence of several such parallels, including (1) their developmental timeline, and (2) their susceptibility to egocentric biases.

Consider first the fact that episodic memory and mindreading share a developmental timeline. According to Tulving (2001), episodic memory retrieval emerges around the age of four years. This is confirmed by Perner & Ruffman (1995), who had children between three and six years of age complete both free-recall and cued- recall memory tasks. These tasks tap different types of memory abilities. In cued-recall tasks, semantic information is quite rich, whereas in free-recall tasks, where no explicit retrieval cues are given, such information is relatively poor. Free-recall tasks cannot be successfully answered without episodic memory. Perner and Ruffman found that only four-to-six-year-old children, not three-year-olds, could succeed on free-recall tasks, supporting the hypothesis that episodic memory retrieval emerges around age four. This corresponds to the traditional timeline for success in advanced mindreading tasks, such as (verbal) false-belief tasks.

Next consider the susceptibility of both high-level mindreading and episodic memory retrieval to egocentric biases. One example of egocentric bias is the "curse of knowledge" (Camerer et al., 1989). This is the tendency to proceed as if other people know what you do, even when you have information to the contrary. In the Camerer et al. study, well-informed people were required to predict corporate earnings forecasts that would be made by other, less-informed people. The better-informed people stood to gain if they disregarded their own knowledge when making predictions about the well-informed people, whom they *knew* lacked the same knowledge. Nonetheless, they failed to disregard their own knowledge completely, letting it "seep" into their predictions. Simulationists would say that the predictors, while attempting to imagine themselves in the shoes of the predictees, allowed their own knowledge to "penetrate" their imaginative construction. In other words, their own genuine mental states were not excluded, or quarantined, from the construction, despite the fact that good (i.e., accurate) simulation requires such quarantining. Quarantine failure is extremely common in (high-level) mindreading. For example, Van Boven & Loewenstein (2003) asked participants to predict states like hunger and thirst in a group of hypothetical hikers lost in the woods with neither food nor water. Their predictions were solicited either before or after they vigorously exercised at a gymnasium. In the case of post-exercise participants, the combined feelings of thirst and warmth were positively associated with their predictions of the hikers' feelings. Here too there is apparent failure to quarantine one's own concurrent states while mindreading hypothetical targets.

Quarantine failure is found in episodic memory. A vivid illustration is from Levine's (1997) study of subjects' memories for their own past emotion states. During the 1992 presidential race, Levine first asked a group of Ross Perot supporters about their emotions immediately after Perot withdrew from the race in July, and later asked them again in November, after they had switched their allegiances to other candidates. Although in July they described themselves as very sad, angry, and hopeless, by November they remembered experiencing much lower levels of emotion (in July). Apparently their November memories were being influenced by their current attitudes toward Perot. Their episodic memories were constructions that were partly influenced, or colored, by the way they felt at the time of memory retrieval.

Shanton argues that the best explanation for these similarities is that the two processes implement the same cognitive strategy, and she argues (based on additional evidence) that this strategy is enactment imagination. This is a different form of simulation than mirroring.

How does enactment imagination differ from mirroring? In the case of mirroring processes, the default upshot is the successful production of a match between the sender state and the receiver state. Disgust in a sender is reproduced, with reasonable accuracy, by disgust in the observer. In the case of enactment imagination, in contrast, the prospects for successful correspondence are much more tenuous. They depend heavily on the vicissitudes of prior information, construction, and/or elaboration. In the case of mindreading, the vicissitudes of prior information are particularly important. If one doesn't have accurate and relevantly complete information about the prior mental states of the target, attempts to put oneself in that person's mental shoes in order to extrapolate some further mental state have relatively shaky prospects for success.

In my view, it isn't entirely clear that the same core brain system described by Buckner, Carroll, and Schacter and colleagues includes ToM, or mentalizing. For example, in describing their core system, Buckner and Carroll say that it extends to lateral parietal regions located within the inferior parietal lobule "near" the temporoparietal junction (TPJ). But being *near* the TPJ may not be sufficient to identify this area as a locus of mentalizing activity. However, my brief for a simulation system that leads to both mindreading and empathizing via *reconstruction* rather than mirroring does not depend essentially on neuroanatomical evidence. If the specific core brain system hypothesized by Buckner, Carroll, and Schacter does not extend to mindreading and empathizing, this would still be compatible with there being a constructive, or reconstructive, species of empathizing. If the core brain system does subserve mindreading and empathizing, that is just gravy.

5. OUTPUT PROFILES OF THE TWO ROUTES TO EMPATHY

The topic of the last three sections has not been states of empathy per se but different *routes* to empathy. Routes to empathy are species of mental activity that (often) lead to empathic states, where empathic states are defined as indicated in section 1 (with possible modifications considered there). The next natural question to ask is how successful or unsuccessful are the two different routes in generating empathic states, that is, states that exemplify substantial isomorphism to those of their targets. How do they fare in comparative terms? Are there characteristic differences in the empathic outputs of the two different routes?

The question of comparative success or accuracy can be decomposed into several sub-questions. First, one can ask about the *reliability* of a route or method. Of the states produced by a given method, how many are genuinely isomorphic to those of the target?[2] Second, one can ask about the *fecundity* of a route or method. For each application of a method, how many isomorphic states (on average) are produced in the empathizer? It should not be assumed that each application generates precisely one output state. Either of the two methods may generate more states (per use) than the other, and such greater fecundity may be important because it is associated with greater intersubjective understanding.

Restricting ourselves initially to the reliability question, the issue resolves into further sub-questions, because each state has more than one dimension, and we can ask with respect to each dimension whether a given method produces output states that resemble the target on that dimension. Vignemont (2010) distinguishes four main dimensions of emotional states: (1) the *type* of state, (2) the *focus* (object) of a state, (3) the *functional role* of a state, and (4) the *phenomenology* of a state. A given route or method of empathizing might be more reliable than another route with respect to some of these dimensions but less reliable with respect to others.

For reasons previously sketched, it seems likely that the mirroring method of empathizing is more reliable than the reconstructive method when it comes to the *type* of emotional state. Mirroring, by its very nature, is a highly reliable method of state generation, one that preserves at least the sameness of mental-state type (e.g., pain, disgust, fear). There is no comparable guarantee (or near guarantee) in the case of the reconstructive method. Outputs of a constructive or reconstructive method depend heavily on the pretend inputs that the empathizer uses, and the accuracy of these inputs can vary widely depending on the quality of her background information. In short, in terms of reliability with respect to type, the mirroring route seems superior to the reconstructive route.

What about the focus dimension of the state: what the emotion or other state is *about*. As Vignemont argues, the mirroring method does not seem to be so helpful in this regard, whereas the reconstructive method is (or might be). Vignemont's example is seeing a smiling stranger on a train. Seeing the smile prompts a happy state in the observer by the mirroring route. But the object of the stranger's happiness remains undisclosed by mirroring. Mirroring reproduces in the observer only happiness, not happiness *about X* or *Y*. On the other hand, argues Vignemont, reconstructive empathizing can be helpful with respect to focus. By adopting the target's perspective, an empathizer can figure out what the object's emotion is about, or directed at, at least when appropriate information is available. Thus, reconstructive empathizing seems to be superior to mirroring in this regard.

Vignemont includes an *intensity* dimension for output states, which she subsumes under phenomenology. Although I agree that an intensity dimension is relevant here, it is not clear that it should be confined to the sphere of phenomenology. As previously indicated, mirroring states often fail to reach the threshold of consciousness, so they may have no phenomenology at all. Does this mean that they have zero intensity? This would be an unsatisfactory inference because unconscious states certainly have important functional properties, including tendencies to influence behavior. On the other hand, what alternative measure of intensity should be selected? Should some measure of neural activity be used? Which one? In any case, once a measure of intensity is chosen, the question is whether the mirroring method or the reconstructive method is more reliable, that is, which tends to produce mirrored states with greater isomorphism? It isn't obvious (to me) what the answer is; this question invites more research.

Vignemont regards the reconstructive method as superior to the mirroring method, but since she herself doesn't draw the reliability/fecundity distinction, it is an open question whether the intended superiority is supposed to hold for both reliability and fecundity, or for fecundity only. She writes:

> Low-level [mirroring] empathy does not meet the condition of isomorphism [because it is limited to the *type* of emotion, and does not go beyond that]. Emotional sharing may be more exhaustive in high-level empathy. (2008)

To support this idea, she considers a case of a woman learning that a friend is pregnant. Because the empathizer knows how much the friend wanted a child, she puts herself in the friend's shoes and realizes how happy she must be. She feels happy with her. The inputs in such a case of reconstructive empathizing are more complex than the inputs to mirroring empathizing, and this allows one to fill out the target's mental states more fully, or in greater detail. Continuing with the pregnancy example, the empathizer pretends that she is pregnant and that she wants a child, which leads her to feel happy. Her emotional state is *about* the pregnancy; it has the same focus as the friend's emotional state. The mirroring method, says Vignemont, "isolates" a mirrored emotion from the rest of the target's mental life. It does not provide a fine-grained sharing of states based on a common network of associated mental states. Reconstructive empathizing does provide this.

Suppose Vignemont means to say that the reconstructive method is superior in both reliability and fecundity. I would be prepared to concede the fecundity part, because the reconstructive method is obviously capable of generating more and more detailed isomorphic states than mirroring. But is it more reliable? I am skeptical. Vignemont ignores two types of error to which only the reconstructive method is liable. The first type of error is one of omission: omitting relevant inputs because of ignorance. If an (attempted) empathizer in the pregnancy case is unaware that her friend is pregnant or is unaware that she wants a child, application of the reconstructive method is unlikely to produce *correct* details involving the target emotion. The second type of error is one of commission. As reviewed above, there is substantial evidence that when people try to simulate the mental states of others, they often fail to quarantine their own genuine states, allowing such states to seep into the simulation process when they don't properly belong there (because the target isn't in them). This results in egocentric biases in the simulation process. Both types of errors can substantially reduce the reliability of the reconstructive method, so I cannot concur with Vignemont's rosy appraisal of it. An assessment of the comparative reliability of the two methods needs more work. Nonetheless, it is good to have this problem placed squarely on the table; it deserves attention.

I have argued that there are two distinct routes to empathy, the mirroring route and the reconstructive route. It is possible, however, that the reconstructive route also involves mirror neurons. This is suggested by Iacoboni and colleagues (Iacoboni, 2011; Uddin et al., 2007). Iacoboni (2011) reports the recent discovery of mirror neurons in several new areas, including the amygdala, hippocampus, parahippocampal gyrus, and entorhinal cortex. He suggests that these mirror neurons may underpin what I earlier called high-level mindreading and empathy (Goldman, 2006), which correspond to what is here called reconstructive empathy.[3] Thus, it is possible that even reconstructive empathy is mediated by neurons with mirror properties. Note, however, that this would not necessarily undercut the distinction between mirroring and reconstructive processes. As standardly conceived, mirror processes are automatic processes generated by observation. In addition, neurons with mirror properties might also participate in such an effortful process as imagination (see Uddin et al., 2007: box 3), a key component of reconstructive empathy. These ideas require further investigation.

NOTES

1. However, it is also not implied that the various routes to empathy are non-overlapping. On the contrary, certain neural centers seem to be involved in the mirroring routes for several different sensations and/or emotions. The anterior insula appears to play a particularly important role in the processing of several such mental states (Singer et al., 2009).
2. The term "reliability" is not used here in exactly the same sense in which it is used in epistemology, because we are not discussing the formation of true or false beliefs. Instead, we are discussing how much, or to what degree, one mental state is isomorphic to (resembles) another. Degrees of reliability are to be computed in terms of proportions of isomorphic versus non-isomorphic features (or something along these lines).
3. Notice that some of the areas containing mirror neurons mentioned by Iacoboni are the same as midline areas mentioned by Okuda et al. (2003) in their study of constructive imagination, specifically the hippocampus and the parahippocampal gyrus.

REFERENCES

Addis, D. R., Wong, A. T., & Schacter, D. L. (2007). Remembering the past and imagining the future: Common and distinct neural substrates during event construction and elaboration. *Neuropsychologia* 45:1363–1377.

Atance, C. M., & O'Neill, D. K. (2005). The emergence of episodic future thinking in humans. *Learning and Motivation* 36:126–144.

Avenanti, A., Bueti, D., Galati, G., & Aglioti, S. M. (2005). Transcranial magnetic stimulation highlights the sensorimotor side of empathy for pain. *Nature Neuroscience* 8:955–960.

Avenanti, A., Paluello, I. M., Bufalari, I., and Aglioti, S. M. (2006). Stimulus-driven modulation of motor-evoked potentials during observation of others' pain. *NeuroImage* 32:316–324.

Baron-Cohen, S. (2003). *The Essential Difference: The Truth about the Male and Female Brain.* New York: Basic Books.

Buccino, G., Binkofski, F., Fink, G. R., Fadiga, L., Fogassi, L., Gallese, V., Seitz, R. J., Zilles, K., Rizzolatti, G., & Freund, H.-J. (2001). Action observation activates premotor and parietal areas in a somatotopic manner: An fMRI study. *European Journal of Neuroscience* 13(2):400–404.

Buckner, R. L., & Carroll, D. C. (2007). Self-projection and the brain. *Trends in Cognitive Sciences* 11:49–57.

Camerer, C., Loewenstein, G., & Weber, M. (1989). The curse of knowledge in economic settings: An experimental analysis. *Journal of Political Economy* 97: 1232–1254.

Cochin, S., Bethelemy, C., Lejeune, B., Roux, S., & Martineau, I. (1998). Perception of motion and qEEG activity in human adults. *Electroencephalography and Clinical Neurophysiology* 107:287–295.

D'Argembeau, A., and Van der Linden, M. (2004). Phenomenal characteristics associated with projecting oneself back into the past and forward into the future: Influence of valence and temporal distance. *Consciousness and Cognition* 13:844–858.

Fadiga, L., Fogassi, L., Pavesi, G., and Rizzolatti, G. (1995). Motor facilitation during action observation: A magnetic stimulation study. *Journal of Neurophysiology* 73: 2608–2611.

Gallagher, H. L., and Frith, C. (2003). Functional imaging of 'theory of mind.' *Trends in Cognitive Sciences* 7:77–83.

Gallese, V., Fadiga, L., Fogassi, L., and Rizzolatti, G. (1996). Action recognition in the premotor cortex. *Brain* 119:593–609.

Gilbert, D.T. (2006). *Stumbling on Happiness*. New York: Knopf.

Goldman, A. I. (2006). *Simulating Minds: The Philosophy, Psychology, and Neuroscience of Mindreading*. New York: Oxford University Press.

Hassabis, D., Kumaran, D., Vann, S. D., & Maguire, E. A. (2007). Patients with hippocampal amnesia cannot imagine new experiences. *Proceedings of the National Academy of Sciences U.S.A.* 104:1726–1731.

Heyes, C. (2005). Imitation by association. In S. Hurley & N. Chater, eds., *Perspectives on Imitation*, vol. 1, *Mechanisms of Imitation and Imitation in Animals*, 157–176. Cambridge, MA: MIT Press.

Hoffman, M. (2000). *Empathy and Moral Development*. New York: Cambridge University Press.

Hume, D. (1939–40/1978). *A Treatise of Human Nature*. Oxford, UK: Clarendon Press.

Iacoboni, M. (2011). Within each other: Neural mechanisms for empathy in the primate brain. In A. Coplan & P. Goldie, eds., *Empathy: Philosophical and Psychological Perspectives*, 45–57. Oxford, UK: Oxford University Press.

Jabbi, M., Swart, M., & Keysers, C. (2007). Empathy for positive and negative emotions in the gustatory cortex. *NeuroImage* 34:1744–1753.

Jackson, P. L., Meltzoff, A. N., & Decety, J. (2004). How do we perceive the pain of others? A window into the neural processes involved in empathy. *NeuroImage* 24: 771–779.

Keysers, C., & Gazzola, V. (2006). Toward a unifying neural theory of social cognition. *Progress in Brain Research* 156:383–406.

Keysers, C., & Perrett, D. (2004). Demystifying social cognition: A Hebbian perspective. *Trends in Cognitive Sciences* 8:501–507.

Keysers, C., Wicker, B., Gazzola, V., Anton, J.-L., Fogassi, L., & Gallese, V. (2004). A touching sight: SII/PV activation during the observation of touch. *Neuron* 42: 335–346.

Klein, S. B., & Loftus, J. (2002). Memory and temporal experience: The effects of episodic memory loss on an amnesic patient's ability to remember the past and imagine the future. *Social Cognition* 20:353–379.

Lamm, C., Batson, C. D., & Decety, J. (2007). The neural substrate of human empathy: Effects of perspective-taking and cognitive appraisal. *Journal of Cognitive Neuroscience* 19(1):42–58.

Levine, L. (1997). Reconstructing memory for emotions. *Journal of Experimental Psychology: General* 126(2):165–177.

Morrison, I., Lloyd, D., di Pellegrino, G., & Roberts, N. (2004). Vicarious responses to pain in anterior cingulate cortex: Is empathy a multisensory issue? *Cognitive, Affective, Behavioral Neuroscience* 4:270–278.

Okuda, J., Fujii, T., Ohtake, H., Tsukiura, T., Tanji, K. Susuki, K., et al. (2003). Thinking of the future and the past: The roles of the frontal pole and the medial temporal lobes. *NeuroImage* 19:1369–1380.

Perner, J., & Ruffman, T. (1995). Episodic memory and autonoetic consciousness: Developmental evidence and a theory of childhood amnesia. *Journal of Experimental Child Psychology* 59:516–548.

Rizzolatti, G., Fadiga, L., Gallese, V., & Fogassi, L. (1996). Premotor cortex and the recognition of motor actions. *Cognitive Brain Research* 3:131–141.

Rizzolatti, G., Fogassi, L., & Gallese, V. (2001). Neurophysiological mechanisms underlying the understanding and imitation of action. *Nature Neuroscience Reviews* 2:661–670.

Rizzolatti, G., & Sinigaglia, C. (2008). *Mirrors in the Brain*. Oxford, UK: Oxford University Press.

Saxe, R. (2006). Uniquely human social cognition. *Current Opinion in Neurobiology* 16: 235–239.

Saxe, R., & Kanwisher, N. (2003). People thinking about thinking people: fMRI studies of theory of mind. *NeuroImage* 19:1835–1842.

Saxe, R., & Powell, L. J. (2006). It's the thought that counts: Specific brain regions for one component of theory of mind. *Psychological Science* 17(8):692–699.

Schacter, D. L., & Addis, D. R. (2007). The cognitive neuroscience of constructive memory: Remembering the past and imagining the future. *Philosophical Transactions of the Royal Society of London, Series B: Biological Sciences* 362: 773–786.

Schacter, D. L., Addis, D. R., & Buckner, R. L. (2007). Remembering the past to imagine the future: The prospective brain. *Nature Reviews Neuroscience* 8:657–661.

Schacter, D. L., Addis, D. R., & Buckner, R. L. (2008). Episodic simulation of future events. *Annals of the New York Academy of Science* 1124:39–60.

Schacter, D. L., & Addis, D. R. (2009). On the nature of medial temporal lobe contributions to the constructive simulation of future events. *Philosophical Transactions of the Royal Society, B* 364:1245–1253.

Shanton, K. (unpublished). Episodic memory and mindreading share a common simulational strategy. Rutgers University, department of philosophy.

Shanton, K., & Goldman, A. I. (2010). *Simulation Theory*. In L. Nadel and S. Nichols, eds., *Wiley Interdisciplinary Reviews: Cognitive Science*,1(4). Published online Feb. 5, 2010, DOI 10:1002/wcs.33.

Singer, T., Critchley, H. D., & Preuschoff, K. (2009). A common role of insula in feelings, empathy and uncertainty. *Trends in Cognitive Sciences* 13:334–340.

Singer, T., Seymour, B., O'Doherty, J., Kaube, H., Dolan, R., & Frith, C. (2004). Empathy for pain involves the affective but not sensory components of pain. *Science* 303:1157–1162.

Singer, T., Seymour, B., O'Doherty, J., Stephan, K. E., Dolan, R., & Frith, C. (2006). Empathic neural responses are modulated by the perceived fairness of others. *Nature* 439:466–469.

Stueber, K. (2006). *Rediscovering Empathy: Agency, Folk Psychology, and the Human Sciences*. Cambridge, MA: MIT Press.

Tulving, E. (1983). *Elements of Episodic Memory*, vol. 2. New York: Oxford University Press.

Tulving, E. (1985). Memory and consciousness. *Canadian Psychologist* 25:1–12.

Tulving, E. (2001). Episodic memory and common sense: How far apart? In Baddeley, A., Conway, M. A., & Aggleton, J. P., eds., *Episodic Memory: New Directions in Research*, 269–288. New York: Oxford University Press.

Uddin, L. Q., Iacoboni, M., Lange, C., & Keenan, J. P. (2007). The self and social cognition: The role of cortical midline structures and mirror neurons. *Trends in Cognitive Sciences* 11:153–157.

Van Boven, L., & Loewenstein, G. (2003). Social projection of transient drive states. *Personality and Social Psychology Bulletin* 29(9):1159–1168.

Vignemont, F. de (2008). Empathie miroir et empathie reconstructive. *Revue Philosophique de la France and de l'Etranger* 133(3):337–345.

Vignemont, F. de (2010). Knowing other people's mental states as if they were one's own. In D. Schmicking & S. Gallagher, eds. (283–299), *Handbook of Phenomenology and Cognitive Science*. Dordrecht, NL: Springer.

Vignemont, F. de, and Singer, T. (2006). The empathetic brain: How, when, and why? *Trends in Cognitive Sciences* 10:435–441.

Wicker, B., Keysers, C., Plailly, J., Royet, J.-P., Gallese, V., & Rizzolatti, G. (2003). Both of us disgusted in my insula: The common neural basis of seeing and feeling disgust. *Neuron* 40:655–664.

CHAPTER 9

Is Social Cognition Embodied?

(WITH FRÉDÉRIQUE DE VIGNEMONT)

INTRODUCTION

A specter is haunting the laboratories of cognitive science, the specter of embodied cognition (EC). For decades, the reigning paradigm of cognitive science has been classicism. On this approach, higher cognitive functions are analogized to the operations of a computer, manipulating abstract symbols on the basis of specific computations. As embodiment theorists tell the story, classical cognitivism (CC) claims that mental operations are largely detached from the workings of the body, the body being merely an output device for commands generated by abstract symbols in the mind (or the "central system" of the mind). Embodiment theorists want to elevate the importance of the body in explaining cognitive activities. What is meant by "body" here? It ought to mean the whole physical body minus the brain. Letting the brain qualify as part of the body would trivialize the claim that the body is crucial to mental life, simply because the brain is the seat of most, if not all, mental events.

Proponents of EC are found in virtually all sectors of cognitive science. They include artificial intelligence [1], psychology [2–5], cognitive neuroscience [6,7], linguistics [8] and philosophy [9–12]. However, embodiment seems to mean widely different things to different EC theorists, and their views range from the radical to the not so radical. In view of this diversity, it is impossible to canvass all varieties of EC (for review, see ref. 13).

We begin by laying out four general constraints on a conceptually satisfying and empirically fruitful definition of EC: (1) A definition should assign central importance to the body (understood literally), not simply to the situation or environment in which the body is embedded. Many theorists more or less equate EC with situated cognition; we focus entirely on the former. (2) The definition should concentrate on the cognizer's own body, not the bodies of others. Perception of another person's body should not automatically count as EC. (3) Any substantial EC thesis should be a genuine rival to CC. (4) It should also make a clear- enough claim that its truth or falsity can be evaluated by empirical evidence. After assessing candidate definitions of EC in terms of the foregoing desiderata, we shall choose our favorite candidate (not necessarily excluding others) and apply it to social cognition. We shall ask how strongly the current empirical evidence supports EC as a thesis about social cognition.

INTERPRETATIONS OF EMBODIMENT

We shall formulate four definitions or conceptions of embodiment (figure 9.1). Because many pre-existing formulations of embodiment are rather opaque, we hope that our proposals will bring increased clarity to this matter. For any conception of embodiment, of course, it could be claimed that "all" of cognition is embodied, or that "90 percent" of cognition is embodied, etc. Nobody is in a position to address this quantificational issue with any precision, and we shall leave this question open here, merely assuming that EC theses want embodiment to have, at least, an "important" role.

Body anatomy interpretation: Parts of the body have an important causal role in cognition in virtue of their distinctive anatomy

This definition is motivated by such obvious facts as this: If we had the bat's system of echolocation instead of human eyes, we would perceive the world differently than we do. Thus, the physical body (distinct from the brain) influences the nature of our perceptions. This characterization of EC does not conflict with CC. No advocate of CC would disagree with this trivial claim. Therefore, constraint (3) is violated, and this interpretation is unacceptable.

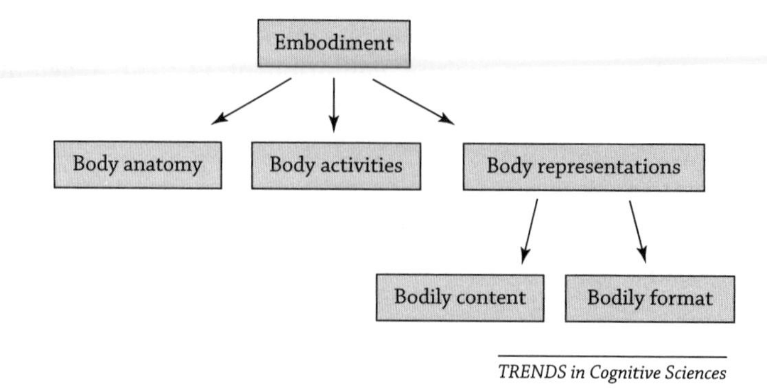

Figure 9.1: Taxonomy of embodiment theses. Classification of alternative interpretations of the notion of embodiment.

Bodily activity interpretation: One's actions and other body-related traits (e.g., posture) have an important causal role in cognition

An example of embodiment under this interpretation is facial feedback. According to the facial-feedback hypothesis, one's own facial musculature activity affects one's mood or emotion [14–16]. The bodily activity interpretation, however, faces a dilemma. On the one hand, cases in which actions or bodily conditions affect cognition are often quite trivial and are recognized by all programs for cognitive science. All theorists, including CC theorists, would cheerfully grant that opening or closing the eyes affects one's perceptions. This is a case in which bodily activity influences cognition (i.e., perception), but it hardly separates EC from CC. On the other hand, there are more ambitious EC theories that are variants of the second interpretation. Some claim that the body and its activities "ground" a variety of our concepts [4,8,17]. Others claim not merely that bodily activity causes perception, but that perception consists of, or is constituted by, sensorimotor contingencies [11]. However, this strong form of an embodied theory of perception (or perceptual experience) is difficult to defend. True, a well-known study found differences of perceptual development in a kitten that was able to explore its environment as contrasted with a kitten required to be passive [18]. And wearing inverting goggles, which changes sensorimotor contingencies, results in reorganization of perception. However, these results only show that sensorimotor contingencies have "effects" on perceptual experience, not that they "constitute" perceptual experience [19].

Not all versions of EC focus on the body per se. Some focus on mental representations of the body—although other EC theorists object to mental representations altogether (e.g., refs. 1,2). How would mental representations of or about the body enter the picture?

Bodily content interpretation: Mental representations with bodily contents have an important causal role in cognition

It is not entirely clear how this interpretation would work in detail. In fact, we prefer to let this third interpretation serve merely as a jumping-off point for a fourth interpretation, which we find more congenial. First, let us introduce the shorthand "B-reps" for bodily representations, a class of mental representations. Next, let us distinguish representations classified as bodily in virtue of their bodily contents (B-contents) from representations classified as bodily in virtue of being encoded in bodily formats (B-formats). B-formats will now become our primary concern.

The idea of a code or format of mental representation is familiar in cognitive science, although there is no consensus about what formats there are or how to individuate them [20]. Some formats are modality-specific: a visual format, an auditory format, and so forth. It is also common to postulate an amodal, or purely conceptual, format. What is the relationship between a particular mental format and the contents of that format's tokens?

We suspect that formats are partly individuated by the contents that their tokens characteristically bear, but that is not the only individuating feature. Representations in different formats can have partly overlapping contents. So, there must be a second factor that has something to do with the neural network that underpins the format. Thus, ventral and dorsal visual pathways are presumed to be different visual codes or formats not only because of presumably different contents, but also because of their different neural pathways. However, this is not the place to say the last word on the question of format individuation.

Let us now revisit the suitability of our third interpretation, the B-content interpretation. Is this suitable to EC? On reflection, it is quite out of the spirit of many forms of EC. If someone represents her own body by means of an amodal, purely conceptual format, then even though the content of the representation is bodily, many proponents of EC might not want to classify it as an instance of EC. This suggests that bodily formats are crucial to embodiment. Thus, we turn to our fourth interpretation of EC:

Bodily format interpretation: Mental representations in various bodily formats or codes have an important causal role in cognition

We regard this interpretation of EC as the most promising one for promoting an embodied approach to social cognition.

Note that, even collectively, our interpretations do not cover all aspects of all EC positions in the literature. For example, some EC position statements feature claims about the inability of standard inferential or computational mechanisms to accommodate the phenomena, appealing instead to the "body" (or the "situation") to do the explanatory work. None of our interpretations of EC explicitly says this. Second, many proponents of EC advance highly global theses about embodiment, whereas our interpretations do not make such claims. Our interpretations are useable in either local or global embodiment claims. For these reasons, both proponents and opponents of EC might criticize our interpretations on the grounds of being excessively tame or "sanitized." There is some justice to this charge, but we regard sanitized variants as scientifically and philosophically fruitful. It makes sense to recognize that selected cognitive tasks might be executed via embodied processes, without ascending to more global claims.

EVIDENCE FOR EMBODIED SOCIAL COGNITION

It is worth noting that the mere fact that most social activities involve the perception of bodily behaviors does not qualify them as embodied. Which types of social-cognitive activities, then, are prime candidates for being embodied? According to proponents of embodied social cognition (ESC) [10,21–23], there are six favorite candidates, including behavior imitation, joint action, emotional contagion, empathy, mindreading, and language understanding (for language understanding, see box 9.1). However, not all of these meet our criteria for embodiment. Imitation and joint action involve an influence by other people's bodily movements on one's own [24,25], so they are not ways that one's own movements and postures affect one's cognitions. The situation is different in emotional contagion via facial mimicry [14,16,26]. When a receiver's facial expression mimics that of the sender, facial feedback is a sub-process of a larger social process of mental contagion [27] (figure 9.2). This is a clear instance of an embodied social process under the bodily activity interpretation.

Yet physical mimicry cannot account for much of social cognition. If social cognition is importantly or pervasively embodied, it must be because of B-reps and their distinctive formats. Thus, the fourth interpretation of

BOX 9.1 EMBODIED LANGUAGE

In discussing the embodiment of language, Gallese [21] distinguishes between phono-articulatory aspects of language and semantic aspects. Concerning the embodiment of the former, a TMS study [64] showed that listening to phonemes induces an increase in amplitude of motor-evoked potentials (MEPs) recorded from the tongue muscles involved in their execution. Embodied simulation at the semantic level concerns the semantic content of a word, verb, or proposition. At the behavioral level, a study [5] asked participants to judge if a read sentence about actions or transfer of information from one person to another was sensible by making a response that required moving toward or away from their bodies. Readers responded faster to sentences describing actions whose direction was congruent with the required response movement, both for sentences describing literal spatial movement and sentences describing abstract movement (information transfer). At the neural level, the authors of ref. 65 found a congruence in the left premotor cortex (a prime mirror area) between effector-specific activations of visually presented actions and actions described by literal phrases. Thus, it seems that conceptual processing of linguistic phrases describing actions re-activates the same cortical sectors activated by observing actions made with the corresponding effectors [7].

However, we would like to emphasize the narrow scope of what these findings show. They reveal that premotor areas are activated when hearing sentences or verbs about motoric actions. However, sentences or verbs about motoric actions are a very limited domain of sentences. If that were all that embodied semantics could establish, it would be a very limited victory.

EC is the most fruitful. Applied to the social realm, it says that representations using B-formats have an important role in social cognition.

Which B-reps have B-formats? It is plausible to posit many such formats. A motoric format is used in giving action instructions to one's hands, feet, mouth, and other effectors. A somatosensory format represents events occurring at the body's surface. Affective and interoceptive representations plausibly have distinctive B-formats, associated with the physiological conditions of the body, such as pain, temperature, itch, muscular and visceral sensations, vasomotor activity, hunger and thirst [28]. If these formats are exploited to represent actions or states of other individuals, these cognitions would be engaged in social cognition via B-formats. For example, in empathy, one might observe another's disgust expression, which in turn elicits a disgust feeling in oneself. If the disgust feeling involves a mental

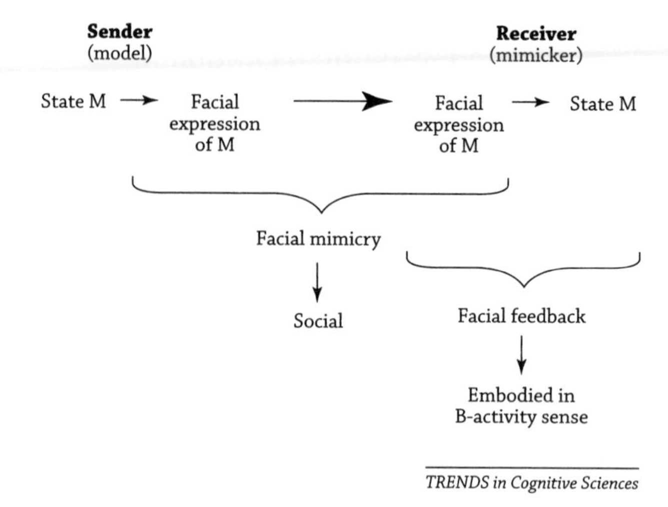

Figure 9.2: Emotional contagion through facial mimicry.
A social process of mental contagion can be broken down into two stages. First, a sender's facial expression is mimicked by that of a receiver; second, the receiver's facial musculature influences his mood or emotion.

representation in a B-format ("the viscera are in such-and-such a state"), and if one labels or assigns this visceral-state representation to the other person (thinking of how she feels from "the inside"), this is a representation of the other in a B-format. The same would happen when one imagines another's disgust.

More widely, representations with B-formats can be exploited for social purposes either during perception or during imagination of another individual in a specific state. The former corresponds to what is called mirroring and has received the most empirical support. We shall thus focus on it, keeping in mind that B-reps can also be activated during imagination [29,30].

The discovery of mirror neurons opened a wide window on the possibility that some social cognition starts at a primitive level of motor planning [31]. The prelude to the discovery was finding a neural vocabulary in monkey premotor cortex in which types of actions (e.g. grasping, holding and tearing) are coded by populations of neurons. Some neurons indicate the goal of an act, others the manner in which a motor act can be performed [32]. This is clearly a motoric code, one also used in some types of social cognition. Cells in area F5 not only send movement instructions to the hand or other effectors, but also echo instructions for the same movements when a monkey merely observes another monkey execute that movement [33] or hears action-related sounds [34]. Networks with mirror-matching properties were also found in humans [35–37].

Whether participants see other individuals acting or hear action-related sounds, they activate effector-specific motor representations involved in the execution of the same action [38]. Mirroring for other experiences—including touch [39,40], pain [41,42], disgust [43], and pleasure [44]—soon followed.

Representations in B-formats are thus activated not only during fundamental motor, somatosensory, affective, and interoceptive functions, but also for some social-cognitive functions. Although the social role of mirror systems is still controversial [45], we believe that they do have such a role, as suggested by further empirical findings that argue in favor of such a social role, in addition to mirroring studies per se. First, mirror activity is often correlated with empathy or mindreading questionnaires [37,41,46]. These correlations indicate that mirroring has a social dimension. More precisely, we suggest that it has a role in action and emotion recognition. It was found that lesions affecting B-reps interfere with action and emotion recognition. Patients with selective impairment in emotion experience have a matching selective impairment in recognizing emotional facial expressions in other people, whereas they have a preserved declarative knowledge about the relevant emotion [47,48]. So, normal subjects, who have no such lesions, must be using their own emotion experience—involving a B-format—in recognizing the emotion of someone they observe. Similarly, apraxic patients with gestural-production impairments can have difficulties in recognizing action-related sounds, although this is not systematic [49]. Whether B-reps have a role in a wider range of social-cognitive functions (e.g., imitation [50], empathy [21], perspective taking [35], mindreading [46,48,51,52], and language production and comprehension [53,54]), as suggested by ESC proponents, is still being debated at the empirical level. Nonetheless, in the next section, we raise some conceptual worries concerning the extent of embodiment in social cognition in general.

BEYOND EMBODIMENT

Our brief literature review makes an empirical case for the causal role of B-formats in a variety of social-cognitive activities. Because of the low-level nature of B-formats, such a pattern was never anticipated by CC. As noted, however, EC enthusiasts often make all-encompassing claims about the extent to which embodiment pervades cognition. Although our review identifies previously unsuspected relationships between B-formats and social cognition, it leaves how far B-reps extend into social cognition quite open. To make a global assessment, prototypical forms of social cognition should be examined.

Three possibilities should be kept in mind: (1) Some social-cognitive tasks could be executed by two or more methods, one involving B-formats and others involving other formats. (2) Even the involvement of B-formats might occur at only one stage of a compound process, in which non-bodily formats predominate. (3) Many social-cognitive activities might involve no B-formats at all.

(1) Empathy illustrates the first possibility: a social-cognitive activity that can be executed with or without B-reps. According to refs. 55 and 56, there are indeed two forms of empathy: "mirror" empathy and "reconstructive" empathy. Mirror empathy involves automatic re-activation of a given affective state in an observer, and this process can be dominated by the mirrored event occurring in a B-format. Reconstructive empathy is a more constructive and controlled process that extracts information from memory in an endogenous production of the empathic event. The final empathy event might be couched in a B-format, but a great deal of non-bodily processing occurs before its production.

(2) Even if some portion of a given simulation heuristic for mindreading (e.g., face-based emotion recognition) involves the mirroring of an emotion, other stages of the process, including the final attribution, could be devoid of any use of a B-format [48,52]. Note the differences here between different proponents of the simulation theory of mindreading. Gallese [21], who defends an explicitly embodied approach, maintains that some mindreading is constituted by mirroring and involves no representation in propositional (non-bodily) format. This posits the deployment of B-formats to the exclusion of any other representational format. In contrast, we admit a role for mirroring as part of (low-level) mindreading, but regard the mirroring stage as only one stage of the mindreading process, a stage that causes but is not constitutive of the subsequent mental-state attribution per se [52]. This leaves more room for non-bodily representational resources to be deployed.

(3) Even defenders of the idea that mirroring has "some" role in mindreading often restrict that role to low-level mindreading, such as emotion recognition or motor-intention prediction. For example, Goldman [52] proposes that face-based attribution of emotions (e.g., disgust) is implemented by a simulation procedure featuring the mirroring of disgust (which is a B-rep), but he equally acknowledges the existence of simulational high-level mindreading that is unlikely to involve any B-formats at all (for discussion, see ref. 57). Simulation, per se, does

BOX 9.2 EMBODIMENT AND SIMULATION

One theme that comes back frequently in EC and ESC is an apparent link between embodiment and reinstantiation, reactivation or simulation [6,23]. Gallese [21] and Niedenthal [23] often speak of "embodied simulation" in one phrase, strongly implying that all simulation is embodied. Why, one might wonder, are reinstantiation or simulation crucial to embodiment? Mental simulation allows (in principle) the simulational mindreading of beliefs and (non-motoric) intentions, as ref. 52 proposes with its notion of "high-level" mindreading. It would be difficult however, to make a case for embodied simulational mindreading of beliefs because the reinstantiation of beliefs would not involve any B-format.

One might reply that only the reenactment of low-level or modality-specific states is embodied. But what exactly makes sensory cognitions embodied? It cannot be their content because the content of sensory cognitions is presumably the external world, not the body. Noë [11] defends the embodiment of perception by arguing that it is tightly linked to action. In the absence of any such argument, however, why should it be presumed that perceptual cognition is embodied?

not entail the use of B-formats (box 9.2). High-level mindreading would standardly involve propositional attitudes like belief and desire. A propositional attitude includes several components, including a type and a propositional content. For example, "Johnny THINKS that 25 + 27 = 52" has THINKING as its type and 25 + 27 = 52 as its content. To be full and accurate, a mindreader must represent both components. Would either representation be a B-rep (occurring in a B-format)? That seems unlikely. Why would a representation of the mathematical content, 25 + 27 = 52 be a B-rep? Equally, why would a representation of the state type THINKING be a B-rep, expressed in a B-code? No evidence known to us supports these hypotheses. Furthermore, evidence that the representation of thinking is especially associated with the right temporoparietal junction (RTPJ) indicates the contrary [58] (but see ref. 59 for a contrary view).

Finally, many other processes involved in social cognition are separate and distinct from mindreading (table 9.1). When interacting with someone, you must recognize who the person is, ascribe personality features to her, categorize her relative to the group to which she belongs, and perhaps apply prejudicial beliefs about her group and prototypical social scripts to guide your behavior.

Table 9.1. SAMPLES OF SOCIAL COGNITION

Person recognition	Face perception, body perception, gait perception, etc.
Personality perception	Trustworthiness, competence, friendliness, fairness, shyness, friendliness, adventurousness, etc.
Membership perception	Gender, race, age, political opinion, religion, etc.
Social beliefs	Prejudice, social scripts, stereotypes, etc.
Social mechanisms	Self/other distinction, gaze detection, perspective-taking, empathy, mirroring, intentionality detector, etc.
Mindreading	Emotion, bodily sensations, intentions, desires, beliefs, etc.
Social attitudes	Envy, pity, contempt, admiration, like/dislike, shame, trust, sympathy, etc.
Social interaction	Imitation, competition, communication, collaboration, punishment, revenge, help, deception, reward, exclusion, etc.

To what extent are these various aspects of social cognition embodied? Here, we highlight a few studies on personality-trait attribution and prejudice-based judgments. It was found that the dimensions of warmth (e.g., friendliness, helpfulness, sincerity, trustworthiness, and morality) and competence (e.g., intelligence, skill, creativity, and efficacy) account almost entirely for how people characterize others [60]. Perceiving someone's warmth and competence, however, is not like perceiving someone's current emotion. At the computational level, personality judgments involve complex calculations requiring the collection of information over time and the drawing of inferences [61]. This computational complexity is hardly compatible with ESC. Even when personality judgment is on-line based on poor perceptual information like short point-light walker body movements, it was found that it activates neural bases distinct from those activated by emotion recognition [62]. Of particular interest here, the authors found greater activity for personality judgments in the medial prefrontal cortex previously associated with propositional-attitude mentalizing. This result is consistent with a recent meta-analysis of brain imaging studies, which found that the attribution of enduring dispositions of others and the knowledge of interpersonal norms and scripts engage the medial prefrontal cortex [63]. Thus, it has not been shown that social cognition is pervasively embodied.

To promote future development of the EC approach, we challenge embodiment theorists (and their critics) to formulate and defend their claims by answering the following questions: First, which interpretation of embodiment do they have in mind? Second, which sectors of cognition, or which cognitive tasks, do they say are embodied; and how fully does each task involve embodiment? Third, how does the empirical evidence support the specific embodiment claims under the selected interpretation(s)? Fourth, how do the proffered claims depart substantially from CC? In our opinion, the findings discussed earlier (in "Evidence for Embodied Social Cognition") do provide support for a limited-scope version of ESC under the B-format interpretation. It is doubtful, however, that such a thesis can be generalized. Researchers who fly their ideas under the banner of EC sometimes advance more far-reaching theses than the evidence warrants. These matters can be clarified only if proponents of EC accompany their writings with detailed answers to the four questions posed earlier.

REFERENCES

1. Brooks, R. (1991). Intelligence without representation. *Artif. Intell.* 47:139–159.
2. Thelen, E., & Smith, L. (1994). *A Dynamic Systems Approach to the Development of Cognition and Action*. Cambridge, MA: MIT Press.
3. O'Regan, J. K., & Noë, A. (2001). A sensorimotor account of vision and visual consciousness. *Behav. Brain Sci.* 24:939–973.
4. Barsalou, L. (1999). Perceptual symbol systems. *Behav. Brain Sci.* 22:577–609.
5. Glenberg, A. M., & Kaschak, M. P. (2002). Grounding language in action. *Psychon. Bull. Rev.* 9:558–565.
6. Gallese, V. (2005). Embodied simulation: From neurons to phenomenal experience. *Phenom. Cogn. Sci.* 4:23–48.
7. Aziz-Zadeh, L., & Damasio, A. (2008). Embodied semantics for actions: Findings from functional brain imaging. *J. Physiol.* 102:35–39.
8. Lakoff, G.. & Johnson, M. (1999). *Philosophy in the Flesh*. Cambridge, MA: MIT Press.
9. Clark, A. (1997). *Being There: Putting Brain, Body and World Together Again*. Cambridge, MA: MIT Press.
10. Gallagher, S. (2005). *How the Body Shapes the Mind*. Oxford, UK: Oxford University Press.
11. Noë, A. (2004). *Action in Perception*. Cambridge, MA: MIT Press.
12. Hurley, S. L. (1998). *Consciousness in Action*. Cambridge, MA: Harvard University Press.
13. Wilson, M. (2002). Six views of embodied cognition. *Psychon. Bull. Rev.* 9:625–636.
14. Larsen, R. J., et al. (1992). Facilitating the furrowed brow: An unobtrusive test of the facial feedback hypothesis applied to unpleasant affect. *Cogn. Emotion* 6:321–338.
15. Strack, F., et al. (1988). Inhibiting and facilitating conditions of the human smile: A nonobtrusive test of the facial feedback hypothesis. *J. Pers. Soc. Psychol.* 54:768–777.

16. Niedenthal, P. M., et al. (2005). Embodiment in attitudes, social perception, and emotion. *Pers. Soc. Psychol. Rev*. 9:184–211.
17. Prinz, J. (2002). *Furnishing the Mind: Concepts and Their Perceptual Basis*. Cambridge, MA: MIT Press.
18. Held, R., & Hein, A. (1963). Movement-produced stimulation in the development of visually guided behavior. *J. Comp. Physiol. Psychol*. 56:872–876.
19. Block, N. (2005). Alva Noë, Action in Perception. *J. Philos*. 102:259–272.
20. Jackendoff, R. (1992). *Languages of the Mind*. Cambridge, MA: MIT Press.
21. Gallese, V. (2007). Before and below 'theory of mind': Embodied simulation and the neural correlates of social cognition. *Philos. Trans. R. Soc. Land. B Biol. Sci*. 362: 659–669.
22. Klin, A., et al. (2003). The enactive mind, or from actions to cognition: Lessons from autism. *Philos. Trans. R. Soc. Lond. Biol. Sci* 358:345–360.
23. Niedenthal, P. M. (2007). Embodying emotion. *Science* 316:1002–1005.
24. Brass, M., & Heyes, C. (2005). Imitation: Is cognitive neuroscience solving the correspondence problem? *Trends Cogn. Sci*. 9:489–495.
25. Sebanz, N., et al. (2006). Joint action: Bodies and minds moving together. *Trends Cogn. Sci*. 10:70–76.
26. Laird, J. D. (1984). The real role of facial response in the experience of emotion: A reply to Tourangeau, Ellsworth, and others. *J. Pers. Soc. Psychol*. 47:909–317.
27. Jabbi, M., & Keysers, C. (2008). Inferior frontal gyrus activity triggers anterior insula response to emotional facial expressions. *Emotion* 8:775–780.
28. Craig, A. D. (2002). How do you feel? Interoception: The sense of the physiological condition of the body. *Nat. Rev. Neurosci*. 3:655–666.
29. Ruby, P., & Decety, J. (2001). Effect of subjective perspective taking during simulation of action: A PET investigation of agency. *Nat. Neurosci*. 4:546–550.
30. Jabbi, M., et al. (2008). A common anterior insula representation of disgust observation, experience and imagination shows divergent functional connectivity pathways. *PLoS ONE* 3:e2939.
31. Gallese, V., et al. (2004). A unifying view of the basis of social cognition. *Trends Cogn. Sci*. 8:396–403.
32. Rizzolatti, G., & Sinigaglia, C. (2007). *Mirrors in the Brain. How Our Minds Share Actions and Emotions*. Oxford, UK: Oxford University Press.
33. Gallese, V., et al. (1996). Action recognition in the premotor cortex. *Brain* 119:593–609.
34. Kohler, E., et al. (2002). Hearing sounds, understanding actions: Action representation in mirror neurons. *Science* 297:846–848.
35. Grèzes, J., & Decety, J. (2001). Functional anatomy of execution, mental simulation, observation, and verb generation of actions: A meta-analysis. *Hum. Brain Mapp*. 12:1–19.
36. Buccino, G., et al. (2001). Action observation activates premotor and parietal areas in a somatotopic manner: An fMRI study. *Eur. J. Neurosci*. 13:400–404.
37. Gazzola, V., et al. (2006). Empathy and the somatotopic auditory mirror system in humans. *Curr. Biol*. 16:1824–1829.
38. de Vignemont, F., and Haggard, P. (2008). What is shared? *Soc. Neurosci*. 3:421–433.
39. Keysers, C., et al. (2004). A touching sight: SII/PV activation during the observation and experience of touch. *Neuron* 42:335–346.
40. Blakemore, S. J., et al. (2005). Somatosensory activations during the observation of touch and a case of vision-touch synaesthesia. *Brain* 128:1571–1583.

41. Singer, T., et al. (2004). Empathy for pain involves the affective but not sensory components of pain. *Science* 303:1157–1162.
42. Jackson, P. L., et al. (2005). How do we perceive the pain of others? A window into the neural processes involved in empathy. *Neuroimage* 24:771–779.
43. Wicker, B., et al. (2003). Both of us disgusted in *my* insula: The common neural basis of seeing and feeling disgust. *Neuron* 40:655–664.
44. Jabbi, M., et al. (2007). Empathy for positive and negative emotions in the gustatory cortex. *Neuroimage* 34:1744–1753.
45. Csibra, G. (2007). Action mirroring and action interpretation: An alternative account. In *Sensorimotor Foundations of Higher Cognition* (Attention and Performance, vol. 22) (Haggard, P., et al., eds.), 435–459. Oxford, UK: Oxford University Press.
46. Iacoboni, M., et al. (2005). Grasping the intentions of others with one's own mirror neuron system. *PLoS Biol.* 3:529–535.
47. Adolphs, R., et al. (2000). A role for somatosensory cortices in the visual recognition of emotion as revealed by three-dimensional lesion mapping. *J. Neurosci.* 20:2683–2690.
48. Goldman, A. I., & Sripada, C. S. (2005). Simulationist models of face-based emotion recognition. *Cognition* 94:193–213.
49. Pazzaglia, M., et al. (2008). The sound of actions in apraxia. *Curr. Biol.* 18:1766–1772.
50. Iacoboni, M., et al. (1999). Cortical mechanisms of human imitation. *Science* 286: 2526–2528.
51. Gallese, V., & Goldman, A. I. (1998). Mirror neurons and the simulation theory of mindreading. *Trends Cogn. Sci.* 2:493–501.
52. Goldman, A. I. (2006). *Simulating Minds. The Philosophy, Psychology and Neuroscience of Mindreading*. New York: Oxford University Press.
53. Rizzolatti, G., & Arbib, M. A. (1998). Language within our grasp. *Trends Neurosci.* 21:188–194.
54. Gallese, V., & Lakoff, G. (2005). The brain's concepts: The role of the sensory-motor system in reason and language. *Cogn. Neuropsychol.* 22:455–179.
55. de Vignemont, F. (In press). Knowing other people's mental states as if they were one's own. In *Handbook of Phenomenology and Cognitive Science* (Schmicking, D. & Gallagher, S. eds.), Dordrecht, NL: Springer.
56. Goldman, A. I. (2011). Two routes to empathy: Insights from cognitive neuroscience. In *Empathy: Philosophical and Psychological Perspectives* (Coplan, A., & Goldie, P., eds.), Oxford, UK: Oxford University Press.
57. de Vignemont, F. (2009). Drawing the boundary between low-level and high-level mindreading. *Philos. Studies* 144:457–466.
58. Saxe, R., & Wexler, A. (2005). Making sense of another mind: The role of the right temporoparietal junction. *Neuropsychologia* 43:1391–1399.
59. Mitchell, J. P. (2008). Activity in right temporoparietal junction is not selective for theory-of-mind. *Cereb. Cortex* 18:262–271.
60. Wojciszke, B., et al. (1998). On the dominance of moral categories in impression formation. *Pers. Soc. Psychol. Bull.* 24:1251–1263.
61. Fiske, S. T., et al. (2007). Universal dimensions of social cognition: Warmth and competence. *Trends Cogn. Sci.* 11:77–83.
62. Heberlein, A. S., and Saxe, R. (2005). Dissociation between emotion and personality judgments: Convergent evidence from functional neuroimaging. *Neuroimage* 28: 770–777.

63. Van Overwalle, F. (2008). Social cognition and the brain: A metaanalysis. *Hum. Brain Mapp.*, DOI:10.1002/hbm.20547.
64. Fadiga, L., et al. (2002). Speech listening specifically modulates the excitability of tongue muscles: A TMS study. *Eur. J. Neurosci.* 15:399–402.
65. Aziz-Zadeh, L., et al. (2006). Congruent embodied representations for visually presented actions and linguistic phrases describing actions. *Curr. Biol.* 16:1818–1823.

CHAPTER 10

A Moderate Approach to Embodied Cognitive Science

1. ELEMENTS OF A WELL-MOTIVATED APPROACH TO EMBODIED COGNITION

There are many research programs in cognitive science that urge a reorientation under the banner of embodied cognition. Most of these programs promote a rather radical alternative to standard or classical cognitivism (for reviews, see Anderson, 2003; Shapiro, 2011). Despite the common label, the programs are very heterogeneous, and I shall make no attempt here to survey them or offer any taxonomy. The present chapter also uses the label of embodied cognition, but it advances a rather moderate conception of embodiment-oriented cognitive science. Although highlighting the pervasiveness in cognition of bodily factors, it does not invoke this as a ground for revolutionizing the methodology of cognitive science.

There are two core elements in my approach. The first element appeals to the idea of bodily representational *codes* (or *formats*) (i.e., hypothesized mental codes that are primarily, or fundamentally, utilized in forming interoceptive or directive representations of one's own bodily states and activities) (Goldman & Vignemont, 2009). The second element of the approach adduces wide-ranging evidence that the brain *reuses* or *redeploys* cognitive processes that have different original uses. If this redeployment idea is applied to bodily formats of representation, they jointly encourage the prospect that body-coded cognition is an extremely pervasive sector of cognition. This seems like a significant hypothesis (though hardly a "totalizing" one, I should stress), and it constitutes the core of the approach to embodied cognition presented here. The approach is compatible with many empirical findings characteristically cited by other embodiment enthusiasts, but it isn't committed to any radical methodological, metaphysical, or architectural theses that some such enthusiasts embrace. In other words, my conception of embodied cognition is fully in sync with existing empirical research and raises no questions, for example, about such staples of traditional cognitive science as mental representation or computational processing.

Actually, what is advanced here is best seen as two distinct proposals. The first is a philosophical, or conceptual, proposal, namely an interpretation of the notion of embodied cognition, a proposed definition of the phrase. The second proposal is an empirically based claim that human cognition in fact realizes or exemplifies this definition to a surprising degree—surprising relative to orthodox or canonical presentations of cognitive science. I support the second thesis with a reasonably wide range of empirical evidence. It is always possible, of course, for readers to take exception to the cited evidence or its probative worth, and there isn't space here to defend that evidence in much detail. But even if only a limited swath of the evidence cited here holds water, embodiment would seem to be realized to a significant degree, a degree quite unanticipated by cognitive science of two or three decades ago and still very far from general acceptance. Future evidence will tell us exactly how extensive embodied cognition is; it would be silly to offer a precisely quantified claim here. This chapter is not a manifesto to the effect that *all* of cognition is embodied. But there is enough existing evidence of its prevalence, I argue, that cognitive science should invest (or continue to invest) a lot of energy and resources into its exploration and documentation.

2. EMBODIMENT AND BODILY REPRESENTATIONAL CODES

What does it mean for cognitive events to be embodied? Or—as there is no prevailing consensus about the meaning of "embodied"—what is a fruitful way to understand this term for purposes of cognitive science? I begin by floating and critiquing several candidate definitions before arriving at a more satisfactory one (cf. Goldman & Vignemont, 2009).

(1) Cognition C (of subject S) is a specimen of embodied cognition if and only if C occurs in S's body.

According to this definition, simply occurring in some subject's body *suffices* for a cognitive event to qualify as being embodied. Is this a happy result? Assuming physicalism, *every* cognition presumably occurs in its subject's body (where "body" includes the brain). This holds even if a mind/body is construed according to the "extended cognition" interpretation (Clark & Chalmers, 1998; Clark, 2008). That interpretation simply expands the size or extent of the body for cognitions to be "in." So, all cognitions would trivially qualify as embodied under definition (1). This is an unhappy result; it should not be such a trivial matter for universal embodiment to obtain.

(2) Cognition C is a specimen of embodied cognition if and only if C is caused (or causally influenced) by one or more parts of S's body (other than the brain).

Like definition (1), statement (2) is an overly lenient definition of embodiment. Assuming that all perceptual experiences are caused by sensory-level inputs involving body parts, (2) implies that every perceptual experience qualifies as an embodied cognition. This follows no matter what detailed account is given of perception. This also seems like an unhappy result, because being (partly) caused by sensory-level inputs is a totally unexceptional property of perceptual events, which no orthodox cognitive scientist would question. Why should it give rise to a special designation associated with embodiment, independent of any new or revelatory story about perception (e.g., the enactive story of perception à la O'Regan & Noe 2001)?

(3) Cognition C is a specimen of embodied cognition if and only if C's representational content concerns S's body; that is, S's body (or some part thereof) is the intentional object of C.

The chief problem with definition (3) centers on the necessity side. It implies that embodied cognitions are restricted to cognitions that have *people* as their intentional objects. This is unduly restrictive and unmotivated. Definition (3) also disallows the possibility that embodied cognitions take people *other* than the subject as intentional objects. Like both (1) and (2), then, condition (3) fails to provide a satisfactory rationale for singling out a special class of cognitions as embodied ones. Nor does it hint at any clear conception of what "embodied cognitive science" might be. I would like to do better, and I think that definition (4) can help.

(4) Cognition C is a specimen of embodied cognition if and only if C uses some member of a special class of *codes* or *formats* for representing and/or processing its content, viz., a *body-related* code or format (*B-code* or *B-format*).

The postulation of multiple codes or formats for mental representation is quite popular in cognitive science. There is no generally accepted treatment of what it is to be such a mental code, and little if anything has been written about the criteria of sameness or difference for such codes. Nonetheless, it's a very appealing idea, to which many cognitive scientists subscribe. Assuming that mental codes are language-like, each code presumably has a distinctive vocabulary, syntax, and set of computational procedures (or some of the foregoing). Each perceptual modality presumably has its own distinctive code. Some modalities—certainly vision—have multiple levels of processing, each with their own code or format. The two visual streams, the ventral and the dorsal, presumably have distinct codes (Milner & Goodale, 1995; Goodale & Milner, 2004). When the brain tries to "convert" information from a visual to a linguistic format, or vice versa, this can be difficult—which attests to the presence of significant differences between such formats (Jackendoff, 1992).

I turn now from representational formats in general to *body-oriented* formats (Goldman & Vignemont, 2009). Many codes in the mind/brain represent states of the subject's own body, indeed, they represent them from an *internal* perspective. Proprioception and kinaesthesis give the brain information—couched, presumably, in distinctive formats—about states of one's own muscles, joints, and limb positions. These interoceptive senses are the basis for B-formats of representation. One's own body, or selected parts thereof, is what they *primarily*, or *fundamentally*, represent. One's own body can also be represented via the external senses, which are not specialized for use in connection with one's own body. One can *see*, for example, that one's right arm is extended. But a token visual representation of one's arm being extended does not qualify—in the stipulated sense being introduced here—as a representation in a bodily code, because vision does not use such a code. (More precisely, it is not assumed from the start that vision uses such a code, although section 4 presents a case for revisiting this question.)

Many bodily codes are tacitly recognized in cognitive science (especially cognitive neuroscience), for example, codes associated with activations in the somatosensory cortex and motor cortex. Stimulation of areas on the surface of the body produces experiential representations, the neural substrates of which comprise a topographically mapped region of somatosensory cortex. This mapping provides a point-for-point representation of the body's surface: a map for the hand, the face, the trunk, the legs, the genitals, and so on (Gazzaniga et al., 2002: 647). Similarly, activation of areas in the motor cortex is topographically organized in order to represent a wide range of bodily effectors and to enable movement commands to be sent to those effectors.

All of the foregoing is "old hat" in cognitive neuroscience; it is introductory-level material. Newer work on internal body representation is illustrated by Craig's (2002) work on body representation that features a hypothesized sense of the physiological condition of the entire body called "interoception." Craig's account focuses on the lamina I spinothalamocortical system. This system conveys signals from small-diameter primary afferents that represent the physiological status of all bodily tissues. Lamina I neurons project to the posterior part of the ventromedial nucleus (or VMpo). Craig calls VMpo the "interoceptive cortex" and argues that it contains representations of distinct, highly resolved sensations, including different types of pain, tickle, temperature, itch, muscular and visceral sensations, sensual touch, and other feelings from (and about) the body.

I propose to classify all mental representations using codes or formats of the sorts just cited as embodied representations. Tokens of such representations qualify as embodied not because their current use, necessarily, is to represent particular body parts or bodily states, but because they belong to an (internal) representational system the primary, or fundamental, function of which is to represent one's own bodily parts and states. In other words, token representations qualify as embodied if and only if they utilize (or belong to) B-formats.

Now suppose it turns out that B-formats are also redeployed or co-opted for representing things *other* than one's own bodily parts or states. These additional representations would also qualify as embodied cognitions. Moreover, if there were extensive application of B-formats to different cognitive tasks, this would be a departure from "business as usual" in cognitive science as traditionally pursued. The more extensive the borrowed or derivative applications are, the greater the departure from traditional cognitive science. This prospect—of extensive derivative applications—is what I shall highlight as the basis for a (somewhat) novel conception of embodied cognition, albeit one with substantial extensional overlap with a number of preexisting programs of embodied cognition. In sections 3 and 4, evidence is presented in support of the contention that wide-ranging derivative use of B-formats is in fact the case in human cognition.

Despite the potentially expansive implications of this approach, it would not necessarily classify as embodied every cognitive event that would be so classified by alternative approaches to embodiment. For example, many proponents of embodied cognition count every perceptual event as embodied. Perceptual cognitions are automatically grouped together with motoric cognitions and assigned the status of embodied cognitions. Barsalou (1999, 2008), for example, seems to class all sensorimotor events as embodied. The present proposal, by contrast, would not automatically (as a matter of definition) treat visual representations as embodied. Why? Because the representational contents of the visual system are (generally) states of affairs *external* to the body. This is the position of almost all representationalist philosophers of mind (e.g., Tye, 1995). These same philosophers, on the other hand, would regard pain cognitions as representational states the contents of which are bodily conditions. So feelings of pain qualify as embodied cognitions. This representationalist approach is clear in the following passage from Tye:

> [P]ains represent... disturbances [in the body].... [A] twinge of pain represents a mild, brief disturbance. A throbbing pain represents a rapidly pulsing disturbance. Aches represent disorders that occur *inside* the body rather than on the surface. These disorders are represented as having volume, as gradually beginning and ending, as increasing in severity and then slowing fading away.... A stabbing pain is one that represents sudden damage over a particular well-defined bodily region.... A racking pain is one that represents that the damage involves the stretching of internal body parts (e.g., muscles). (Tye 1995: 113)

If such representations are elements of one or more bodily formats, as I assume, and if they are used for secondary, or derived, purposes, then according to definition (4), they would still be classified as embodied cognitions, just like B-formatted cognitions when used in their primary, fundamental role.[1]

3. THE MASSIVE REDEPLOYMENT HYPOTHESIS

"Massive redeployment hypothesis" is the name that M. Anderson (2007a, 2008, 2010) gives to a postulated principle of the mind. Another label he uses for this principle is "neural reuse." Whichever phrase is used, the underlying idea is that reuse of neural circuitry for a variety of cognitive purposes is a central organizing principle of the brain. In other words, it is common for neural circuits originally established for one purpose to be exapted (exploited, recycled, or redeployed) during evolution or normal development and put to different uses, without necessarily losing their original functions. An initial type of datum from neuroscience that motivates this idea is that many neural structures are activated by different tasks across multiple cognitive domains. Broca's area, for example, is not only involved in language processing, but also in action-related and imagery-related tasks, such as movement preparation, action sequencing, action recognition, and imagery (Anderson, 2010: 245). Accordingly, rather than posit a functional architecture for the brain in which regions are dedicated to large-scale cognitive domains like vision, audition, and language, respectively, neural reuse theories posit that low-level neural circuits are used and reused for various purposes in different cognitive tasks and domains.

Anderson identifies four theories of this kind in the literature: (1) Gallese and Lakoff's "neural exploitation" hypothesis (Gallese, 2008; Gallese & Lakoff, 2005); (2) Hurley's "shared circuits model" (Hurley, 2005, 2008); (3) Dehaene's "neuronal recycling" theory (Dehaene, 2005, 2009; Dehaene & Cohen, 2007); and (4) his own "massive redeployment" theory (Anderson, 2007a, 2007b, 2010). The core idea of the massive redeployment theory is that evolutionary considerations favor the brain's reusing existing components for new tasks as opposed to developing new circuits de novo. This implies that we should expect a typical brain region to support numerous cognitive functions that are used in diverse task domains. Also, more recent functions should generally use a greater number of widely scattered brain areas than evolutionarily older functions, because the later a function comes onboard, the more likely it is that there will already be useful neural circuits that can be incorporated in the service of the new function (2010: 246). In several publications, Anderson reports an assortment of evidence that supports these and related predictions (Anderson, 2007a, 2007c, 2008). Here I shall review just a smattering of the evidence he adduces.

The massive redeployment hypothesis, argues Anderson, implies the falsity of anatomical modularity, the notion that each functional module in the brain is implemented in a dedicated and fairly circumscribed piece of neural hardware (Bergeron, 2007). Instead, brain regions will not be dedicated to single high-level tasks ("uses"). Different cognitive functions will be supported by putting many of the same neural circuits together in different arrangements (Anderson, 2010: 248). This approach to cognitive architecture is attractively different from various classical alternatives. It is attractive, argues Anderson, because neural reuse is in fact a significant and widespread feature of the brain, inadequately accounted for by other classical architectures (such as massive modularity theory, J. R. Anderson's (2007d) "Act-R" theory, and classical parallel distributed processing theory).

What empirical evidence is cited for this claim about the superiority of the neural reuse approach, and in what tasks or domains is the theory best exemplified? M. L. Anderson gives six primary examples, of which I shall report three. The first example is the use of circuits associated with motor control functions in higher-level tasks of language comprehension. For example, Pulvermuller (2005) found that listening to the words "lick," "pick," and "kick" activates successively more dorsal regions of primary motor cortex (M1). This is consistent with the idea that understanding these verbs relies on motor activation. Indeed, the action concepts may be stored in a motoric code, and understanding the verbs might involve partial simulations of the related actions.

Another example comes from Glenberg & Kaschak's (2002) research. Participants were asked to judge whether a sentence does or does not make sense by pressing a button that required movement either toward or away from the body. The sentences of interest described actions that would also require such movement, and the main finding was an interaction between conditions such that it took longer to respond to a sentence that makes sense when the action described runs counter to the required response motion. So the simple comprehension of a sentence apparently activated action-related representations. More striking yet was that even sentences describing abstract transfers, such as "he sold his house to you," which involves no directional motor action, elicited an interaction effect.[2]

A second category of examples is the reuse of motor control circuits for memory. An example of the motor system's involvement in memory is reported by Casasanto & Dykstra (2010) who found bidirectional influence between motor control and autobiographical memory. Participants were asked to retell memories with either positive or negative valence while moving marbles either upward or downward from one container to another. They retrieved more memories and moved marbles more quickly when the direction of movement was congruent with the valence of the memory (upward for positive memories, downward for negative memories).

A third category of examples is the reuse of circuits that mediate spatial cognition for a variety of higher-order cognitive tasks. One such mediation is the use of spatial cognition for numerical cognition. There is substantial evidence, for example, that response effects observed during number processing feature the reuse of a particular circuit in the left inferior parietal sulcus that plays a role in shifting spatial attention (Hubbard et al., 2005). The idea is that there is a "number line"—a spatial cognitive construct—on which numerical magnitudes are arrayed from left to right in order of increasing size. Another study, reported by Andres et al. (2007), found that hand motor circuits were activated during adults' number processing in a dot-counting task. These activations play a functional role in both domains, as was confirmed by Roux et al. (2003), who found that direct cortical stimulation of a site in the left angular gyrus produced both acalculia and finger agnosia (a disruption of finger awareness).

Anderson recognizes that other theoretical approaches in the embodied cognition family also make appeals to the idea of reuse or redeployment.

He gives considerable attention, for example, to the *conceptual metaphor* approach originating with Lakoff & Johnson (1980, 1999) and to the *concept empiricist* approach of Barsalou (1999, 2008; Barsalou et al., 2003) and Prinz (2002). The conceptual metaphor approach holds that cognition is dominated by metaphor-based thinking whereby structures and logical protocols used in certain domains guide or structure thinking in another domain. Thus, Lakoff and Johnson argue that metaphorical mapping is used to borrow concepts from the domain of war to understand events occurring in the domain of love. Concept empiricism focuses on the content of cognitive representations—symbols, concepts, and other vehicles of thought. The issue here is the degree to which such vehicles (our mental carriers of meaning) are ultimately tied to sensory experience. Concept empiricists endorse a highly modal approach, according to which "the vehicles of thought are reactivated perceptual representations" (Weiskopf, 2007: 156). They are "perceptual symbols" (i.e., "record[s] of the neural activation that arises during perception" (Barsalou, 1999: 578, 583). This contrasts with a rationalist or amodal approach in which the vehicles of cognition are nonperceptual, or abstract, structures (Fodor, 1975; Fodor & Pylyshyn, 1988).

My own original interest in theories of reuse or redeployment arose from work on theory of mind. According to the simulation approach to mindreading, characteristic attempts to read another person's mind are executed by running a simulation of the target in one's own mind and seeing what mental state emerges (Gordon, 1986; Goldman, 1989, 1992, 2006; Gallese & Goldman, 1998; Heal, 1986; Currie & Ravenscroft, 2002). For example, the mindreading task of figuring out what decision another person will make is executed by piggybacking on the capacity to make decisions of one's own. Rival accounts of mindreading make no comparable appeal to the redeployment of one's own mental activity in predicting (or retrodicting) others' mental states. If neural reuse or massive redeployment is an organizing principle of the brain, however, this would render mental redeployment in mindreading tasks an unsurprising method for the human brain to seize upon.

An interesting side question arises here. Simulationists about mindreading have long said that people use their capacity for pretense or imagination to generate stretches of mental activity intended to (mentally) imitate a target. But what is this capacity for imagination? How should it be viewed from the perspective of the massive redeployment hypothesis? The massive redeployment hypothesis says that the brain adapts certain preexisting "uses" of a neural circuit to new types of uses. In each case, a particular circuit C is deployed for use U and then gives way—some of the time—to a new use, U′, of C. But much of this adaptation, Anderson implies, takes place at a biological level, rather than a cognitive level, whereas the imagination presumably does what it does at the cognitive level. It seems to be a general-purpose redeployment device that allows people to select routines from a seemingly unlimited variety of mental activities and allow the selected routines to be used for a wide swath of novel applications. Given the capacity for visual and auditory experience, the imagination allows one to generate faux visual experiences or faux auditory experiences (i.e., visual or auditory images) and then employ them in a wide variety of tasks. The tasks include planning a novel sequence of behavior, recalling a past experience, and understanding what someone else is experiencing. How did this remarkably flexible cognitive tool—a "universal" tool for reuse or redeployment—manage to evolve? I throw this question out as an interesting side issue, but assessment of the massive redeployment hypothesis does not hinge on it. It certainly does not threaten that hypothesis.

The most striking and pervasive redeployment phenomenon discovered in the last two decades is the family of mirroring phenomena, which include not only motor mirroring but also the mirroring of emotions and sensations. Mirror neurons have been discovered in monkeys (Rizzolatti et al., 1996; Rizzolatti & Craighero, 2004; Rizzolatti & Sinigaglia, 2010), humans (Keysers et al., 2010), and most recently in birds (Prather et al., 2008; Keller & Hahnloser, 2009). The fundamental phenomenon, first found in the ventral premotor cortex of macaque monkeys (area F5), is that a specific class of neurons discharge during both the execution of a given behavior and its observation. More recently it has been shown that neurons in the posterior parietal area LIP involved in oculomotor control fire both when the monkey looks in a given direction and when it observes another monkey looking in the same direction (Sheperd et al., 2009). Several studies in humans demonstrated that observing someone else performing a given motor act recruits the same parieto-premotor areas involved in executing that act (Rizzolatti & Sinigaglia, 2010). Brain imaging experiments in humans have shown that witnessing someone else expressing a given emotion or sensation (e.g., disgust, pain, and touch) recruits some of the same visceromotor and sensorimotor brain areas as are activated when one experiences the same emotion or sensation (Wicker et al., 2003; Keysers et al., 2010).

How do these findings mesh with our B-format framework? Very well indeed. Early in their work, the Parma group identified what they call a "motor vocabulary" in the monkey premotor area, where individual cells or populations of cells code for particular hand actions such as holding, grasping, breaking, etc. (Rizzolatti & Sinigaglia, 2006). This obviously qualifies as a B-format, because selected hand actions—a species of bodily activity—are being represented as "to be executed." Of course, these premotor activations (or premotor-parietal activations) would not be *redeployments* of B-formatted cognitions unless they are recruited to perform *different* cognitive tasks than commanding their own effectors to move. However, there is strong evidence that mirroring activity in the observation mode does indeed recruit different tasks, although it is (somewhat) controversial exactly which tasks are recruited via observation-driven mirroring. Most of the candidates for these additional tasks are interpersonal or social in nature, but the specifics of the tasks are still being debated.

Early publications of the Parma group already interpreted mirroring activity (at the observational end) as involving more than the mere reoccurrence of an action instruction directed at one's own effectors. Mirroring at the observational end is interpreted as including something like the following representation: "The individual over there is doing *this*," where "this" indexes the motor command associated with the neural activation in question (e.g., "grasp object X with such-and-such a grip"). Because the "ordinary" (non-observational) activation of the relevant cells includes no reference to another individual, if these cells are activated as part of a representation of what another person is doing, or planning to do, the latter activity would be a different cognitive task than the fundamental one. Thus, even an observer's fairly minimal interpretation of what a target actor is doing constitutes a redeployment of the motoric format in a novel, cognitively interpersonal, task. Stronger interpretations are possible, of course. The observer might engage in the task of *understanding* the target's action (specifically, understanding it "in" the motor vocabulary of his own action). This need not include an attribution of a *goal-state* to the target. On the other hand, if mirroring is part and parcel of an act of *mindreading* by the observer, as Gallese & Goldman (1998) conjectured, then the observer utilizes his motoric vocabulary to perform a very different task, a theory-of-mind task, which is not a fundamental use of the motoric B-format. As we see, however, even "deflationary" interpretations of an observer's cognitions can still warrant the conclusion that he engages in redeployment of the motoric code (which, of course, is a bodily code).[3]

The idea that mirroring prompts mindreading is controversial. There are studies by proponents of mirror-based mindreading that purport to support this hypothesis (Fogassi et al., 2005; Iacoboni et al., 2005), but these may be open to different interpretations.[4] Once again, however, even if those experiments don't decisively demonstrate mirror-based mindreading, they still provide excellent evidence for *some* kinds of reuse of the motoric format to perform cognitive acts beyond the fundamental use of the motoric format. For example, a further cognition might be an expectation that the actor will perform a specified *action*, that is, a piece of behavior (Goldman, 2009). A prediction-of-action cognition is not an ascription of a mental state (e.g., an intention), but it is very different from commanding one's own effectors to act, so it qualifies as a secondary, derivative application of the motor code in question.

In any event, the case for mirror-based realizations of the redeployment principle does not hinge on motoric mirroring alone. Indeed, in my view, evidence for the mindreading species of reuse is more clear-cut for sensation and emotion mindreading than for mindreading of motor intentions.[5] Challenges to the suggestion that embodied redeployment is exemplified in these cases, however, might be predicated on doubts that the relevant formats are *bodily* formats. However, it seems straightforward that pain representations and touch cognitions are fundamentally, or primarily, representations in bodily formats. This may be less clear for disgust or fear, at least initially; but a strong case for this is made by Jabbi et al. (2007).

Thus, there is substantial evidence in support of the pervasive occurrence of embodied cognition. Not only does embodied cognition occur in fundamental uses of B-formats—as all cognitive scientists will presumably acknowledge—but there is massive redeployment of B-formats for other cognitive tasks. Section 4 will present yet different types of B-format redeployment—not based on mirroring or any other social interactions. Thus, our initial bodily-format interpretation of embodied cognition not only has intuitive appeal in itself, but it also paves the way for a strong empirical case for the widespread incidence of embodied cognition. It is noteworthy that Gallese, a long-time proponent of the importance of embodiment (see Gallese, 2005, 2007), has recently signed on to the B-format interpretation of embodiment (Gallese & Sinigaglia, 2011).[6]

However, I have yet to offer a specific formulation of an embodied cognition thesis. The core thesis, containing many admittedly vague terms, is stated below. Greater exactness should hardly be expected in this kind of generalization, especially at the present stage of inquiry.

> (Core thesis): Embodied cognition is a significant and pervasive sector of human cognition both because
>
> (1) B-formats in their primary uses are an important part of cognition, and
>
> (2) B-formats are massively redeployed or reused for many other cognitive tasks, including tasks of social cognition.

A corollary of this factual thesis is research "advice" to cognitive science: It should devote considerable attention to the two forms of embodied cognition (primary and derived). Obviously, this is not a terribly revolutionary proclamation. Substantial sectors of cognitive science are already doing this. The important take-home message, therefore, is not the advice per se but the theoretical unification of the empirical findings that makes systematic sense of these assorted findings. Moreover, this unification highlights features of human cognition that were nowhere on the horizon twenty years ago and are still ignored, doubted, or denied by many cognitive scientists. A wider acceptance of the entire "ball of wax" would mark a major shift in cognitive science as a whole.

Some of the aforementioned skeptics level their criticisms at allegedly excessive claims about the influence of mirroring. Notice, however, that mirroring is just one strand in a much broader landscape of embodied cognition. Similarly, some criticisms (e.g., Jacob & Jeannerod, 2006) are aimed at arguably excessive claims about the significance of motoric phenomena. The critics may see these claims as incipient attempts to provide a global "reductive" explanation of all cognitive phenomena to the motoric domain. But motoric phenomena are just a slice of the body-related phenomena that get exploited for supplementary cognitive uses. Section 4 examines a family of redeployed bodily representations that are, at most, only minimally motoric. So the breadth of the present conception of embodiment—and the breadth of its empirical support—should be carefully weighed before accusing it of being some narrow form of reductionism. Finally, no *universal* claims are made here. Nobody is saying that *all* cognition is embodied.

4. PERCEPTION AND EMBODIED COGNITION

Although perception has figured in a number of programs for embodied cognitive science, no previous program, to my knowledge, has appealed to redeployment or reuse to frame the case for embodied cognition in perception. In this section, I add weight to the unifying value of the B-format conception of embodiment by showing how recent evidence from vision science fits snugly under this umbrella. The research program summarized in this section is that of Proffitt and colleagues. Their research shows that representations of one's own body are tacitly at work in executing tasks that are ostensibly far removed from the perceiver's own bodily state, viz., estimating properties of the distal environment.

A good introduction to Proffitt's thinking about perception can profitably begin with his reflections on how the brain functions vis-à-vis the body as viewed from the perspective of behavioral ecology. A principal function of the brain, says Proffitt (2008), is to control the body in order to achieve desired states in both the body's external environment and its internal environment. Studies in behavioral ecology (e.g., Krebs & Davies, 1993) show that the behavior of organisms is primarily governed by energetic and reproductive imperatives:

> [O]rganisms have been shaped by evolution to follow behavioral strategies that optimize obtaining energy (food), conserving energy, delivering energy to their young, and avoiding becoming energy for predators. To meet these ends, species have evolved behavioral strategies for achieving desired outcomes in the external physical environment while concurrently maintaining desired states in the internal environment of the body. (Proffitt, 2008: 179–180)

Maintaining desired states, it might be added, requires a regular monitoring of, or representation of, what bodily states currently obtain.

Proffitt is a vision scientist, of course. How do these reflections bear on vision? Like any vision scientist, Proffitt assumes that optical factors (e.g., light impinging on the retinas) play a crucial role in determining visual experience. What is distinctive to his approach, however, is that non-optical factors—specifically bodily factors—also play a role in determining visual experience. (So far, of course, this says nothing about B-formats or their reuse in visual perception. That will come later.) Here are some of his principal findings concerning the relation of bodily influences on vision.[7]

(A) Visual representations of object size are scaled by reference to one's own bodily parts. In experiments reported by Linkenauger et al. (2010), objects were either magnified or "minified" by the wearing of goggles. Following magnification, when a subject's hand was placed next to the magnified object, the object appeared to shrink to near-normal size. Similarly, following minification, when a subject's hand was placed next to the previously minified object, it appeared to grow to near-normal size. The compelling inference is that objects appeared to shrink or grow when placed next to one's hand because they were *rescaled* to the magnified or minified hand. (This would be an especially natural thing to do for graspable objects.) It is noteworthy that rescaling did not occur when familiar *objects* were placed next to the target object. Nor did rescaling occur when someone else's hand was placed next to the object. Only the proximity of one's own hand had the indicated effect.

(B) Visual judgments of environmental layout (e.g., distance or steepness) are influenced by physiological or energetic states of the body. Proffitt's early research showed a pattern of error in judgments of hill inclines viewed straight on. Angles were systematically overestimated. Five-degree hills were judged to be 20-degrees steep and 10-degree hills were judged to be

30-degrees steep. Experimental studies pertaining to bodily states were done by Bhalla & Proffitt (1999). They presented four studies that showed that hills appeared steeper when people were fatigued, encumbered, of poor physical fitness, elderly, and in declining health. In these studies, three dependent measures were taken: a verbal report, a visual matching task, and a manual adjustment of a tilting palm board. Across these four studies, the measures of explicit awareness (i.e., verbal report and visual matching) were affected by the factors listed above, whereas the implicit visual guidance of action measure (i.e., palm board adjustment) was not. When subjects wore heavy backpacks while making slant judgments, "explicit" judgments showed increased overestimations. Bhalla and Proffitt studied students with varying fitness levels (including varsity athletes) while riding a stationary bicycle that measured oxygen uptake and recovery time. When subjects made slant judgments of hills, the greater their fitness, the shallower were their judgments of hill inclines.

(C) When intended or anticipated actions are more difficult in terms of required effort, distance judgments are affected. Witt & Proffitt (2008) found that when a subject expected to walk to a target as contrasted with expecting to throw a beanbag to it, the apparent distance to the target differed. In other words, subjects made distance judgments from an "actional" perspective. Their distance judgments reflected their being a "thrower" or a "walker," or expecting to be one or the other. If a person was to be a thrower, the estimate was influenced by the effort required to throw, and analogously if he/she was to be a walker.

What internal mechanism or mechanisms are responsible for these influences? Witt & Proffitt (2008) posit a process of *internal motor simulation* to explain the influences. Although they don't spell out the steps of such a motor simulation, or how exactly it influences distance judgments, it presumably runs something like this: The subject tries to reenact the cognitive activity that would accompany the motor activity in question—without actually setting any effectors in motion. During this series of steps—or perhaps at the end—the energetic or physiological states of the system are monitored. Distance judgments are arrived at partly as a function of the detected levels of these states. If this reconstruction is correct, then the processes of detecting and representing the indicated bodily-state levels qualify as cognitive activities that utilize one or more B-formats. Moreover, the output B-representations generated by these cognitive activities either during the simulation or at its end are then used or deployed for a *non-bodily* cognitive task, namely estimating the (external) distance between self and target object. This, then, would be a clear case of reusing or redeploying B-formats to execute a fundamentally non-bodily cognitive task. Obviously, if this account is correct, it exemplifies the massive redeployment hypothesis applied to embodied cognitions.

Independent of this motor simulation hypothesis, Proffitt has in mind a general hypothesis that comports well with the ideas I have been advancing. Modifying the maxim of the Greek philosopher Protagoras, who famously held that "man is the measure of all things," Proffitt propounds the maxim that one's body is the measure of all things. In other words, one's own body is used to scale physical judgments about other (non-bodily) subject matters. As we have seen, dramatic demonstrations of this maxim are presented in the form of the visual shrinkage or growth of external objects when one's own hand—but not another person's hand—is brought into sight. As Linkenauger et al. express it, "the perceptual system uses the body as a perceptual ruler, and thus the sizes of graspable objects are perceived as a proportion of the hand's size. This proportion directly indicates for the perceiver how large objects are with respect to his or her hand's grasping capabilities" (Linkenauger et al., 2010).

The body-based scaling idea is also articulated in the following passage:

> [W]e argue that visual information is not [merely] combined with, but rather *is scaled by*, non-visual metrics derived from the body. . . . We do not perceive visual angles, retinal disparities, and ocular-motor adjustments, which are the stuff of visual information; rather, we perceive our environment. The angular units of visual information must be transformed into units appropriate for the specification of such parameters of surface layout as extent, size, and orientation. We propose that these scaling units derive from properties of the body in a way that makes perception, like all other biological functions, a phenotypic expression. (Proffitt & Lingenauger 2012)

Clearly, Proffitt and colleagues are not simply saying that the body and its parts are causally responsible for certain (antecedently) surprising effects. They are saying that *representations* of bodily parts are used to influence representations of non-bodily objects. This point is noteworthy because Proffitt was strongly influenced by J. J. Gibson, a major force in the embodied cognition movement, who resisted *representations* as unhelpful theoretical tools for cognitive science. Yet Proffitt's writing on this topic seems to side more with orthodox cognitive science.

One line of criticism of Proffitt's work challenges his conclusions about bodily effects on perception. In particular, it questions whether subjects who made steeper slope estimates while wearing backpacks were genuinely influenced by the backpacks' weight, via physiological effects on the wearers. Durgin et al. (2009) argue that experimental demands of the situation might have led subjects to elevate their cognitive estimates of slope. In other words, perhaps the experimenters' hypotheses in Proffitt's studies were transparent to participants and their reported response differences reflected biases in judgment in compliance with experimental demand characteristics. If a new experiment were conducted that manipulated experimental demand characteristics and if it were to produce changes in judgment similar in magnitude to those previously attributed to backpacks, this would undermine the argument that the physical burden of the backpack affects perception directly. Durgin et al. conducted an experiment that purported to have exactly these results.

Proffitt (2009) offers a very ample reply to this challenge. First, there were enormous differences between the Bhalla-Proffitt experimental setup and that of Durgin et al. In the former, hills were always of very considerable length, and their crests were well above eye level. Thus, subjects could easily see that there would be a real energy cost to climbing the hill. In the Durgin et al. experiment, by contrast, the "hill" was a 1 m × 2 m ramp, which was viewed indoors. Proffitt concedes that Durgin's clever demand manipulations might have produced response differences in their experiment, especially because there was no serious energy cost involved. But it does not follow that the prospect of energy costs had no impact in Bhalla and Proffitt's very different experimental setup. In fact, Proffitt had anticipated the kind of worry Durgin et al. tested, and had therefore used converging measures and manipulations in which the anticipated outcome would not be intuited by participants. For example, in the third experiment reported in Bhalla & Proffitt (1999), physical fitness was assessed using a cycle ergometer test, and they found that fitness was negatively correlated with slant judgments (for the two explicit measures). There was no experimental manipulation in this experiment; all participants were treated the same, and the experimenter was blind to the participants' fitness status. More recently, other experiments have been conducted using blood glucose as an indicator of bioenergetic condition. Schnall et al. (2009) found that hills appear steeper to those with depleted levels of blood glucose. It would seem, then, that this line of research is methodologically very well grounded and extremely sound.

5. CONCLUSION

With only a few conceptual resources, I have articulated a comprehensible and rather natural conception of embodied cognition. Moreover, drawing on a wide-ranging body of research, much of which has already attracted high levels of attention, a straightforward case has been presented for the very considerable role of embodiment in human cognition. Finally, it has been shown how the proposed conception of embodied cognition makes important points of contact with a number of other programs in embodied cognition. Perhaps it is time to converge on a single approach—the B-format approach—as a unifying and comprehensive one, rather than persist with the dispiriting balkanization of embodiment theory.[8]

NOTES

1. A more complex representationalist analysis of one class of feelings (i.e., emotions) is offered by Prinz (2004). He holds that emotions are perceptions of the body, specifically, of bodily changes. He argues that emotions represent organism-environment relations, but they do so by perceiving bodily changes. In different terminology, organism-environment relations are said to be the "real" contents of emotions, and bodily changes are said to be their "nominal" contents (2004: 68). This is close enough, for our purposes, to treat emotions as embodied under our proposal; and Prinz certainly presents himself as an embodiment theorist.
2. For additional supportive evidence of motor involvement in language understanding, see Pulvermuller & Fadiga (2010), Jirak et al. (2010), and Glenberg & Gallese (2011).
3. Vignemont & Haggard (2008) pose the question of what specific representations are "shared" between sender and receiver in a mirror transaction. At what level of the hierarchical structure of the motor system do shared representations occur? This is a good question, to which they offer a complex array of interesting possible answers. However, the central question for present purposes does not concern the specific motoric level at which the observer's representations replicate those of the actor, but whether any different cognitive tasks *at all* are undertaken by the observer that are distinct from those of the actor. And that question can be addressed without settling the one that concerns Vignemont and Haggard.
4. In the experiment by Iacoboni et al. (2005), subjects observed a person in three kinds of conditions: an action condition, a context condition, and an intention condition. The intention condition was one that may have suggested an intention beyond that of merely grasping a cup: an intention to either drink tea or clean up (after tea). This condition yielded a significant signal increase in premotor mirroring areas where hand actions are represented. This was interpreted by the researchers as evidence that premotor mirroring areas are involved in understanding the *intentions* of others. There is room for doubt about this interpretation. However, even if the enhanced mirroring activity during the intention condition did not constitute an *intention* attribution by the observer, it very plausibly did constitute a prediction or expectation of a future *action* by the portrayed individual. Because an action is not a mental state, predicting an action would not qualify as mindreading.

5. For details of evidence about pain, see Avenanti et al. (2006); Shanton & Goldman (2010). For details about emotions like disgust and fear, see Goldman & Sripada (2005); Goldman (2006: chap. 6); Jabbi et al. (2007). The main "special" evidence for mirror-based mindreading in the case of emotions is evidence involving patients with paired deficits in experiencing and attributing the same emotion. For example, Calder et al. (2000) found such a pairing in patient NK, who had suffered insula and basal ganglia damage. On a questionnaire to probe the experience of disgust, NK's score was significantly lower than that of controls. Similarly, in tests of his ability to recognize emotions from faces, NK showed a marked deficit in recognizing disgust, but not other emotions. The natural inference (when conjoined with the Wicker et al., 2003, study) is that a normal subject who sees someone else's disgust-expressive face undergoes a mirrored experience of disgust and uses it to recognize disgust in the other. This is why an impaired disgust system leaves a subject (selectively) unable to mindread disgust normally based on a facial (or vocal) expression (i.e., because he does not undergo a mirrored disgust experience).
6. Gallese had previously advanced the concept of reuse, one of the core elements of the B-format approach, as the linchpin of an account of embodiment (Gallese, 2007, 2008, 2010). Note, however, that reuse by itself is neither necessary nor sufficient for embodiment. It isn't necessary because primary, or fundamental, uses of a B-format—which do not constitute reuses—still qualify as instances of embodiment. It isn't sufficient because there may be many cognitions that reuse non-bodily formats. Such reuses are not instances of embodied cognition. Gallese and Sinigaglia take notice of the latter point, writing, "The notion of reuse, however, is not sufficient to explain the MM [mirror mechanism]" (2011: 513).
7. Proffitt and colleagues often speak of influences on visual "experience," and some might question whether this is fully supported by the evidence. The issue is whether their findings are merely post-perceptual phenomena, rather than genuine perceptual phenomena. This matter has been tested and addressed in Witt et al. (2010). What they say seems quite re-assuring on this point, namely that the findings do pertain to genuinely perceptual phenomena. The reader can judge for him/herself.
8. I thank Dennis Proffitt, Michael Anderson, Vittorio Gallese, Frederique de Vignemont, and Lucy Jordan for helpful feedback on some of the main ideas in this chapter. An earlier version of the material on Proffitt was presented to a workshop at the Australian National University. A talk based on the original paper as a whole was presented at the Rutgers University Center for Cognitive Science. I am grateful to members of these groups for instructive criticisms.

REFERENCES

Anderson, J. R. 2007d. *How Can the Human Mind Occur in the Physical Universe?* New York: Oxford University Press.

Anderson, M. L. 2003. Embodied cognition: A field guide. *Artificial Intelligence* 149(1):91–103.

Anderson, M. L. 2007a. Evolution of cognitive function via redeployment of brain areas. *The Neuroscientist* 13:13–21.

Anderson, M. L. 2007b. Massive redeployment, exaptation, and the functional integration of cognitive operations. *Synthese* 159(3):329–345.

Anderson, M. L. 2007c. The massive redeployment hypothesis and the functional topography of the brain. *Philosophical Psychology* 21(2):143–174.

Anderson, M. L. 2008. Circuit sharing and the implementation of intelligent systems. *Connection Science* 20(4):239–313.

Anderson, M. L. 2010. Neural reuse: A fundamental organizational principle of the brain. *Behavioral and Brain Sciences* 33:245–266.

Andres, M., X. Seron, & E. Olivier. 2007. Contribution of hand motor circuits counting. *Journal of Cognitive Neuroscience* 19:563–576.

Avenanti, A., I. M. Paluello, I. Bufalai, & S. Aglioti. 2006. Stimulus-driven modulation of motor-evoked potentials during observation of others' pain. *NeuroImage* 32:316–324.

Barsalou, L. W. 1999. Perceptual symbol systems. *Behavioral and Brain Sciences* 22:577–660.

Barsalou, L.W. 2008. Grounding cognition. *Annual Review of Psychology* 59:617–645.

Barsalou, L. W., W. K. Simmons, A. K. Barbey, & C. D. Wilson. 2003. Grounding conceptual knowledge in modality-specific systems. *Trends in Cognitive Sciences* 7(2):84–91.

Bergeron, V. 2007. Anatomical and functional modularity in cognitive science: Shifting the focus. *Philosophical Psychology* 20(2):175–195.

Bhalla, M., & D. R. Proffitt. 1999. Visual-motor recalibration in geographical slant perception. *Journal of Experimental Psychology: Human Perception and Performance* 25:1076–1096.

Calder, A. J., J. Keane, F. Manes, N. Antoun, & A. W. Young. 2000. Impaired recognition and experience of disgust following brain injury. *Nature Neuroscience* 3:1077–1078.

Casasanto, D., & K. Dykstra. 2010. Motor action and emotional memory. *Cognition* 115(1): 179–185.

Clark, A. 2008. *Supersizing the Mind: Embodiment, Action, Cognitive Extension*. Oxford, UK: Oxford University Press.

Clark, A., & D. Chalmers. 1998. The extended mind. *Analysis* 58(1):7–19.

Craig, A. D. 2002. How do you feel? Interoception: The sense of the physiological condition of the body. *Nature Reviews Neuroscience* 3:655–666.

Currie, G., & I. Ravenscroft. 2002. *Recreative Minds*. Oxford, UK: Oxford University Press.

Dehaene, S. 2005. Evolution of human cortical circuits for reading and arithmetic: The 'neuronal recycling' hypothesis. In *From Monkey Brain to Human Brain*, eds. S. Dehaene, J.-R. Duhamel, M. D. Hauser, & G. Rizzolatti, 131–157. Cambridge, MA: MIT Press.

Dehaene, S. 2009. *Reading in the Brain*. New York: Viking.

Dehaene, S., & L. Cohen. 2007. Cultural recycling of cortical maps. *Neuron* 56:384–398.

Durgin, F. H., J. A. Baird, M. Greenburg, R. Russell, K. Shaughnessy, & S. Waymouth. 2009. Who is being deceived? The experimental demands of wearing a backpack. *Psychonomic Bulletin and Review* 16(5):964–969.

Fodor, J. A. 1975. *The Language of Thought*. Cambridge, MA: Harvard University Press.

Fodor, J. A., & Z. Pylyshyn. 1988. Connectionism and cognitive architecture: A critical analysis. *Cognition* 28:3–71.

Fogassi, L., P. F. Ferrari, B. Gesierich, S. Rozzi, F. Chersi, & G. Rizzolatti. 2005. Parietal lobe: From action organization to intention understanding. *Science* 308:662–667.

Gallese, V. 2005. Embodied simulation: From neurons to phenomenal experience. *Phenomenology and Cognitive Science* 4:23–48.

Gallese, V. 2007. Before and below 'theory of mind': Embodied simulation and the neural correlates of social cognition. *Philosophical Transactions of the Royal Society, B* 362: 659–669.

Gallese, V. 2008. Mirror neurons and the social nature of language: The neural exploitation hypothesis. *Social Neuroscience* 3(3–4):317–333.

Gallese, V., & A. I. Goldman. 1998. Mirror neurons and the simulation theory of mindreading. *Trends in Cognitive Sciences* 2:493–501.

Gallese, V., & G. Lakoff. 2005. The brain's concepts: The role of the sensory-motor system in conceptual knowledge. *Cognitive Neuropsychology* 22(3–4):455–479.

Gallese, V., & C. Sinigaglia. 2011. What is so special about embodied simulation? *Trends in Cognitive Sciences* 15(11):512–519.

Gazzaniga, M. S., R. B. Ivry, & G. R. Mangun. 2002. *Cognitive neuroscience,* 2nd ed. New York: Norton.

Glenberg, A. M., & M. F. Kaschak. 2002. Grounding language in action. *Psychonomic Bulletin and Review* 9:558–565.

Goldman, A. I. 1989. Interpretation psychologized. *Mind and Language* 4:161–185.

Goldman, A. I. 1992. In defense of the simulation theory. *Mind and Language* 7(1–2):104–119.

Goldman, A. I. 2006. *Simulating Minds: The Philosophy, Psychology, and Neuroscience of Mindreading.* New York: Oxford University Press.

Goldman, A. I. 2009. Mirroring, simulating, and mindreading. *Mind and Language* 24(2): 235–252.

Goldman, A. I., & C. S. Sripada. 2005. Simulationist models of face-based emotion recognition. *Cognition* 94:193–213.

Goldman, A. I., & F. Vignemont. 2009. Is social cognition embodied? *Trends in Cognitive Sciences* 13(4):154–159.

Goodale, M., & D. Milner. 2004. *Sight Unseen.* Oxford, UK: Oxford University Press.

Gordon, R. M. 1986. Folk psychology as simulation. *Mind and Language* 1:158–171.

Heal, J. 1986. Replication and functionalism. In *Language, Mind, and Logic,* ed. J. Butterfield (135–150). Cambridge, UK: Cambridge University Press.

Hubbard, E. M., M. Piazza, P. Pinel, & S. Dehaene. 2005. Interaction between number and space in parietal cortex. *Nature Reviews Neuroscience* 6(6):435–448.

Hurley, S. L. 2005. The shared circuits hypothesis: A unified functional architecture for control, imitation, and simulation. In *Perspectives on Imitation: From Neuroscience to Social Science,* eds. S. Hurley & N. Chater, 76–95. Cambridge, MA: MIT Press.

Hurley, S. L. 2008. The shared circuits model: How control, mirroring and simulation can enable imitation, deliberation, and mindreading. *Behavioral and Brain Sciences* 31(1): 1–58.

Iacoboni, M., I. Molnar-Szakacs, V. Gallese, G. Buccino, J. C. Mazziotta, & G. Rizzolatti. 2005. Grasping the intentions of others with one's own mirror neuron system. *PLoS Biology* 3:529–535.

Jabbi, M., M. Swart, & C. Keysers. 2007. Empathy for positive and negative emotions in the gustatory cortex. *NeuroImage* 34:1744–1753.

Jackendoff, R. 1992. *Languages of the mind: Essays on mental representation.* Cambridge, MA: MIT Press.

Jacob, P., & M. Jeannerod. 2006. The motor theory of social cognition: A critique. *Trends in Cognitive Sciences* 9(1):21–25.

Jirak, D., M. M. Menz, G. Buccino, A. M. Borghi, & F. Binkofski. 2010. Grasping language: – A short story on embodiment. Consciousness and Cognition 19: 711–720.

Keller, G. B., & R. H. Hahnloser. 2009. Neural processing of auditory feedback during vocal practice in a songbird. *Nature* 457:187–190.

Keysers, C., J. H. Kaas, & V. Gazzola. 2010. Somatosensation in social perception. *Nature Reviews Neuroscience* 11:417–428.

Krebs, J. R., & N. B. Davies. 1993. *An Introduction to Behavioral Ecology,* 3rd edition. Oxford, UK: Blackwell.

Lakoff, G., & M. Johnson. 1980. *Metaphors We Live By*. Chicago: University of Chicago Press.

Lakoff, G., & M. Johnson. 1999. *Philosophy in the Flesh: The Embodied Mind and Its Challenge to Western Thought*. New York: Basic Books.

Linkenauger, S. A., V. Ramenzoni, & D. R. Proffitt. 2010. Illusory shrinkage and growth: Body-based rescaling affects the perception of size. *Psychological Science* 21(9): 1318–1325.

Milner, A. D., & M. A. Goodale. 1995. *The Visual Brain in Action*. Oxford, UK: Oxford University Press.

O'Regan, J. K., & A. Noe. 2001. A sensorimotor approach to vision and visual consciousness. *Behavioral and Brain Sciences* 24(5):939–973.

Prather, J. F., S. Peters, S. Nowicki, & R. Mooney. 2008. Precise auditory-vocal mirroring in neurons for learned vocal communication. *Nature* 451:249–250.

Prinz, J. J. 2002. *Furnishing the Mind: Concepts and Their Perceptual Basis*. Cambridge, MA: MIT Press.

Prinz, J. J. 2004. *Gut Reactions: A Perceptual Theory of Emotion*. New York: Oxford University Press.

Proffitt, D. R. 2008. An action-specific approach to spatial perception. In *Embodiment, Ego-Space, and Action*, eds. R. L. Katzky, B. MacWhinney, M. Behrmann. Psychology Press.

Proffitt, D. R. 2009. Affordances matter in geographical slant perception. *Psychonomic Bulletin and Review* 16(5):970–972.

Proffitt, D. R., & S. A. Linkenauger. 2012. Perception viewed as a phenotypic expression. *Tutorials in Action Science*, eds. W. Prinz, M. Beisert, & A. Herwig. Cambridge, MA: MIT Press.

Pulvermuller, F. 2005. Brain mechanisms linking language and action. *Nature Reviews Neuroscience* 6:576–582.

Pulvermuller, F., & L. Fadiga. 2010. Active perception: Sensorimotor circuits as a cortical basis for language. *Nature Reviews Neuroscience* 11:351–360.

Rizzolatti, G., & L. Craighero. 2004. The mirror neuron system. *Annual Reviews of Neuroscience* 27:169–192.

Rizzolatti, G., & C. Sinigaglia. 2006. *Mirrors in the Brain: How Our Minds Share Actions and Emotions*. Oxford, UK: Oxford University Press.

Rizzolatti, G., & C. Sinigaglia. 2010. The functional role of the parieto-frontal mirror circuit: Interpretations and misinterpretations. *Natural Reviews Neuroscience* 11:264–274.

Rizzolatti, G., L. Fadiga, V. Gallese, & L. Fogassi. 1996. Premotor cortex and the recognition of motor actions. *Cognitive Brain Research* 3:131–141.

Roux, F.-E., S. Boetto, O. Sacko, F. Chollet, & M. Tremoulet. 2003. Writing, calculating, and finger recognition in the region of the angular gyrus: A cortical stimulation study of Gerstmann syndrome. *Journal of Neurosurgery* 99:716–727.

Schnall, S., J. R. Zandra, & D. R. Proffit. 2009. Direct evidence for the economy of action: Glucose and the perception of geographical slant. *Perception* 39:464–482.

Shanton, K., & A. I. Goldman. 2010. Simulation theory. *Wiley Interdisciplinary Reviews, Cognitive Science*, doi: 10.1002/wcs.33.

Shapiro, L. 2011. *Embodied Cognition*. London: Routledge.

Sheperd, S. V., J. T. Klein, R. O. Deaner, & M.L. Platt. 2009. Mirroring of attention by neurons in macaque parietal cortex. *PNAS* 106:9489–9494.

Tye, M. 1995. *Ten Problems of Consciousness*. Cambridge, MA: MIT Press.

Vignemont, F., & P. Haggard. 2008. Action observation and execution: What is shared? *Social Neuroscience* 3 (3–4):421–433.

Weiskopf, D. 2007. Concept empiricism and the vehicles of thought. *Journal of Consciousness Studies* 14:156–183.

Wicker, B., C. Keysers, J. Plailly, J.-P. Royet, V. Gallese, & G. Rizzolatti. 2003. Both of us disgusted in *my* insula: The common neural basis of seeing and feeling disgust. *Neuron* 40:655–664.

Witt, J. K., & D. R. Proffitt. 2008. Action-specific influences on distance perception: A role for motor simulation. *Journal of Experimental Psychology: Human Experimental Psychology* 34:1479–1492.

Witt, J. K., D. R. Proffitt, & W. Epstein. 2010. When and how are spatial perceptions scaled? *Journal of Experimental Psychology: Human Perception and Performance* 36(5): 1153–1160.

PART THREE

The Metaphysics of Action

CHAPTER 11

The Individuation of Action

The problem of act individuation was introduced into current philosophical discussion by G. E. M. Anscombe,[1] who asked the question:

> Are we to say that the man who (intentionally) moves his arm, operates the pump, replenishes the water-supply, poisons the inhabitants, is performing *four* actions? Or only one? (45)

This is an example of a larger problem, the problem of when, in general, act A is the same as act A′. Anscombe offers no solution to this general problem, no general criterion of act identity. But her remarks concerning this example can be examined with profit. She answers her question as follows:

> In short, the only distinct action of his that is in question is this one, A. For moving his arm up and down with his finger round the pump handle *is*, in these circumstances, operating the pump; and, in these circumstances, it *is* replenishing the house water-supply; and, in these circumstances, it *is* poisoning the household (46).

The "is" here is clearly the "is" of identity; what Anscombe intends to assert is that the man's moving his arm up and down = the man's operating of the pump = the man's replenishing of the water supply = the man's poisoning of the inhabitants.

The order in which these putative identities are presented is significant. Anscombe herself is aware of this, for she speaks of the four descriptions as forming

> ...a series, A–B–C–D, in which each description is introduced as dependent on the previous one, though independent of the following one.[2]

The existence of a natural ordering here is confirmed by considering a rearrangement of the elements. The following arrangement is clearly jumbled: "Replenishing the water supply *is* (in these circumstances) operating the pump; and it *is* (in these circumstances) poisoning the inhabitants; and it *is* (in these circumstances) moving his arm up and down." Note, in particular, an interesting asymmetry. Although it is fairly natural to say that moving his arm up and down *is* replenishing the water supply, it is less natural to say that replenishing the water supply *is* (in these circumstances) moving his arm up and down.

The existence of a natural ordering is an embarrassing fact for Anscombe's position. The ordering is readily explained if we grant that the descriptions designate distinct acts that stand in a certain asymmetric relation to one another. But if they all designate the same act, whence arises the ordering? Admittedly, there are natural orderings among descriptions that name the same object. A natural ordering exists among the descriptions 1 + 1, 2 + 0, 3 – 1, 4 – 2, etc., though they all designate the number 2. In Anscombe's case, however, unlike the numerical case, the ordering is imposed not by the descriptions themselves, but rather by the world. If the man's behavior had been different, a different ordering would be required. Had his arm been resting on the pump, he might have moved his arm up and down *by* operating the pump (with his other arm). And to express this fact, a different sort of ordering among the descriptions would be appropriate. There is reason to suspect, therefore, that Anscombe's descriptions are not in fact descriptions of the very same action, that the "is" it seems natural to employ here is not the "is" of identity.

The use of the preposition "by" provides additional, and more conclusive, evidence for the same point. It is true to say that the man poisons the inhabitants *by* replenishing the water supply, that he replenishes the water supply *by* operating the pump, and that he operates the pump *by* moving his arm up and down. As used here, the preposition "by" seems to express a relation that holds between acts (e.g., between an act of replenishing the water supply and an act of operating the pump, in that order). Now if Anscombe is right in claiming that the man's operating of the pump is *identical* to his replenishing of the water supply, then any relation that holds between these acts in one direction must hold between them in the opposite direction. Moreover, the relation must hold between each of these acts and itself. But though it is true to say that the man replenishes the water

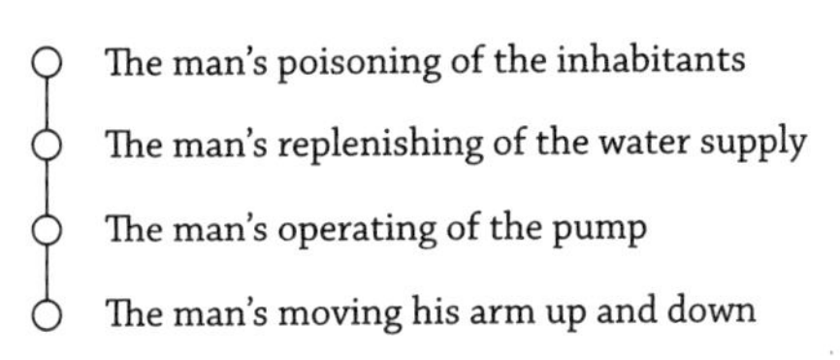

Figure 11.1: Four actions linked via the by-relation

supply *by* operating the pump, it is false to say that he operates the pump *by* replenishing the water supply. (Similarly, it is false to say that he moves his arm up and down *by* operating the pump, or that he replenishes the water supply *by* poisoning the inhabitants.) Moreover, it would be odd to say that the man operates the pump *by* operating the pump. The by-relation, then, seems to be an asymmetric and irreflexive relation. Because no such relation can hold between a given thing and itself, we must conclude that the acts in this example are not identical.

There may be an initial tendency to confuse the by-relation with the causal relation, for in many cases, the by-relation holds between a pair of acts in virtue of a causal relation holding between one of the acts and one of its consequences. This confusion must be avoided. Suppose that I push the button and this causes the bell to ring. There is a causal relation that holds between my pushing of the button and the bell's ringing, but no by-relation holds between this act and this consequence. In virtue of the causal relation, though, it is true to say that I ring the bell by pressing the button. Thus, the by-relation holds between my ringing of the bell (i.e., my *causing* the bell to ring) and my pushing of the button. The causal relation, however, does not hold between these two acts of mine (in either direction).

Although many instances of the by-relation depend upon causal relationships, many do not. The by-relation holds between John's signaling for a turn and John's extending his arm out the window, and it holds between John's fulfilling of his promise and John's returning of the book. In neither of these cases, though, does the by-relation obtain in virtue of a causal consequence. The existence of a certain convention makes it true that John signals *by* extending his arm, and the existence of a certain background condition (viz., that John promised to return the book) makes it true that he fulfills his promise *by* returning the book.

It is instructive to think of the by-relation in diagrammatic terms. In Anscombe's case, we can capture the force of the by-relation by drawing a column of circles, the lowest of which represents the man's moving his arm up and down and the higher of which represent the other acts. (See figure 11.1.) A vertical line (or, later, a diagonal line) indicates that the

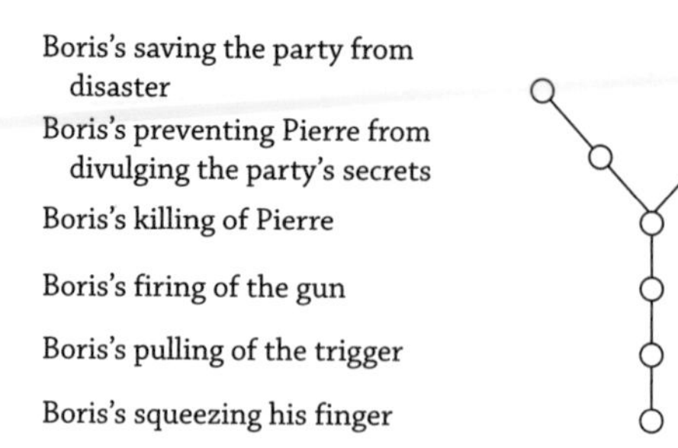

Figure 11.2: By-relations with branching

upper circle represents an act that bears the by-relation to the act represented by the lower circle. As the diagram suggests, the by-relation is not only asymmetric and irreflexive, but also transitive.

Although Anscombe's example can be represented with a single column of circles, a complete treatment of the sorts of relations that hold between acts requires more complex diagramming. Specifically, we must admit the possibility of tree-shaped diagrams, with branches going off in different directions. Suppose that Boris squeezes his finger, thereby pulling the trigger, thereby firing the gun, thereby killing Pierre, thereby preventing Pierre from divulging the party's secrets, thereby saving the party from disaster. By killing Pierre, he also drives Pierre's lover to suicide. The by-relationships here do not permit a single column of acts. Although Boris drives the lover to suicide by killing Pierre and saves the party from disaster by killing Pierre, he does not save the party from disaster by driving the lover to suicide; nor does he drive the lover to suicide by saving the party from disaster. This case is shown in figure 11.2.

These simple examples place in clear relief the most important kinds of relationships that are ignored by Anscombe's view on act identity. Not only does her treatment of such acts neglect the asymmetry of the by-relation, but it also ignores the distinctiveness of the relation that obtains between a pair of acts that bear the by-relation to a common act, though not to each other.

The problem confronting Anscombe's view on act individuation is equally a problem for other writers. Donald Davidson's treatment of particular cases, for example, is substantially the same as Anscombe's. In "Actions, Reasons and Causes," Davidson writes:

> I flip the switch, turn on the light, and illuminate the room. Unbeknownst to me I also alert a prowler to the fact that I am home. Here I do not do four things, but only one, of which four descriptions have been given.[3]

Unlike Anscombe, however, Davidson has buttressed his position with considerable argumentation. Moreover, in another paper, "The Individuation of Events,"[4] Davidson proposes a general criterion of act individuation. (Strictly speaking, it is a criterion of event individuation, but it is intended to hold for acts as well, because actions are a species of events.) The general criterion is this:

(1) Events are identical if and only if they have exactly the same causes and effects.

I do not believe that sameness of causes and effects is a sufficient condition for act identity. But I have no quarrel with the claim that it is a necessary condition; indeed, assuming that being a cause of *E* and being an effect of *E* are genuine properties, its being a necessary condition follows from the indiscernibility of identicals. What I wish to stress, however, is that Davidson's treatment of particular cases of putative act identity founders on his own general criterion. Apparently, many pairs of acts he would class as identical do not have exactly the same causes or do not have exactly the same effects.

Let us look first at effects. Consider Boris's act of pulling the trigger and his act of killing Pierre. Then consider the event that consists of the gun's firing. Clearly, Boris's pulling of the trigger causes this event. But does Boris's killing of Pierre cause it? Certainly

(2) Boris's killing of Pierre caused the gun to fire

seems to be false; at any rate, it sounds extremely odd. If it is, indeed, false, then by appeal to (1) we must conclude that Boris's pulling of the trigger ≠ Boris's killing of Pierre.

A possible reply to this difficulty has already been hinted at: Although (2) sounds odd, it might be argued it is nonetheless true. But how can this reply be justified? A clue is provided in a passage where Davidson says: "To describe an event as a killing is to describe it as an event (here an action) that caused a death."[5] This suggests the view that

(3) Boris's killing of Pierre

is to be paraphrased as

(4) the action of Boris's that caused Pierre's death.

If this paraphrase were correct, then presumably (3) and (4) would be co-referential. It could next be argued that the referent of (4), and hence of (3) as well, is Boris's squeezing his finger. Because Boris's squeezing his finger caused the gun to fire, it would then be concluded that (2) is true.

Can it be established that (4) is a correct paraphrase of (3)? It might be noted, with this in mind, that the sentences

(5) Boris killed Pierre

and

(6) There is an action of Boris's that caused Pierre's death

are "virtually" equivalent. (I say *virtually* equivalent because one can cause Pierre's death without killing him, e.g., by getting someone else to shoot him. But for our purposes this point can be ignored.) Now if (3) and (4) were derived from (5) and (6) by the same transformation, there would be reason to think that they are paraphrases of each other, or at least co-designative. However, they are not so derived: Although (3) is a gerundial nominalization of (5), (4) is not a gerundial nominalization of (6). Moreover, even the suggested principle is unsound. Not all gerundial nominalizations of equivalent sentences are co-designative. For example, although "Effect *E* occurred" and "Something caused *E* to occur" are equivalent, their respective nominalizations ("Effect *E*'s occurring" and "Something's causing *E* to occur") are not co-designative. There is not much reason to believe, then, that (4) is a good paraphrase of (3), or that (3) and (4) are co-referential. Of course, if (3) cannot be construed as (4), one wants to know how it should be construed. I shall return to this question later.

Let me turn now to the requirement that identical actions have the same causes. Davidson clearly wants to regard the following simultaneous actions as identical: John's singing, John's singing loudly, and John's singing off-key. This seemingly innocuous view, however, runs into difficulties. John's singing loudly, let us suppose, is partly caused by his being angry, but his being angry does not at all cause his singing off-key. Similarly, his singing off-key is partly caused by his having a sore throat, but his having a sore throat is not at all a cause of his singing loudly. According to (1), therefore, John's singing off-key ≠ John's singing loudly.

Two replies are open here. First, one might say that, contrary to appearances, John's singing loudly *is* caused by his having a sore throat, and his singing off-key *is* caused by his being angry. How can these contentions be supported, however, without relying on the disputed identity of the acts in question (or on the dubious assumption that John's being angry = his having a sore throat)? Secondly, one might reply that the effect of John's having a sore throat is not the "entire" action of singing off-key, but simply one aspect, feature, or property of it, viz., its being off-key. Similarly, the effect of his being angry is not the entire action, but simply its being loud. This reply raises important questions concerning the conception of an action. What things are "constitutive" of an action and what things are mere aspects or properties of it? If something causes the direction of an arm movement, does it cause the action itself, or merely one feature of it? However one answers this question, one is forced to admit that some of the terms in causal relations are things that consist in *something's having a certain property*: Either *John's* having the property of singing loudly, or his *action*'s having the property of being loud. Once the need for such particulars is recognized, the conception of an act I wish to recommend receives additional support. But this runs ahead of the story.

Let me mention one final problem for the Anscombe-Davidson mode of act individuation. This is the *temporal* problem that has been raised by Lawrence Davis[6] and Judith Jarvis Thomson.[7] Suppose that John shoots George at noon but George does not die of his wounds until midnight. It is true that John killed George and that he killed him *by* shooting him; but is the killing the same as the shooting? Though it is clear that George's death occurs twelve hours after John shoots him, it seems false to say that George's death occurs twelve hours after John *kills* him. But if the death follows the shooting, but not the killing, by twelve hours, the shooting and the killing must be distinct.

Davidson discusses this sort of problem in "The Individuation of Events." He points out that we may easily know that an action is a pouring of poison without knowing that it is a killing; so we may be unprepared to describe the action as a killing until the death occurs. But the problem is not an epistemic one. Even after the death has occurred, and we *know* it has occurred, it still sounds wrong to say that it occurred twelve hours after the killing. Davidson goes on to suggest that there may be a tendency to confuse an event described in terms of an effect with an event described in terms of a "terminal state" (e.g., "his painting the barn red," the referent of which is not over until he has finished painting the barn red). But what grounds does one have for supposing that we are *confused* when we think that John's killing of George is not over until George dies? There is a conflict here between a pre-analytic intuition and a theory. Though it is possible to protect the theory by calling the intuition confused, it is just such intuitions that must be used in testing the theory. Moreover, the intuition that John's killing of George does not precede George's death by twelve hours is just the sort of intuition that sheds doubt on the view that phrases like (3) can be paraphrased according to the model of (4). If "John's killing of George" just *means* "the act of John's that caused George's death," why does it strike us as false to say that John's killing of George preceded the death by twelve hours? Why does this seem false even when we know that the shooting caused the death and that it preceded the death by twelve hours?

Sufficient reasons have now been given for regarding the Anscombe-Davidson pattern of act individuation as inadequate. But although their particular identity claims are mistaken, or at least highly questionable, it does not follow that Davidson's general criterion for act identity is incorrect. I have indicated my doubts that (1) provides a sufficient condition for act identity; but even if (1) were correct, it would not be wholly satisfactory. Ideally, a criterion of act identity should not merely be "correct"—if that were enough, why not be content with Leibniz's law as a criterion of act-identity? A further desideratum of such a criterion is that it lay bare the nature, or ontological status, of an act. Although Davidson's causal criterion tells us something of ontological importance about actions—viz., that they are terms in causal relations—there is more to be said.

The formulation of a fully adequate criterion of identity, then, requires a better understanding of the ontological status of an act. To attain this understanding, I suggest we begin by considering ordinary action sentences. A satisfactory analysis of action sentences, I believe, requires the notion of an *act type* (roughly, what others have called a "generic" action). The usefulness of this notion derives from various sources. Frequently we wish to say that the *same* act is performed on more than one occasion (e.g., that John did the same thing as Oscar, or that John performed the same act today as he did yesterday). The act in question here cannot be an "individual" or "concrete" act, for an individual act cannot occur at different times, nor can two persons be agents of the same individual act (neglecting the possibility of collective actions). The notion of an act type is also needed in talking about ability. If I say that there is an act that John, but not Oscar, has the ability to perform, I do not assert the existence of any "concrete" action, for I do not imply that John has performed an act that Oscar is unable to perform. My statement is best construed as asserting the existence of an act *type*, which John, but not Oscar, has the ability to perform.

An act type, as I construe it, is simply an act *property*, something that an agent exemplifies. When we say, "John weighed 170 pounds" or "John was bald," we ascribe to John the property of weighing 170 pounds or the property of being bald. Similarly, when we say "John signaled for a turn" or "John killed George," we ascribe act properties or act types to John: the property of signaling for a turn or the property of killing George. To ascribe an act type to someone is to say that he *exemplified* it. If John and Oscar perform the same act, they exemplify the same act type. If John has the ability to wiggle his ears, he has the ability to exemplify the property of wiggling one's ears. Admittedly, there is a difference between exemplifying a property, in general, and performing an act. This difference, I believe, is to be analyzed in terms of what *causes* the exemplifying of the property. If I sneeze as a result of the usual causes, I exemplify the property of sneezing, but I do not perform an act. If sneezing is under my voluntary control, however, and if I exercise this control by sneezing on purpose, then I have *performed* an *act* of sneezing.

How adequate is the approach to action sentences that I am recommending? Davidson has argued that an analysis of the logical form of action sentences must account for the fact that "Sebastian strolled through the streets of Bologna" entails "Sebastian strolled."[8] According to my analysis, however, each of these sentences ascribes an act property to a person. This seems to imply that these sentences are to be parsed, respectively, as *Bs* and *Ss*, where *s* stands for Sebastian, *Bx* for *x* strolls through the streets of Bologna, and *Sx* for *x* strolls. If these are unstructured predicates, however, how is the entailment ensured?

Recent works by Romane Clark[9] and Terence Parsons[10] suggest how to handle the problem. (A related suggestion is made by Roderick Chisholm.[11]) On the Clark-Parsons approach to predicate modification, a predicate modifier is construed as an operator that maps a property expressed by one predicate onto a new property. "Slowly," for example, is an operator that, when attached to the predicate "*x* drives," yields a new predicate, "*x* drives slowly," that expresses the property of driving slowly. Clark suggests that for standard modifiers (excluding "negators" such as "nearly"), a principle of *predicate detachment* holds. Let an initial segment of a predicate be either a "core" predicate or the result of prefixing a modifier operator to an initial segment. Then any predicate *P* implies any initial segment of itself *Q*—that is, any statements P^* and Q^* that result from *P* and *Q* by uniformly filling in the term positions are such that P^* entails Q^*. With this rule of detachment, it is easy to see that the predicate "*x* strolled through the streets of Bologna" implies "*x* strolled," hence that "Sebastian strolled through the streets of Bologna" entails "Sebastian strolled."

Given this technique for representing the logical form of sentences with predicate modifiers, we may continue to regard such sentences as ascribing act types to persons. Of course, because strolling through the streets of Bologna is a different property from the property of strolling (though perhaps in some sense "constructed from" the latter property), the two sentences in question ascribe different act types to Sebastian. However, nobody can exemplify the first of these act types without also exemplifying the second at the same time.

We have introduced the notion of an act type, but our main interest is in "individual" acts (i.e., acts that have a particular agent, that occur at a particular time [or during a stretch of time], and that serve as terms in causal relations). Obviously these cannot be act types, which are universals. Let us call these individual, or particular, acts "act tokens." Because an act token is standardly designated by a nominalized form of an action sentence, and because an action sentence associated with such a nominalization asserts that a person exemplifies a certain act property, it is natural to view the designatum of such a nominalization as an exemplifying of an act property by a person. Thus, John's jumping ten feet is an exemplifying by John of the property of jumping ten feet. And Boris's killing of Pierre is an exemplifying by Boris of the property (act type) of killing Pierre. Moreover, as the act type of killing Pierre is distinct from the act type of pulling the trigger, it seems natural to say that Boris's exemplifying of the act type of killing Pierre is distinct from Boris's exemplifying of the act type of pulling the trigger.

Because, in general, the same person can exemplify the same act type on different occasions, and because we want to count these exemplifyings as distinct, we should incorporate the temporal element into our characterization of an act token. We may say, then, that *an act token is an exemplifying of an act type by a person at a time* (or during a stretch of time).[12] Actually, even this characterization is not quite complete, because a person may be the agent of two or more exemplifyings of the same act type at the same time. If, at time t, John points with his right hand and points with his left, then he is the agent of two simultaneous act tokens of pointing. This problem can be dealt with by specifying the *way* in which an act token is performed, in this case, either with the right hand or with the left. For simplicity, however, this refinement will be ignored in what follows.

We are now in a position to state a criterion for the individuation of act tokens that reflects their ontological status:

(7) For any act token A and any act token A', where A is the exemplifying of φ by X at t and A' is the exemplifying of ψ by Y at t', $A = A'$ if and only if $X = Y$, $\varphi = \psi$, and $t = t'$.

This criterion obviously has the consequence that many of the acts equated by Anscombe and Davidson are classed as distinct. In light of the problems that were raised for the Anscombe-Davidson approach, this is a welcome consequence.

Although our criterion yields a multiplicity of acts where the Anscombe-Davidson approach yields just one, there can still be different descriptions of the same act token. Like any entity, an act token can be referred to by a variety of non-synonymous expressions. Just as an act token may have many descriptions, it may exemplify many properties. John's operating of the pump (at *t*) exemplifies the property of being caused by a certain desire and the property of causing the inhabitants to be poisoned. This act token does not exemplify the property of operating the pump, however. We must distinguish between the property expressed by "*x* operates the pump" and the property expressed by "*x* is an operating of the pump." The former is exemplified by John, not by John's operating of the pump (at *t*). The latter is exemplified by John's act, but not by John. In my terminology, the latter is equivalent to "*x* is a *token* of the type, *operating the pump*," and this predicate can be true only of an act token, not of a person.

The distinction between exemplifying a property and being a token of a property is very important. To say of an act token that it is a killing of George is not, on my view, to assert that it *exemplifies* the property of causing George's death; rather, it is to say that it is a *token of* the act type killing George. Now, although John's shooting of George *exemplifies* the property of causing George's death, it is not a *token of* the act type, causing George's death. There is just one act type of which it is a token, viz., shooting George.[13] Thus, although John's shooting of George is an action of John's that causes George's death, it is not a killing of George.

Our criterion of individuation has the virtue of distinguishing acts in a manner that accommodates the by-relation, as well as the causal and temporal properties of acts. The multiplicity of acts that it countenances, however, may seem ontologically objectionable. Because an agent may exemplify indefinitely many act types at one time, he may be the agent of indefinitely many simultaneous act tokens. Is this an unacceptable consequence? This multiplicity of acts may appear to credit a person with too many accomplishments. But the degree of accomplishment of an agent is not to be measured by the number of his acts. Whether someone "gets a lot done" in a day is determined by *what* he does, not by the number of his act tokens. A different sort of objection to the criterion is that it fails to provide an explicit method for *counting* actions: It implies the existence of indefinitely many act tokens, no particular number of them. This concern betrays a confusion. A criterion of individuation need not provide a method of counting. One can provide a criterion of identity for patches of red or for pieces of wood (e.g., "occupies exactly the same place at the same time") without providing a principle for counting patches of red or pieces of wood. We do not need a criterion of numerosity to heed Quine's precept "No entity without identity."[14]

Although the ontological commitments of our analysis may look unattractive at first glance, its ontological virtues must not be overlooked. Instead of treating actions (or events) as a primitive or irreducible category, our account reduces act tokens to persons, act properties, and times. This supports the Aristotelian-Strawsonian ontology in which substances are primary, and events and states of affairs are derivative. If ontological parsimony is, as I believe, primarily a matter of the number of *kinds* of entities countenanced, rather than the number of instances of a given kind that are countenanced, then the view of action I propose fares very well on the dimension of parsimony. Ultimately, of course, an analysis of action cannot be assessed purely in terms of ontological parsimony, but must be judged by its general fruitfulness in theory construction. On this count too, I believe, our approach proves its mettle.[15]

The mere multiplicity of acts countenanced by our theory is not so troublesome. It would be a serious flaw, however, if we were unable to account for the important unity among the acts in, say, the pumping case, a unity that is rightly stressed by Anscombe and Davidson. But our own theory is perfectly capable of capturing the unity that one senses here. Instead of conceiving of it in terms of the relation of identity (which has the disadvantage of being symmetrical and reflexive), we capture it in terms of the by-relation (or a relation that closely approximates the by-relation). Using our diagrammatic conception, we think of this unity in terms of a single *act tree*, where each of the nodes on the tree represents either a basic act or an act that bears the by-relation to a single basic act.

Strictly speaking, the set of distinct acts we want on a single act tree cannot be ordered by the ordinary "by" locution. According to our criterion of individuation, John's singing (at *t*) and John's singing loudly (at *t*) will be distinct acts; but we would not ordinarily say either that John sings "by" singing loudly or that John sings loudly "by" singing. This problem can be handled by introducing a slightly broader relation, which I have called *level generation*, under which the ordinary by-relation is subsumed. I have tried to analyze the notion of level generation elsewhere and cannot review it here.[16] I believe, however, that the inclusion of the additional cases does not upset our diagrammatic conception. For example, the pair of acts consisting in John's extending his arm and John's extending his arm out the window can be neatly and naturally fitted onto an act tree as shown in figure 11.3.

The analysis I have sketched, then, satisfies two fundamental desiderata. First, it slices the units of action thinly enough to accommodate the

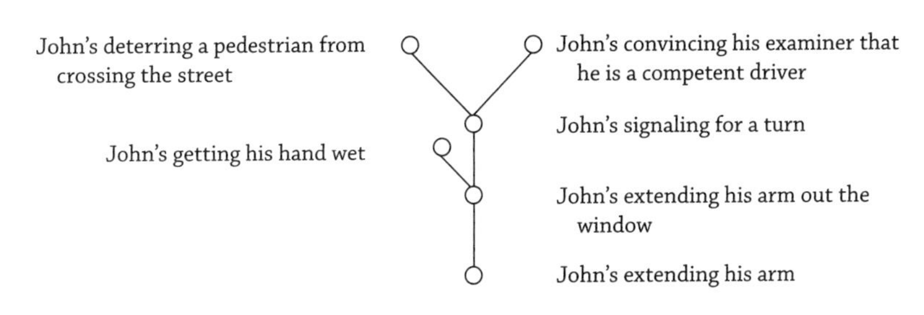

Figure 11.3: An action tree

by-relation. This need has been felt by a number of philosophers who have drawn a distinction between basic and non-basic actions, but my theory allows for a more discriminating ordering of acts than a mere dichotomy. Second, we make use of this ordering to introduce the notion of an act tree, and we use this notion to explicate the unity among the diverse acts in, say, Anscombe's pumping case or Davidson's switch-flipping case.

NOTES

1. G. E. M. Anscombe, *Intention* (Oxford, UK: Basil Blackwell, 1957).
2. Ibid., 45. Anscombe seems to associate this series with the man's intentions. But the same series would be appropriate even if the man did not have all the intentions Anscombe ascribes to him.
3. D. Davidson. (1963). Actions, reasons and causes. *Journal of Philosophy* 60:685–700, 686.
4. In N. Rescher et al., eds. (1969). *Essays in Honor of Carl G. Hempel* (Dordrecht, Neth.: Reidel), 216–234.
5. Ibid., 229. Also see Davidson's paper "Agency," in R. Binkley, R. Bronaugh, & A. Marras, eds. (1971), *Agent, Action, and Reason* (Toronto: University Press), 3–25.
6. L. Davis. (1970). Individuation of actions. *Journal of Philosophy* 67:520–530.
7. J. J. Thomson. (1971). The time of a killing. *Journal of Philosophy* 68:115–132.
8. This point was first raised by Davidson in N. Rescher, ed. (1967), The logical form of action sentences, in *The Logic of Decision and Action* (Pittsburgh, PA: University of Pittsburgh Press).
9. R. Clark. (1970). Concerning the logic of predicate modifiers. *Noûs* 4:311–335.
10. T. Parsons. (1970). The logic of grammatical modifiers. *Synthese* 21:320–334.
11. R. Chisholm. (1971). States of affairs again. *Noûs* 5:179–189.
12. My conception of an act token bears a close resemblance to the account of an (individual) event given by Jaegwon Kim (1966), On the psycho-physical identity theory, *American Philosophical Quarterly* 3:227–235; (1971), Causes and events: Mackie on causation,*Journal of Philosophy* 68: 426–441; and (forthcoming), Causation, nomic subsumption, and the concept of event; and R. M. Martin (1969), On events and event-descriptions, in J. Margolis, ed., *Fact and Existence* (Oxford, UK: Basil Blackwell). The distinction between an act type and an act token is also

considered by Charles Landesman (1969), Actions as universals: An inquiry into the metaphysics of action, *American Philosophical Quarterly* 6: 247–252. Although Landesman rejects the term "act-token," his concept of an "action-exemplifying fact" is very close to my notion of an act token.My theory might be brought even closer to that of Kim and Martin by construing act types not as relational properties (e.g., killing George) with objects or patients "built in," but rather as two-term (or *n*-term) relations (e.g., killing) that an agent and patient jointly exemplify (or that an agent exemplifies "with respect to" a patient).

13. We might relax our requirement to allow John's singing off-key to be a token of the type *singing* as well as of the type *singing off-key*. This relaxation would not imply, however, that John's killing of George is a token of the type *shooting George*.
14. This point is made by Davidson in "The Individuation of Events" (see footnote 4).
15. See my *A Theory of Human Action* (Englewood Cliffs, NJ: Prentice-Hall, 1970).
16. Cf. ibid., chap. 2.

CHAPTER 12

A Program for "Naturalizing" Metaphysics, with Application to the Ontology of Events

1. METAPHYSICS AND COGNITIVE SCIENCE

I wish to advance a certain program for doing metaphysics, a program in which cognitive science would play an important role.[1] This proposed ingredient is absent from most contemporary metaphysics. There are one or two local parts of metaphysics where a role for cognitive science is commonly accepted, but I advocate a wider range of application. I begin by laying out the general program and its rationale, with selected illustrations. Then I explore in some detail a single application: the ontology of events. I do not push hard for any particular ontological conclusion, about either events or any other topic. The focus is methodology, not a particular output of the methodology.

Here is a recently published characterization of the metaphysical enterprise, one that probably captures orthodox practice pretty well and to which I take no exception:

> Metaphysical investigations begin with initial appearances.... In everyday life, these appearances are seldom questioned. In metaphysics, we investigate further. As we pursue a metaphysical topic, we seek to get beyond appearances. We consider arguments about how things really are. We seek to learn the reality of the situation. Reality may confirm initial appearances or it may undercut them. Either way, our goal is to find the ultimate reality. This suggests that the subject matter distinctive of metaphysics is ultimate reality. (Conee & Sider, 2005: 200)

The author of this passage, Earl Conee,[2] does not endorse the stated view unqualifiedly. He cavils at the claim that metaphysics is concerned with "ultimate" reality, suggesting that the term "ultimate" adds nothing of importance. He also worries that if metaphysics is simply concerned with reality, it doesn't differ from other factual investigations. Leaving these quibbles aside, Conee seems to find this characterization of metaphysics fairly satisfactory, as would other metaphysicians, I suspect.

Although Conee doesn't explicitly say this, "reality" is normally understood to refer to what exists (in the broadest sense) in a *mind-independent* way. Metaphysics seeks to understand the nature of the world as it is independently of how we think of it. The suggestion that we should study the *mind* to understand reality would therefore strike many metaphysicians as wrongheaded. They would readily concede that a *portion* of metaphysics—the metaphysics of mind—has mind and mental states as its proper province. But the mind is just a fragment of reality. Most sectors of metaphysics are concerned with extra-mental reality. It would be grossly misdirected for those other sectors of metaphysics to aim their inquiries at the mind. Nonetheless, this is what I propose. I don't mean that a study of the mind is the *final* aim of all metaphysical inquiries; but it should be studied *in the course of* other metaphysical investigations. It should be a contributing part of metaphysical investigation, even those investigations for which the mind is not the primary target.

To clarify the proposal, it must be explained that the referent of "the mind" is (or includes) the aggregate of *organs* or *mechanisms of cognition*. Cognitive organs or mechanisms play a critical role in the causal production of appearances, including metaphysical appearances (whatever exactly we take that to connote). In considering whether such metaphysical appearances should be accepted at face value or, alternatively, should be superseded through some sort of metaphysical reflection, it obviously makes sense to be as informed as possible about how these mechanisms of cognition work. That is why cognitive science is relevant.

Let me expand on this proposal. As Conee indicates, metaphysical inquiries usually start with default metaphysical assumptions (i.e., naïve, intuitive, or unreflective judgments). These correspond to what he calls "appearances." We intuitively judge that objects are colored, that people have free will, that some events cause others, that time passes (always flowing in the same direction), and that some possibilities are unactualized. Metaphysical inquiry starts from such default judgments, but it is prepared to analyze or interpret them in alternative ways, or even to abandon them altogether. They are all up for critical scrutiny, of one sort or another. How should we proceed in this critical, reflective activity? To what degree should precedence, or priority, be given to our naïve metaphysical convictions?

Virtually all metaphysicians agree that our default metaphysical views are subject to philosophical refinement. If there are inconsistencies among these views, some must be abandoned. In addition, most contemporary metaphysicians would agree that science should sometimes override our naïve metaphysics. Physics might give us reason to conclude that time doesn't "pass" at all; that it has no asymmetrical directedness; or, indeed, that there is no such thing as time, only space-time. Again, physics might give us reason to abandon certain assumptions about causal relations. Most existing appeals to science in defense of metaphysical refinements (or revolutions) are appeals to *physical* science. This is understandable, given that most of metaphysics is concerned with ostensibly non-mental targets (e.g., color, causation, time, possibilia). I argue, however, that even in these sectors of metaphysics, evidence from mental science (i.e., cognitive science) can and should be part of metaphysical inquiry.

There are at least two ways in which a metaphysician might try to "revise" naïve experience or naïve belief in matters metaphysical. First, a metaphysician might advance an *eliminative* thesis. This or that ontological phenomenon, assumed to exist on the basis of common sense or naïve experience, might be denied any sort of existence at all. This transpires when a metaphysician denies the existence of time or free will, for example, yet commonsense affirms their existence.

Second, a metaphysician might propose a more nuanced kind of ontological revision of commonsense or naïve experience. Starting with a naïve conception of a certain property, a metaphysician might suggest that the property is really different in crucial ways from the way commonsense or experience represents it. The proposal does not deny the phenomenon's existence (*some* phenomenon worthy of the name). It merely suggests that the property's ontological status is importantly different from the way it is ordinarily represented. One such move is to claim that, although the property is ordinarily represented as being intrinsic to its bearers, it is in fact a relation between members of the class of bearers and another class of objects or events. For example, it's a response-dependent property, a relation between the presumed bearers of the property and some class of mental responses that occur in the subject (perceptual responses, cognitive responses, or emotional responses). Although we think of ourselves as detecting a non-relational property, one that "resides" in the observed objects themselves, the claim is that there is no such "well-behaved" intrinsic property of the bearers. (The critic's conception of well-behavedness may vary across revisionary theories. Some might require the property to be a "natural kind," others a physical property, etc.) Instead, what unites the objects or events in question is only a relation between them and a class of subjective responses to them. This yields an anthropocentric property, rather than a well-behaved one. If it is to be revisionary, a response-dependence thesis must also show that there is a divergence between its true response-dependent character and its naïve "appearance."

Cognitive science-influenced metaphysics need not be restricted to revisionary theories. Cognitive science might also introduce evidence that favors commonsense views. I concentrate, however, on how cognitive science might support revisionary claims. Here is an example.

In speech perception, we seem to hear boundaries of words. It sounds as if there are silences between words that enable us to distinguish their beginnings and ends. But this is an illusion. Steven Pinker (1994), a leading cognitive scientist, describes the matter as follows:

> All speech is an illusion. We hear speech as a string of separate words, but unlike the tree falling in the forest with no one to hear it, a word boundary with no one to hear it has no sound. In the speech sound wave, one word runs into the next seamlessly; there are no little silences between spoken words the way there are white spaces between written words. We simply hallucinate word boundaries when we reach the edge of a stretch of sound that matches some entry in our mental dictionary. This becomes apparent when we listen to speech in a foreign language: it is impossible to tell where one word ends and the next begins. (1994: 159–160)

Although we seem to detect word boundaries in the speech of our native language, there are no such boundaries in the acoustic stimulus.[3] The experienced boundaries are supplied or constructed by the hearer's speech-perception system, building on its knowledge of words in the language that is likely to be spoken. Moreover, even the sequences of sounds we think we hear within a word are an illusion. Information about each component of a word is smeared over the entire word, and this information is reconstructed or reassembled by the listener's system.

So what shall we say about the metaphysics of word boundaries? One could adopt eliminativism. That is the view suggested by Pinker's term "hallucinate." But there is also another option: a response-dependence maneuver.[4] Perhaps word boundaries consist in relations between the spoken sequences of sounds and the auditory experiences of hearers who understand the language in question. Word boundaries are tendencies for such sound sequences to produce "boundary" experiences in (suitable) hearers. These tendencies are genuine relational properties, actually instantiated.

So the response-dependence move is not equivalent to eliminativism. Nonetheless, it is clearly a revisionary view relative to naïve experience, because auditory experience represents word boundaries as being features intrinsic to the speech signal.

The evidence that might support such a response-dependence maneuver includes the evidence Pinker adduces from cognitive science. The acoustic properties of speech signals are studied in the laboratory. These are the kinds of studies on which Pinker bases his conclusions. The same holds for other studies that support the notion of a specialized sense for speech, which is responsible for the mental creation of our phonetic experience. Conjectures on these matters might be made on the basis of casual observation rather than scientific experimentation, especially of felt differences when hearing one's native language versus a foreign language. But really strong evidence against naïve realism and toward either eliminativism or response dependence comes from research in cognitive science. Cognitive science cannot adjudicate between the two revisionary theses. Philosophical reflection must assess the advisability of eliminativism versus respondence dependence. But evidence from cognitive science tilts strongly against realism (or intrinsicalism) about audible word boundaries.

The ontological status of word boundaries is not a very significant metaphysical topic. So let me turn to a more salient example in metaphysics: color. Here I focus on a slightly different issue than response dependence, but one that again shows how cognitive science can be relevant.

Since ancient times, philosophers and physical scientists have offered "secondary quality" accounts of color. These are response-dependence or dispositional theories, which treat colors as ("merely") dispositions to produce certain visual experiences in normal perceivers. On this type of approach, redness is the disposition to look red to standard perceivers in standard conditions of visual observation. This is a revisionary approach because color doesn't "present itself" as a disposition to look a certain way. Redness presents itself (in visual experience) as being "in" objects themselves, or on their surfaces. Thus, in place of dispositional theories, several contemporary theorists propound the view that colors are the *categorical bases* of such dispositions. Brian McLaughlin characterizes redness (in part) as "that property which disposes its bearers to look red to standard visual perceivers in standard conditions of visual observation" (2003: 100). As McLaughlin explains, this is not a dispositional theory of color properties. A dispositional property is a functional-role property. On McLaughlin's theory, by contrast, a color property is the *basis or ground* of a functional role. Redness is the *occupant* of the redness role. More fully, McLaughlin says that being a basis for the disposition of objects to look red doesn't suffice for being redness; to be redness, "a property must be a basis *common* to all things so disposed" (2003: 102, emphasis added). He also implies that, to be redness, a property must be the *unique* property that occupies the redness role.

McLaughlin proceeds to pose three problems for such an occupant, or role-player, theory, two of which I'll mention here. First is the problem of the "common ground." Is there a (physical) categorical property that all red things have in common? The phenomenon of metamerism, identified by color science, seems to stand in the way. What is seen as a single color can be produced by a single light-wave frequency or by many different combinations of light-wave frequencies. As C. L. Hardin notes: "[I]f an observer's unique yellow [yellow that is not at all greenish or reddish] were at 575 nm on the spectrum [in a certain observational setting], an appropriate mixture of spectral 550 nm and 650 nm light would match it exactly" (1988/1993: 198). So there doesn't seem to be any common basis of the disposition to produce an experience of unique yellow. Second is the problem of "multiple grounds." If there is more than one common basis of the disposition to look red in circumstances C, then it's not the case that there is some property that is *the* property that disposes its bearers to look red to standard perceivers in circumstances C.

The problems of the common ground and of multiple grounds flow directly from the demands imposed by the role-occupant theory. The question is how serious these problems are and whether they can be satisfactorily solved. Cognitive science is relevant here, first in creating at least one of the problems and second in helping to solve it. The problem of the common ground is made particularly salient by metamerism, and metamerism is a phenomenon discovered by cognitive science. Although McLaughlin thinks that the problem of multiple grounds can be adequately solved by purely philosophical maneuvers (2003: 125–126), the problem of the common ground is viewed as one that can only be solved with the help of vision science. McLaughlin appeals specifically to the opponent-processing theory of color vision (Hurvich & Jameson, 1957) (see McLaughlin, 2003: 128–131). Opponent-processing theory is also the product of color science (indeed, the cognitive science branch of color science). Whether McLaughlin's considerations succeed in solving the problem of the common ground I shall not try to settle. The point is simply to illustrate how detailed theories of color properties can readily encounter problems when confronted with empirical facts of the sort cognitive science uncovers. At the same time, cognitive science can often help solve some of those problems.

As things go with color, so they go with other properties that invite either a dispositional treatment or a categorical basis treatment. Another example of this type is disgustingness. Now it's not entirely clear how our ordinary conception of disgustingness should be elucidated. But let us mainly focus on a property that, on reflection, seems worthy of being the referent of the term "disgusting" (just as redness is the referent of "red").

If disgustingness is the ground of the disposition to elicit feelings of disgust, mustn't it be a property common to all disgusting things, as in the case of redness? Is there any such common ground? Cognitive science findings make such a thesis problematic. The emotion of disgust has a remarkably diverse range of elicitors, which range from the concrete to the abstract. These are summarized by Kelly (2007). At the concrete end of the spectrum of disgust elicitors are feces, vomit, urine, and sexual fluids (Rozin et al., 2000). Other likely candidates are corpses and signs of organic decay (Haidt et al., 1994). Bodily orifices and, via contamination, things that come in contact with them are also powerful elicitors (Rozin et al., 1995). All of these elicitors are disgust universals. Finally, disease and parasitic infection provide another set of disgust universals (Curtis et al., 2004). At the abstract end of the spectrum, violations of norms having nothing to do with the types of elicitors just mentioned can also trigger disgust. Flouting a norm central to a cultural in-group, for example, can produce an emotion of disgust.

What could possibly be a common ground among these diverse elicitors? It seems unlikely that there is any such property. If there is one, it surely isn't a natural kind, but at most, a wildly heterogeneous disjunction of properties. If disgustingness is either an intrinsic property shared by all disgust elicitors or nothing at all, eliminativism looms.

Let us consider the alternative approach: dispositionalism or response dependence. On this view, disgustingness is simply the disposition to produce a disgust response in humans. But what, precisely, is the response? Kelly (2007) divides the disgust response into three parts, of which I'll emphasize two. The first part concerns the so-called affect program for disgust, where "affect program" is a theoretical term for reflex-like emotional responses, a set of coordinated physiological and behavioral elements. In the case of disgust, the responses include (1) a behavioral response involving immediate withdrawal, (2) a distinctive facial expression, known as the "gape face," and (3) qualitative responses including revulsion and the physiological concomitants of nausea (Ekman, 1992). The second part is what Kelly calls, following Rozin et al. (2000), "core disgust." Its features are distinct from those of the disgust affect program; they are less reflexive and more cognitive in nature. Two of its central features are a sense of offensiveness and contamination sensitivity. Offensiveness is a type of aversion that views the disgusting entities as somehow dirty, tainted, or impure. Contamination sensitivity refers to the fact that once an item is marked as offensive by the disgust system, the item can contaminate otherwise pure and un-disgusting entities. Entities so contaminated are then treated as disgusting. Moreover, there is an asymmetric relationship between disgustingness and purity. It is far easier for something pure to be contaminated than it is to purify something contaminated: A single drop of sewage can spoil a jug of wine, but a single drop of wine doesn't much help in purifying a jug of sewage (Hejmadi et al., 2004).

Returning now to the dispositional approach to disgustingness, exactly which responses are constitutive of this disposition? Are they *all* constitutive of it? Here is a different way to approach the problem. As we are already dealing with revisionary accounts of disgustingness, why assume there is a single property? Why not distinguish different disgustingness properties, in order to cut nature better at its joints? One way to do this is to bifurcate disgustingness in terms of the two families of responses: the affect-program responses and the core responses. We would then have DISGUSTINGNESS_1 and DISGUSTINGNESS_2, where the former is a disposition to produce affect-program responses and the latter is a disposition to produce core-disgust responses.

No matter which choice we make in this territory, the choice should reflect knowledge of the types of functional (and evolutionary) relations that I have been presenting from cognitive science research. Any natural slicing of the terrain should be informed by the relevant empirical facts, not restricted to what can be known about disgust and disgustingness a priori or via commonsense alone.

Finally, I turn to the case of moral concepts. There is powerful evidence from recent cognitive science that moral judgment is intimately associated with emotion. This might naturally lead to a response-dependence account of our moral concepts, as recently argued by Prinz (2006). Of course, a number of moral philosophers in recent decades have advanced "sensibility" accounts of moral concepts, in the general tradition of Hume. These include McDowell (1985), McNaughton (1988), Johnston (1989), Dreier (1990), Wiggins (1991), and Wright (1992). The point is not that it takes cognitive science to think up such a position in metaphysics; rather, cognitive science is uniquely able to contribute certain kinds of compelling evidence, evidence that might tilt meta-ethical or metaphysical arguments in one way or another. Here is some of the relevant psychological evidence that supports a response-dependence view in meta-ethics.[5]

Moll et al. (2003) measured brain activity as subjects evaluated moral sentences. They found that when subjects made moral judgments (but not when they made factual judgments), brain areas associated with emotions were active. Sanfey et al. (2003) measured brain activity as subjects played an ultimatum game. When a division of a monetary sum was deemed inequitable, a player had brain activity in areas associated with emotion. Berthoz et al. (2002) found similar engagement of brain areas when subjects considered violations of social rules. Again, Greene et al. (2001) found emotion activation as subjects considered moral dilemmas. In a different kind of study, Schnall et al. (2008) found that negative emotion can lead people to make a more negative moral appraisal than they would otherwise. Subjects were given a questionnaire with a series of vignettes and asked to rate the wrongness of the actions described. Half of the subjects who read the vignettes were seated at a nice clean desk; the other half was seated at a filthy desk. The latter rated the vignettes as more wrong than the former. Finally, there is considerable evidence that psychopaths, who seem incapable of making genuine moral judgments, are profoundly deficient in negative emotions (Blair et al., 2001).

Again, there is room for debate over which meta-ethical/metaphysical theories of moral concepts or moral properties should be inferred from these studies. The present point is the fairly mild one that empirical studies are profoundly important in helping us (as theorists) understand the ordinary grasp of moral matters, part of the task of doing metaphysics in the ethical realm. What is especially noteworthy is that affective entanglement with moral judgment is not evident in conscious experience. It presumably occurs, for the most part, at a sub-threshold level, which is why cognitive science and neuroscience are needed to tease out what takes place. Astute philosophers (such as Hume and his intellectual descendents) can tender such hypotheses, but their suspicions are not conclusive and, indeed, are hotly disputed by other philosophers. One needs a cleaner methodology—one not beholden to conscious awareness—to get sounder and more decisive evidence on the matter.

For evidence on the role of emotion even in everyday "factual" judgment, consider another phenomenon discovered by cognitive science: the fascinating delusional symptom called *Capgras syndrome*. In Capgras syndrome, people report that their acquaintances (spouse, family, friends, and so on) have been replaced by "body doubles" (Capgras & Reboul-Lachaux, 1923). They acknowledge that their husband or wife looks like their husband or wife (respectively). They can pick the person out from a lineup as resembling their husband or wife, but they steadfastly maintain that he or she is an impostor. Apparently, they consciously recognize the person's face, but they lack an emotional response to it.[6] They therefore conclude that the person observed isn't in fact the husband, wife, friend, etc.[7] Of special interest here is what Capgras syndrome tells us about so-called normal people. Although we are quite unaware of this, we apparently use emotional associations, at a sub-threshold level, in identifying persons. I am not gearing up, on this basis, to defend a response-dependence account of personal identity. I add this item to the mix only to underline the point that what transpires in judgment-making and concept possession is not introspectively transparent. This is why metaphysics needs cognitive science—at least insofar as metaphysics proceeds by starting from our naïve worldview and proceeding "outward" from there.

In the remainder of this chapter, I illustrate the cognitivist program with a rather different example: the metaphysics of events. This angle again turns on the fact that the ways we conceptualize many types of entities of metaphysical interest are not available to introspection. We need the help of cognitive science to illuminate these conceptualizations.

2. EVENTS AND THE PROBLEM OF EVENT INDIVIDUATION

Let us approach the metaphysics of events while looking over our shoulders at the metaphysics of time and disgustingness. In the case of time, a variety of scientific theorizing, especially in physics, suggests that there is no such thing as time, at least as naïvely experienced. The flow of time, the existence of something called "the present," and the directionality of time are all problematic from the standpoint of physics. But if time is anything at all, it cannot be something merely in our minds. So what external "thing" can be nominated by the word "time"? Metaphysicians struggle with this problem. It will not do to pick out *any* old metric or dimension that is countenanced by contemporary space-time physics and *call* it time. Metaphysicians look for a property or dimension of the space-time manifold that comports as closely as possible to our naïve understanding or grip on time (i.e., our time experience). But when we try to look at what that understanding or experience consists in, we need help from cognitive science. To identify a suitable *physical* dimension to identify with time, we need a better grip on our *anthropocentric* conception of time; and that may require an understanding of temporal *psychology*.

The case of disgustingness has parallels. If we want to understand what disgustingness is, we need to understand the psychology of disgust (the emotion). Disgustingness may prove to have an additional parallel with events. As we have seen, once the relevant facts are learned about disgust and its family of elicitors, we see that a good case can be made for *bifurcating* the phenomenon of disgustingness. Instead of positing a single kind of state, it may make better sense to posit two types of disgustingness. A strong case for such a proposal, however, hinges partly on empirical findings about the internal responses associated with disgust. Similarly, I argue that certain philosophical preconceptions about events should give way once we have a better grip on the way we (psychologically) conceptualize items that can be considered events. Instead of positing a single ontological category of events, it is better to posit a *pair* of such categories.

The preceding paragraph describes where we are headed. Now let me set the stage for the discussion by making some preliminary clarifications. There are narrower and wider uses of the term "event," such as colloquial uses and technical uses. These different uses indicate that different writers about events have different targets in their crosshairs. Colloquially, events are equated with *happenings:* things that *occur,* or *take place*; this can't be said of objects like books or statues. A variety of theories conceive of events rather differently, however (Casati & Varzi, 2008). There are scientific theories that treat events as qualified points in space-time (general relativity), or as sets of outcomes (probability theory); and some philosophical theories treat events as properties of cross-world classes of individuals (Lewis, 1986). These technical uses are put aside for present purposes. Another distinction is between event-*types* and event-*tokens*. The term "crash" can be used to refer to either a general type of event or individual events that occur on particular occasions. Our principal interest is event-tokens.

In the 1960s and 1970s, a lively debate emerged about the problem of event individuation and the nature of events. Some philosophers, most prominently Donald Davidson, championed a view that came to be called the coarse-grained, or unifier, approach. This was initially motivated by the theory of action, where actions, of course, are a subcategory of events. Davidson's unifier approach is succinctly illustrated by the following passage about actions:

> [I]t is hard to imagine a satisfactory theory of action if we cannot talk literally of the same action under different descriptions. Jones managed to apologize by saying "I apologize"; but only because, under the circumstances, saying "I apologize" *was* apologizing. (1969/1980: 164–165)

There aren't two distinct actions of Jones, his saying "I apologize" and his apologizing; these actions are one and the same. Similarly, if Oliver moves his finger, flips a switch, turns on a light, and alerts a prowler, all these actions are identical with one another.

Multipliers contend that Oliver's actions are four distinct actions. The multiplier position was developed (semi-)independently by Jaegwon Kim (for events generally) and me (for actions) (Kim, 1966, 1969, 1973, 1976;

Goldman, 1965, 1970, 1971). We each defended a general account of events or actions as exemplifications of properties. In Kim's well-known formulation, event E is identical with event E′ only if they have as "constituents" the same substance(s), the same property or relation, and the same time (Kim, 1976). For example, the dyadic event consisting of the collision of the *Titanic* with the iceberg has as its two substances the ship and the iceberg, as its relation *colliding with*, and as its time a period of around ten seconds on the fateful night (Simons, 2003). In the Oliver case, the switch-flipping action is not identical to the prowler-alerting action because they involve both different substances (the switch versus the prowler) and different relations (flipping versus alerting). No comparably agreed-upon principle of individuation is associated with the unifier theory.

The unifier approach commanded a majority of adherents in the 1960s and 1970s, and this probably remains true today. Because the property-exemplification approach has been a minority view, I begin the present discussion with a limited defense of it. The purpose is not to defend it as *the* correct theory but only as a serious contender. So let me review some considerations in its favor and some of the dialectic that has characterized the debate.

Here are some arguments from causal relations that I advanced in the early period of the debate (Goldman, 1970, 1971)[8]. Consider these actions:

(B1) Boris's pulling (of) the trigger
(B2) Boris's firing (of) the gun
(B3) Boris's killing (of) Pierre

According to the unifier approach, B1 = B2 = B3. If this is correct, every effect of B1 must equally be an effect of B2 and B3. Now consider the following event:

(F) the gun's firing

Is F an effect of B1? Surely it is. The pulling of the trigger causes the gun to fire. Is F an effect of B3? Apparently not; it certainly sounds wrong to say that Boris's killing (of) Pierre caused the gun to fire. Hence, some effect of B1 is not an effect of B3, a counterexample to the unifier approach. This assumes, of course, that if X and Y are identical, then there are no properties possessed by X that Y lacks. This principle holds for causal-relational properties among others.

Davidson tried to ward off this kind of difficulty by arguing that the singular term "Boris's killing (of) Pierre" should be translated (roughly) as the definite description "The action of Boris that caused Pierre's death." The denotation of this expression is said to be B1, and as F is an effect of B1, it is also an effect of B3. But the proposed translation is contentious. If the translation is correct, why does it seem intuitively incorrect to say that Boris's killing (of) Pierre causes the gun to fire?

Here is another example (Goldman, 1971[9]): John sings; indeed, he sings loudly and off-key. Are the following identity statements true: John's singing (J1) = John's singing loudly (J2) = John's singing off-key (J3)? The unifier theory affirms these identity statements. But consider John's being angry (A) as a candidate cause. Suppose that state A was a contributing cause of J2, because the loudness of John's singing was an inadvertent effect of his anger. Although state A is a contributing cause of J2, it is implausible that it's a contributing cause of J1, because the decision to sing was independent of the angry mood. Because one of the causes of J2 is not a cause of J1, we must conclude that J2 ≠ J1, and another counterexample confronts the unifier theory.

This kind of difficulty has resurfaced in more recent literature on causation. L. A. Paul (2000) gives the example of Suzy who falls and breaks her right wrist while skiing. The next day she writes a philosophy paper. Simplifying Paul's example, note that Suzy's breaking her right wrist is plausibly a contributing cause of her writing the paper *with her left hand*, but it isn't a contributing cause of her writing the paper. Now Paul does not regard this point (or the rest of her example) as grounds for endorsing the property exemplification view of *events*. She introduces the term "aspect" to refer to property instances, but doesn't identify them with events. An aspect can be a property instance *of* an event; it doesn't have to *be* an event. Paul proceeds to argue that aspects, rather than events, are causes and effects. This position has substantial affinity with the multiplier theory because it views property-individuated entities (viz., aspects) as the bearers of causal properties, a role usually assigned to events.

Notice that the parties to the debate agree on two things. First, they agree that there are such things as events. Second, they agree that exactly one of the two approaches is correct; the task is to determine which one that is.[10] For purposes of this debate, then, eliminativism isn't an option.[11] It also isn't an option to consider *both* theories to be correct, as one view explicitly denies what the other asserts, namely the identity of several enumerated items. It is premature, however, to exclude this possibility. Perhaps there are two viable concepts or conceptions of events: Under one the unifier view is correct, and under the other the multiplier (property-exemplification) view is correct. I now try to show how cognitive science lends support to this compromise position.

3. TWO SYSTEMS OF MENTAL REPRESENTATION AND THEIR RELEVANCE TO THE ONTOLOGY OF EVENTS

The thesis I wish to advance is that the mind has two ways of representing events, or two cognitive formats for event representation. I don't mean that there are two representational systems dedicated exclusively to events, but two different systems that can be used to represent events *among other things*. Crucially, they represent events using different representational principles, and this gives rise to different intuitive principles of identity or individuation.

It is arguable that Jonathan Bennett's theory of events (Bennett, 1988) recognizes this kind of point. His position might be encapsulated by saying that people get us to feel multiplier intuitions by using one set of event names ("the killing of the man") and the unifiers use another set of event names ("the murder"). If Bennett's theory approximates the one I'll offer, that's fine from the present perspective. But the one I shall present has a more explicit basis in cognitive science, at a level below the level of language.[12]

The general idea of multiple systems or formats of mental representation for (roughly) the same target domain is extremely common in cognitive science. One example is the two systems of visual representation: the ventral and dorsal systems (Milner & Goodale, 1995). Conscious visual experience of the world is subserved by the ventral visual stream, which generates conscious visual perception of objects. In addition, the recently discovered dorsal visual system unconsciously mediates the visual control of motor behavior (e.g., reaching for a cup that is in view). That there are two distinct systems is evident from neurological dissociations in certain patients, in which one system, but not the other, is impaired. These systems deploy different types of representation, although each is activated by visual inputs.

The study of cognitive processes directed at events and actions is fairly recent. Some of this research is specifically focused on the *individuation* of actions and events, although the particular individuation problem that psychologists study is a bit different from the one metaphysicians concern themselves with. We will have to tread carefully. Nonetheless, here is a passage by the psychologist Karen Wynn that indicates what motivates her interest in the subject:

> [V]irtually nothing is known about how people individuate actions. The individuation of actions is likely to be complex because actions are not definable purely in terms of objective properties. Frege's... observation that number is not an inherent property of portions of the world applies as much to actions as it does to physical matter. Just as a given physical aggregate may be an instance of many different numbers (three forks is also an instance of 12 tines, 2 decks is an instance of 104 cards), so may a given portion of activity (one dance may equally be an instance of 437 steps, one speech may be an instance of 72 utterances). There is no objective fact of the matter as to where, in the continuously evolving scene, one "action" ends and the next "action" begins. The individuation of discrete actions from this continuous scene is a *cognitive imposition*. (Wynn, 1996: 165, emphasis added)

Of special interest to developmental psychologists (including Wynn) is how infants manage to parse the ongoing activity of the world into distinct actions (or events). Infants' parsing abilities should interest us as well, because they bear on the question of what representational capacities are also available to adults for individuating actions and events.

Sharon & Wynn (1998: 357) proceed to enumerate several reasons why an ability to individuate actions and events is essential to the infant. First, such parsing is required to perceive causality. For example, to recognize causality in a "launching episode"—say, where a billiard ball causes movement in another billiard ball by striking it—it wouldn't be possible to recover the causality if the whole motion of the two balls was perceived as one undifferentiated swoosh. Second, individuating actions is a necessary precursor of learning to produce them. A child who wants to throw a ball should realize that pulling back one's arm is a necessary component of the throw, but jumping up and down in excitement afterward is not. A third major motivation for parsing motion into discrete actions lies in the need to interpret the behaviors of others. Parsing motion into discrete acts provides the behavioral units that are explainable in terms of desires and beliefs. Finally, an ability to parse actions into the same units as other people do is necessary for verb acquisition and hence linguistic communication. For language to be comprehensible, two people must have in mind the same bounded pattern of motion when they refer to a "jump," a "hug," or a "hit."

What cues might infants use to individuate actions? Ideas are suggested by studies of individuation in other domains. Spelke (1988) found that infants construe spatially separated surfaces in the visual layout as belonging to distinct objects. Similarly, Bregman (1990) demonstrated that portions of acoustic energy that are temporally separated by silence are interpreted by adults as arising from independent acoustic events, whereas energy that is temporally continuous and judged to originate from the same location is commonly interpreted as arising from the same acoustic event. These findings suggest that infants may use spatiotemporal discontinuities to segment scenes of motion into distinct units. Portions of motion separated by "gaps" in space, time, or both may make different actions. This hypothesis was tested and corroborated by Wynn (1996). In a first test, infants were able to count—hence must have individuated—the jumps of a puppet when there were brief pauses between jumps. In a second test, motionless pauses were replaced by having the puppet gently wag its head back and forth between jumps. Infants were still able to detect boundaries between jumps, though not as strongly as in the first test. Sharon & Wynn (1998) extended these findings, but they also showed that infants are not able to parse heterogeneous actions from a continuous, nonpatterned stream of motion.

The possibility of two distinct systems for action or event individuation was suggested by findings on the individuation of objects. Another study by Spelke and colleagues supported the notion that young infants analyze spatiotemporal continuity in establishing representations of one object or two. Spelke, Breinlinger, Macomber, & Jacobson (1992) arranged a scene with two screens separated by a gap. Young infants who saw objects emerge one at a time from opposite sides of the two screens, but never appear in the middle, interpreted this as two objects, not one. This was another example of the significance of spatiotemporal continuity. (If there were one object, continuous motion would have required it to appear in the gap between the screens.) However, by twelve months of age, infants bring *kind membership* to bear on the problem of object individuation. Xu & Carey (1996) found that when twelve-month-old infants were shown objects of two different kinds emerging one at a time from opposite sides of a single screen, they established representations of two numerically distinct objects. This could not have been inferred on the basis of spatiotemporal continuity considerations alone, because such considerations would be compatible with a single object. Shipley & Shepperson (1990) also used counting as a wedge into studying preschool children's criteria for object individuation. Children were shown displays of objects, some of which were broken. For example, there might be three intact cars and one car broken into two distinct pieces. Even when asked to count "the cars," children as old as five years would often count each separate piece of a car, along with each intact car (yielding a total count of five, in the foregoing example). This is despite the fact that five-year-olds would have known the kind word "car" for some time. Still, when considering a scene consisting of whole cars and split cars, children analyzed the scene in terms of spatiotemporally discontinuous units, not units determined by the "basic-level" kind term "car." This tends to show that although five-year-olds have the capacity to individuate on the basis of kinds, or sortals, they have a preference for the apparently more primitive spatiotemporal criterion of object individuation.

What about event individuation? Wagner & Carey (2003) compared children and adults on individuation tasks concerning both objects and events. Shown a series of slides on a computer screen, subjects were asked for each slide, "How many Xs are here?" The X term was either an appropriate kind label for the object (e.g., "car" or "fork") or else a kind-neutral word (e.g., "thing"). Following the object task, children were told that the game was going to change and now they would count what happened in movies (the event individuation experiment). Here the analogous term X was an appropriate event description, targeting either the goal of the event or its temporally discrete sub-actions. The latter was considered a spatiotemporal parse of the observed activity. In both the object and event cases, children showed a spatiotemporal bias in individuation. They often ignored the kind term (or goal-based term) that was specified and counted in terms of purely spatiotemporal criteria. Nonetheless, even by age three and definitely by age five, they showed some sensitivity to kind-based individuation. Adults, by contrast, were at or near ceiling in their use of kind-based (or goal-based) individuation.

All the studies and theoretical interpretations we have reported are from developmental psychology, which focuses on the growth or unfolding of cognitive competences. The development angle is not our primary interest, however. We are concerned with the cognitive competences of adults, because commonsense ontological judgments issue mainly from adults (philosophers themselves, in most cases). The main point I wish to make is that adults apparently have two systems for individuating events, one system based on spatiotemporal discontinuities and a second system based on kinds. The first is a very early developing system, and the second is not quite so early developing. I report evidence from developmental psychology not because I am intrinsically interested (here) in infants and children, but because it happens to be the branch of cognitive science where the most relevant work on event individuation has been done.

The psychological research just reviewed is the main, already existing, empirical work most relevant to the metaphysics of events, but admittedly it isn't as directly relevant as one might wish. It isn't fully and squarely relevant because the individuation issue addressed in cognitive science is different from the individuation issue discussed in section 2. Psychologists are concerned with event *segmentation*, or *parsing*. This is decomposing the world's spatiotemporal flux into disjointed units. Philosophers are concerned with events occurring at one and the same time and place.[13] Arguably the two questions are connected, but they aren't the same.

Consider the example from section 2 in which Boris pulls a gun's trigger and thereby fires the gun. Both the trigger-pulling and the gun-firing occur at the same time and place. But we can still ask whether additional ontological slicing should be done. Are the trigger-pulling and the gun-firing two *distinct* actions, or are they one and the same action, simply described in terms of different effects? I call this new question the *slicing* question. It isn't one that the psychology literature addresses. Although the psychological work doesn't squarely address the slicing question, it may be indirectly relevant. Recognition of the two disparate systems can help us diagnose the emergence and persistence of the dispute between unifiers and multipliers.

The property-exemplification view assigns a pivotal role to the kind or property of which a token action or event instantiates. According to this approach, if action *a* is a token of action-type *A* and action *b* is a token of action-type *B*, then *a* is not identical to *b*, even if *a* and *b* are performed by the same agent at the same time. Boris's pulling the trigger (at *t*) and Boris's firing the gun (at *t*) are not the same token action, because pulling the trigger and firing the gun are not the same property or kind. This conceptualization of actions or events comports with the second, late-developing system of event representation, in which kinds (or sortals) play a crucial role. By contrast, the unifier view of actions or events assigns no comparably pivotal role to kind-differences in action or event individuation. According to the coarse-grained approach, Boris's pulling of the trigger and Boris's firing of the gun are one and the same action. Is it plausible that this conceptualization of actions is associated with the first, early-developing system of event representation, a system focusing on spatiotemporal factors without attention to (other) kinds? I think it is plausible.

If purely spatiotemporal representations are used to represent actions, it is natural to represent them as bodily movements only, not as entities with higher-order, non-geometric or non-kinematic properties, such as firing the gun, poisoning the king, or writing a check. Interestingly, Davidson, the leading proponent of the coarse-grained theory, explicitly says that actions are, really, just bodily movements:

> We must conclude, perhaps with a shock of surprise, that our primitive actions, the ones we do not do by doing something else, [are] mere movements of the body—these are all the actions there are. We never do more than move our bodies; the rest is up to nature. (1971/1980: 59)

Thus, what many philosophers of action call "basic actions" are the only actions there are. Davidson does not mean to deny, of course, that grander-sounding things like sending spaceships to Mars are also included among our actions. He certainly doesn't deny that these types of descriptions can be applied to our actions. But these "other" actions are just identical to the basic ones, which are movements of the body (e.g., the finger movement that depresses the button that launches the spaceship). This fits a purely spatiotemporal construal of action, which conforms comfortably to the constraints of the conjectured early-developing system of event representation.

A different take on event individuation emerges if one focuses on event-referring expressions such as "Quisling's betrayal of Norway" or "Quisling's betraying Norway," both of which are derived from the indicative sentence "Quisling betrays Norway." Verbs pick out kinds or sortals, which typically abstract from the spatiotemporal properties displayed by their particular instances. If one mentally represents an event by concentrating on the abstract property by which it is specified, this conceptual mode of mental representation would naturally be processed by the later-developing system for event representation.

My hypothesis, then, is that the two groups of partisans in the event-slicing debate are theorizing about the referents of two different types of mental representations, either event representations in the spatiotemporal system or event representations in the kind-based system. Both groups of partisans, of course, have access to both systems of representation, but when they engage in philosophical reflection about events, they implicitly focus on representations from one system rather than the other; and they differ on which system of representations they select. Because Davidson did not dispute the existence of "grander" actions—he did not deny that bodily movements have "higher-order" properties in addition to geometric or kinematic ones—he did not mentally represent actions *exclusively* by means of the spatiotemporal system. He used the kind-based system as well. However, he might have privileged, or concentrated on, the deliverances of the spatiotemporal system as compared with the kind-based system. This is my conjecture. In other words, I don't suggest that unifiers turn off their kind-based system when thinking (during everyday business) about events. Nor do I suggest that property-exemplification theorists turn off their spatiotemporal system when thinking about events. It is only a matter of which representational format is given greater weight or preference in one's theorizing moments. Each representational format can be a source of intuitions about events. A theorist merely chooses to accentuate one family of intuitions rather than another. Nor do I mean to deny that the two types of theorists are unmoved by the various philosophical principles that are proposed and debated. Obviously they are so moved. Nonetheless, I suspect that theorists of different persuasions *dwell* more persistently (in their minds) on one family of intuitions compared with another, and that's a non-negligible factor in what influences their theoretical judgments.

Let me return now to the comparisons drawn earlier (at the beginning of section 2) with the metaphysics of time and disgustingness. In both cases the (initial) question is: What is the property in the world that makes the best fit with a term used uncritically and unreflectively to characterize objects or relations in the world? As the example of time illustrates, finding a suitable fit may not be an easy matter. There may be "less" in the world than what our cognitive structures incline us to suppose. As the disgustingness example illustrates, we may also find that things in the world—and in our cognitive and emotional repertoire—offer a more complex array of alternatives than offered by our language. (The latter offers just a single predicate "disgusting" and its approximate synonyms.) In the case of disgustingness, I proposed that the best solution is to reject the assumption that there is a single property of disgustingness, even understood in response-dependent terms. Instead, there is DISGUSTINGNESS_1 and DISGUSTINGNESS_2. Similarly, the metaphysical moral that may emerge (in part) from the findings of cognitive science is that we should abandon the original assumption of both unifiers and multipliers that exactly *one* of their views is right (the only interesting question being "which one?"). Instead we should conclude that both are right. The best solution is to countenance two metaphysical categories of events, EVENTS_1 and EVENTS_2. This is how cognitive science can play a role in the conduct of metaphysicalizing.[14]

NOTES

1. The theme of the present chapter is very close to that of an earlier paper (Goldman, 1989). In terms of the earlier paper's division between "descriptive" and "prescriptive" metaphysics, this chapter emphasizes the prescriptive enterprise.
2. Although the passage appears in a co-authored book by Conee and Sider, Conee is listed as the lone author of the chapter containing this passage.
3. Instead of describing our auditory experience of word boundaries as featuring "silences" (as Pinker puts it), we might describe the matter in the language of "categorical perception." The boundary between two words is perceived as being discrete and discontinuous, whereas in fact it is as non-discrete and continuous as what is perceived within a spoken word. Even in this formulation of the phenomenon, there is an illusion or error in our naïve representation of word boundar-

ies that is open to correction by cognitive science. Thanks to Roberto Casati for emphasizing this alternative formulation.

4. S.-J. Leslie suggested this as an example of response dependence.
5. I culled this summary from Prinz (2006).
6. Evidence supporting this explanation was first produced by Ellis and Young (1990). They predicted that Capgras patients would not show the normally appropriate skin conductance responses to familiar faces, and this prediction was confirmed. Subsequent research has added further confirmation.
7. For an instructive novelistic account of the syndrome, see Powers (2006).
8. Chap. 11 (this volume).
9. Ibid.
10. I assume that the unifier and multiplier approaches exhaust the territory. That assumption may appear problematic, because there is, for example, the "part" or "componential" view of events (Thalberg, 1971; Thomson, 1971, 1977), which treats some members of a candidate list of identical events as spatiotemporal parts of other members. For present purposes, however, the part theory will be considered a species of the multiplier view. It does, after all, answer the standard question of identity, in relevant types of cases, by saying "different" rather than "same."
11. For a defense of eliminativism about events, see Horgan (1978).
12. Thanks to Dean Zimmerman for suggesting the parallel with Bennett. I am not convinced, however, that the two-conceptions-of-events thesis quite fits Bennett's considered view. In chapter 8 of his book, Bennett says that a good semantics of event names should be intermediate between Kim's view and Quine's, where Quine defended a "zonal" view that identifies any two happenings in the same spatiotemporal zone. Bennett does not say that *both* kinds of semantics are right. On the contrary, he searches for a single intermediate position between them. He despairs of finding one, saying, "The truth lies between Kim and the Quinean, but there is no precise point between them such that it lies there" (Bennett, 1988: 126). In searching for a single intermediate position, Bennett appears to disavow a two-conceptions-of-events view in which both conceptions are "correct."
13. This characterization is slightly contentious, because many multipliers would deny that the events in question take place at the same time and place. For example, the time of killing a person might include the time of the victim's death, making it longer than the time of pulling the trigger. The exact times and places of complex actions and events is a highly vexing issue. Nonetheless, the formulation in the text is a useful approximation.
14. Thanks to Roberto Casati, Holly Smith, and Dean Zimmerman for valuable comments on an early draft of this chapter. Zimmerman was particularly helpful in persuading me that certain material contained in that draft might be profitably deleted (although those were not his words).

REFERENCES

Bennett, J. (1988). *Events and Their Names*. Indianapolis, IN: Hackett.
Berthoz, S. et al. (2002). An fMRI study of intentional and unintentional (embarrassing) violation of social norms. *Brain* 125:1696–1708.

Blair, R. J. R., et al. (2001). A selective impairment of the processing of sad and fearful expressions in children with psychopathic tendencies. *Journal of Abnormal Child Psychology* 29:491–498.

Bregman, A. S. (1990). *Auditory Scene Analysis: The Perceptual Analysis of Sound*. Cambridge, MA: MIT Press.

Capgras, J., & J. Reboul-Lachaux. (1923). L'illusion des 'sosies' dans un délire systématisé chronique. *Bulletin de la Société Clinique de Médicine Mentale* 2:6–16.

Casati, R., & A. Varzi (2008). "Event concepts," in T. F. Shipley & J. Zacks, eds., *Understanding Events: How Humans See, Represent, and Act on Events*. Oxford, UK: Oxford University Press.

Conee, E., and T. Sider (2005). *Riddles of Existence: A Guided Tour of Metaphysics*. Oxford, UK: Clarendon Press.

Curtis, V., R. Aunger, & T. Rabie. (2004). Evidence that disgust evolved to protect from risk of disease. *Proceedings of the Royal Society: Biological Science Series B*, 271(4):S131–S133.

Davidson, D. (1969). The individuation of events. In N. Rescher, ed., *Essays in Honor of Carl G. Hempel*, 216–234, Dordrecht, Neth.: Reidel; reprinted in Davidson, *Essays on Actions and Events*, 163–180. Oxford, UK: Oxford University Press.

Davidson, D. (1971). Agency. In R. Binkley, R. Bronaught, & A. Marras, eds., *Agent, Action, and Reason*. Toronto: University of Toronto Press. Reprinted in Davidson, *Essays on Actions and Events*, 43–61. Oxford,UK: Oxford University Press.

Dreier, J. (1990). Internalism and speaker relativism. *Ethics* 101:722–748.

Ekman, P. (1992). An argument for basic emotions. *Cognition and Emotion* 6:169–200.

Ellis, H. D., & A W. Young (1990). Accounting for delusional misidentifications. *British Journal of Psychiatry* 157:239–248.

Goldman, A. I. (1965). *Action*. Ph.D. diss., Princeton University, Princeton, NJ, department of philosophy.

Goldman, A. I. (1970). *A Theory of Human Action*. Englewood Cliffs, NJ: Prentice-Hall.

Goldman, A. I. (1971). The individuation of action. *Journal of Philosophy* 68:761–774.

Goldman, A. I. (1989). Metaphysics, mind, and mental science. *Philosophical Topics* 17:131–145. Reprinted in A. Goldman, *Liaisons: Philosophy Meets the Cognitive and Social Sciences*, 35–48. Cambridge, MA: MIT Press (1992).

Greene, J., et al. (2001). An fMRI investigation of emotional engagement in moral judgment. *Science* 293:2105–2108.

Haidt, J., C. McCauley,,& P. Rozin (1994). Individual differences in sensitivity to disgust: A scale sampling seven domains of disgust elicitors. *Personality and Individual Differences* 16:701–713.

Hardin, C. L. (1988). *Color for Philosophers: Unweaving the Rainbow*. Indianapolis, IN: Hackett.

Hejmadi, A., P. Rozin, & M. Siegal (2004). Once in contact, always in contact: Contagious essence and conceptions of purification in American and Hindu Indian children. *Developmental Psychology* 40(4):467–476.

Horgan, T. (1978). The case against events. *Philosophical Review* 87:28–47.

Hurvich, L. M., & D. Jameson (1957). An opponent-process theory of color vision. *Psychological Review* 64:384–408.

Johnston, M. (1989). Dispositional theories of value. *Proceedings of the Aristotelian Society* 63 (Supplement):139–174.

Kelly, D. (2007). Projectivism psychologized: The philosophy and psychology of disgust. Ph.D. diss. Rutgers University, New Brunswick, NJ, department of philosophy.

Kim, J. (1966). On the psycho-physical identity theory. *American Philosophical Quarterly* 3:227–235.

Kim, J. (1969). Events and their descriptions: Some considerations. In N. Rescher, ed., *Essays in Honor of Carl G. Hempel*, 198–215. Dordrecht, Neth.: Reidel.

Kim, J. (1973). Causation, nomic subsumption and the concept of event. *Journal of Philosophy* 70:217–236.

Kim, J. (1976). Events as property exemplifications. In M. Brand & D. Walton, eds., *Action Theory*. Dordrecht, Neth.: Reidel.

Lewis, D. (1986). Events. In *Philosophical Papers*, vol. 2, 241–269. New York: Oxford University Press.

McDowell, J. (1985). Values and secondary qualities. In T. Honderich, ed., *Morality and Objectivity*. London: Routledge & Kegan Paul.

McLaughlin, B. (2003). Color, consciousness, and color consciousness. In Q. Smith and A. Jokic, eds., *Consciousness*, 97–154. Oxford, UK: Oxford University Press.

McNaughton, D. (1988). *Moral Vision: An Introduction to Ethics*. Oxford, UK: Blackwell.

Milner, A. D., and M. Goodale (1995). *The Visual Brain in Action*. Oxford, UK: Oxford University Press.

Moll, J., R. De Oliveirra-Souda, and P. J. Eslinger (2003). Morals and the human brain: A working model. *Neuroreport* 14:299–305.

Paul, L. A. (2000). Aspect causation. *Journal of Philosophy* 97:235–256.

Pinker, S. (1994). *The Language Instinct: How the Mind Creates Language*. New York: William Morrow.

Powers, R. (2006). *The Echo Maker*. New York: Farrar, Strauss and Giroux.

Prinz, J. (2006). The emotional basis of moral judgments. *Philosophical Explorations* 9(1):29–43.

Rozin, P., et al. (1995). The borders of the self: Contamination sensitivity and potency of the body apertures and other body parts. *Journal of Research in Personality* 29:318–340.

Rozin, P., J. Haidt, & C. McCauley (2000). Disgust. In M. Lewis & J. M. Haviland-Jones, eds., *Handbook of Emotions*, 2nd ed. New York: Guilford.

Sanfey, A. G., et al. (2003). The neural basis of economic decision-making in the ultimatum game. *Science* 300:1755–1758.

Schnall, S., J. Haidt, G. Clore & A. H. Jordan (2008). Disgust as embodied moral judgment. *Personality and Social Psychology Bulletin* 34:1096–1109.

Sharon, T., and K.Wynn (1998). Individuation of actions from continuous motion. *Psychological Science* 9:357–362.

Shipley, E. F., and B. Shepperson (1990). Countable entities: Developmental changes. *Cognition* 34:109–136.

Simons, P. (2003). Events. In M. Loux & D. Zimmerman, eds., *The Oxford Handbook of Metaphysics* (357–385). New York: Oxford University Press.

Spelke, E. (1988). Where perceiving ends and thinking begins: The apprehension of objects in infancy. In A. Yonas, ed., *Perceptual Development in Infancy*, vol. 20 (219–234). Hillsdale, NJ: Erlbaum.

Spelke, E., K. Breinlinger, J. Macomber, and K. Jacobson (1992). Origins of knowledge. *Cognition* 99:605–632.

Thalberg, I. (1971). Singling out actions, their properties and components. *Journal of Philosophy* 68:781–786.

Thomson, J. J. (1971). Individuating Actions. *Journal of Philosophy* 68:771–781.

Thomson, J. J. (1977). *Acts and Other Events*. New York: Cornell University Press.

Wagner, L., and S. Carey (2003). Individuation of objects and events: A developmental study. *Cognition* 90:163–191.

Wiggins, D. (1991). A sensible subjectivism. In *Needs, Values, Truth: Essays in the Philosophy of Value*. Oxford, UK: Blackwell.

Wright, C. (1992). *Truth and Objectivity*. Cambridge, MA: Harvard University Press.

Wynn, K. (1996). Infant's individuation and enumeration of actions. *Psychological Science* 7:164–169.

Xu, F., and S. Carey (1996). Infants' metaphysics: The case of numerical identity. *Cognitive Psychology* 30:111–153.

CHAPTER 13

Actions, Predictions, and Books of Life

I

Are actions determined? Because it is difficult to tell "directly" whether or not actions are governed by universal laws, some philosophers resort to the following "indirect" argument:

If actions are determined, then it is possible in principle to predict them (with certainty).

It is not possible in principle for actions to be predicted (with certainty).

Therefore, actions are not determined.

I shall call a defender of this argument an "anti-predictionist," and his position will be called "anti-predictionism." In this chapter, I shall try to rebut anti-predictionism.

Both premises of the anti-predictionist argument will come under attack here. The first premise, affirming that determinism entails predictability, is often accepted without adequate scrutiny. Some writers not only assume that determinism entails predictability but even *define* determinism as the thesis that every event is predictable in principle.[1] I believe, however, that it is essential to distinguish between determinism and predictability. We must first notice that there are various kinds or senses of "possibility" that may be involved in the "possibility of prediction." Moreover, it can be shown that in many of these senses, determinism does *not* entail the possibility of prediction. Many anti-predictionists have failed to notice this, however. Therefore, upon discovering some unpredictability in the arena of human action, they have wrongly concluded that actions must be undetermined. This error will be avoided only if we carefully distinguish between determinism and predictability. Hence, an important aim of this chapter is to differentiate various senses of "possibility of prediction" and to ascertain how they are related to determinism.

Let us assume now that we can find some suitable sense of "possibility of prediction" that is closely related to, if not entailed by, determinism. The second premise of the anti-predictionist argument asserts that, in such a sense, it is impossible for actions to be predicted. Various arguments have been offered in support of this premise. One that I shall consider concerns the possibility of writing a complete description of an agent's life—including his voluntary actions—even before he is born. According to anti-predictionism, if actions were determined, it would be possible to write such books. Indeed, it would be possible for such a "book of life" to be written even if the agent were to read its prediction of a given action before he was to perform that action. It seems clear to the anti-predictionist, however, that such books of life are impossible. Predictions of my actions cannot be made with certainty; for when I read these predictions, I can easily choose to falsify them; so argues the anti-predictionist. But it is far from clear that he is right. I think, on the contrary, that it may well be possible (in a suitable sense) for books of life to be written. And thus it seems to me that the anti-predictionist is unable to establish the truth of his second premise.

In general, anti-predictionists support their second premise by contrasting the predictability of human behavior with that of physical events. It is alleged that special difficulties of a purely conceptual sort arise for the prediction of action, and that these difficulties are unparalleled in the realm of merely physical phenomena. I shall claim, however, that there are no essential differences between actions and physical events with respect to the problem of prediction. More precisely, I shall claim that *conceptual* reflection on the nature of human behavior (as opposed to investigation by the special sciences) does not reveal any peculiar immunity to prediction.

It must be emphasized that I offer no proof of the thesis that actions *are* determined; I merely wish to show that the anti-predictionists' arguments fail to prove that they are *not* determined. It is conceivable, of course, that actions are not determined. And if actions are not determined, then I would admit that they are not perfectly predictable (in any sense at all). What I contend, however, is that the arguments of philosophers, based on familiar, commonsense features of human action and human choice, do not prove that actions are undetermined or unpredictable. The basic features of human action are quite compatible with the contention that actions are determined and susceptible to prediction. In other words, my aim here is not to establish the *truth*, but merely the *tenability*, of the thesis that actions are determined.

II

Let us begin with some definitions. I shall define determinism as the view that every event and state of affairs is determined in every detail. An event is determined (in a given detail) if and only if it is deducible from some set of antecedent conditions and laws of nature. A law of nature is, roughly, any true non-analytic universal statement of unlimited scope that supports counterfactual conditionals.[2] Both "low-level" empirical connections, like all metals expand when heated, and "theoretical" connections, like F = ma, are included. Antecedent conditions can be either events, like moving at 10 m.p.h., or states of affairs, like having a specific gravity of 1.7. (Throughout I shall be concerned both with events and states of affairs, but for brevity I shall often omit reference to states of affairs.) Negations of events, like a ball's *not* moving at 10 m.p.h., are also included. Antecedent conditions may be directly observable phenomena, but they need not be. Theoretical, hypothetical, and dispositional states—like being brittle or being intelligent—can serve as antecedent conditions.

Notice that my definition of determinism is in terms of a formal relationship (i.e., the relationship of deducibility holding between events and sets of laws and antecedent conditions). In particular, this definition makes no explicit reference to the ability of anyone to predict these events, and thereby leaves open the question of the connection between determinism and predictability.

If determinism is true, human actions are determined. But determinism alone does not tell us what laws or kinds of laws take human actions as their dependent variables. I shall assume, however, that these laws would include ones with psychological states like desires, beliefs, intentions, etc., as their independent variables. This presupposes—correctly, I think—that statements connecting actions with, for example, wants and beliefs are not purely analytic.[3] Rather, their logical status would correspond to quasi-analytic, quasi-empirical generalizations like many theoretical statements of science. If determinism is true, wants, beliefs, intentions, etc., are themselves determined by prior events of various sorts. The determinants of these mental states are quite diverse, however, so I shall make no attempt to delineate them.

In ordinary language, not all determining factors of an event are called its "causes." A body's having a certain mass may be a determining antecedent condition of that body's moving at a certain velocity after being struck by another object, but its having that mass would not be called a "cause" of its velocity. Similarly, although a person's having a certain intention or desire would not ordinarily be termed a "cause" of his action, it may be an antecedent condition of the relevant sort. Because determinism is often connected with what philosophers call "causal necessity," I shall use the technical term *causally necessitate* to apply to antecedent conditions that, together with laws of nature, determine a given event. Thus, I shall say that desires and beliefs (together with other conditions) "causally necessitate" a given action, even though ordinary language would not condone such an expression.

In our discussion of predictability we need a sense of "prediction" distinct from mere lucky guesses or precognition. We must be concerned with predictions made on the basis of laws and antecedent conditions. I shall call a prediction a *scientific prediction* if and only if it is made by deducing the predicted event from laws and antecedent conditions. A scientific predictor may learn of the laws and antecedent conditions in any number of ways. (On my definition, most predictions made by actual scientists are not "scientific predictions," for real scientists seldom, if ever, *deduce* what will occur from laws and prior conditions. Nevertheless, scientific prediction as defined here may be regarded as an ideal of prediction to which scientists can aspire.)

As indicated above, it is important to identify different senses of the phrase "possibility of prediction." I shall now distinguish four relevant species of possibility, though further distinctions will be made later within some of these categories. The four species are (1) *logical possibility*, (2) *logical compossibility*, (3) *physical possibility*, and (4) *causal compossibility*.

An event is *logically possible* if and only if it is not self-contradictory, and logically impossible if and only if it is self-contradictory. Drawing a square circle is a logically impossible event, whereas jumping 90 feet is a logically possible event. *Logical compossibility* is defined for two or more events. A set of two or more events is logically compossible if and only if the conjunction of the members of the set is logically consistent. A set is logically incompossible (i.e., not logically compossible) if and only if each of the events is logically possible but their conjunction is logically inconsistent. Thus, the two events, (a) *x*'s being a pumpkin from 11 o'clock to 12 o'clock, and (b) *x*'s turning into a pumpkin at 12 o'clock, are logically incompossible.

An event is *physically possible* if and only if it is not inconsistent with any law or laws of nature; an event is physically impossible if and only if there are laws of nature with which it is inconsistent. Traveling faster than the speed of light, for example, is physically impossible. I shall speak not only of events being physically impossible *in general*, but also as being physically impossible *for* certain kinds of entities. Thus, the act of lifting a ten-ton weight is not, in general, physically impossible; but it is physically impossible for (normal) human beings. Given the physical constitution of human beings, laws of nature make it impossible for them to lift such weights.

Causal compossibility differs from physical possibility in attending to groups of events rather than events taken singly. Roughly, a set of events is causally compossible just in case laws of nature allow each of them to occur singly and allow them to occur as a group. More precisely, consider a set of events $\{e_1,\ldots,e_n\}$ each of which is logically possible and physically possible, and which are jointly logically compossible. Then $\{e_1,\ldots,e_n\}$ is a causally compossible set if and only if there is no set of laws of nature such that the conjunction of these laws with $\{e_1,\ldots,e_n\}$ is logically inconsistent.[4] I shall say that the set as a whole is causally compossible or that each member is causally compossible "with" or "relative to" the other members.

A set of events $\{e_1,\ldots,e_n\}$ is causally *in*compossible (i.e., not causally compossible) if and only if there are some laws of nature $L_1,\ldots,L_k$ such that the conjunction of $L_1,\ldots,L_k$ with $e_1,\ldots,e_n$ is logically inconsistent. Assuming, as we do, that $e_1,\ldots,e_n$ satisfy the other three species of possibility, the set $\{e_1,\ldots,e_n\}$ will be causally incompossible if and only if the negation of (at least) one member of the set is entailed by the conjunction of the other members of the set conjoined with $L_1,\ldots,L_k$. Thus, if the negation of a given member of the set is causally necessitated by the other members of the set, then the set is causally incompossible.

III

The most interesting questions concerning the prediction of action are best handled in terms of the notion of causal compossibility. The reflexivity of predictions—the fact that a prediction often has an effect that bears on its own truth—can be understood properly with the use of this notion. But the question of the causal compossibility of predictions of action cannot arise unless the other three species of possibility are satisfied. Our definition of causal compossibility makes a set causally compossible only if its members are logically possible, physically possible, and (jointly) logically compossible. For example, if it is physically impossible to make scientific predictions of actions, the question of causal compossibility does not ever arise. Therefore, before turning to the questions of reflexivity, including the question of whether "books of life" can be written, we must focus on certain problems connected with the logical compossibility and the physical possibility of predicting actions.

The logical possibility and compossibility of predictions can be discussed together, because the distinction between them is somewhat blurred. This is because a correct prediction is not really a single event, but a pair of events—a prediction and an event predicted. Two different examples of logical incompossibility have been uncovered in connection with the prediction of behavior. I shall discuss these examples briefly and argue that, contrary to what their authors suppose, they do not prove that actions are undetermined, and they do not prove that actions have a peculiar immunity to prediction unparalleled by physical phenomena.

The first logical incompossibility, as discussed by Maurice Cranston,[5] can be summarized as follows: Suppose that Sam invents the corkscrew at time *t*. In the intended sense of "invent," this means (a) that Sam thinks of the corkscrew at *t*, and (b) that no one ever thought of the corkscrew before *t*. Cranston argues that no one could have predicted Sam's inventing the corkscrew. In order for him to make this prediction, he would himself have to think of the corkscrew. And had he thought of the corkscrew, it would be false to say that Sam "invented" the corkscrew. Yet, ex-hypothesi, Sam *did* invent the corkscrew. Using the terminology of "logical incompossibility," we can formulate Cranston's problem by saying that the three events, (a) Sam thinks of the corkscrew at *t*, (b) no one ever thought of the corkscrew before *t*, and (c) someone predicted Sam's inventing the corkscrew, are logically incompossible.

The second example poses a problem for predicting not actions, but decisions. However, as the concept of a voluntary action is so closely tied to that of a decision, an unpredictability connected with decisions is very important for us to discuss. Carl Ginet claims that it is impossible ("conceptually" impossible) for anyone to predict his own decisions.[6] The argument begins by defining "deciding to do *A*" as *passing into* a state of knowledge (of a certain kind) that one will do *A*, or try to do *A*.[7] Suppose now that Sam, at *t* decides to do *A*. Had Sam predicted that he would make this decision—and had this prediction involved *knowledge*—he could not have decided later to do *A*. For if, before *t*, he had known that he would decide to do *A*, he would have known then that he would do *A*, or try to do *A*. But if, before *t*, he had known that he would do *A* (or try to do *A*), then he could not, at *t*, have *passed into* a state of knowing that he would do *A*. Thus, according to Ginet, Sam could not have predicted that he would make this decision.

Of course, Sam might make his prediction and then forget it. If so, he can still decide, at *t*, to do *A*. However, if Sam not only knows, before *t*, that he will decide to do *A*, but also *continues* to know this up until *t*, then Sam cannot, at *t*, decide to do *A*. In other words, the following three events are logically incompossible: (a′) Sam decides, at *t*, to do *A*, (b′) Sam predicts (i.e., knows) that he will decide to do *A*, and (c′) Sam continues to know this until *t*.

What do these two logical incompossibilities prove? Do they prove that decisions and inventions are undetermined? Do they prove that voluntary actions, including the decisions that lead to them, have a special immunity to prediction? The answer to both questions is "No," I believe.

Our examples of logical incompossibilities do not establish any special status for human behavior, for precisely analogous incompossibilities can be produced for physical phenomena. Let the expression "a tornado strikes *x by surprise*" mean (1) a tornado strikes *x* at a certain time, and (2) before that time nobody ever thought of a tornado striking *x*. Now suppose that, as a matter of fact, a tornado strikes Peking by surprise. Then it is logically incompossible for this event to have been predicted. That is, the set consisting in the tornado striking Peking by surprise and a prediction of the tornado striking Peking by surprise is a logically incompossible set. In general, it is logically incompossible for tornadoes striking places by surprise to be predicted. For if anyone were to predict these events, they could no longer be described as "tornadoes striking places *by surprise*." Nevertheless, there certainly are (or could be) events correctly describable as "a tornado striking *x* by surprise."

I wish next to argue that the invention and decision incompossibilities do not show that these human phenomena are undetermined. Notice first that the tornado case, though it has the same logical structure, does not bear on the question of determinism. Although it is logically incompossible for anyone to predict the tornado striking Peking by surprise, I am in no way inclined to suppose that this event is not determined. Similarly, our logical incompossibilities fail to show that inventions and decisions are undetermined. How could such logical incompossibilities demonstrate that these events are not governed by laws of nature? The notion of a law is in no way involved in the concept of logical incompossibility. And hence the presence of logical incompossibilities sheds no light on the question of whether there are laws and antecedent conditions that entail inventions or decisions.

The critical error here is the assumption that if an event is determined (under a given description), it must be possible to predict it (under that description).[8] The falsity of this proposition should be adequately clear from the invention case. Suppose that Sam's thinking of the corkscrew at *t* is deducible from laws and antecedent conditions. And suppose that the fact that no one ever thinks of the corkscrew before *t* is also deducible from laws and antecedent conditions. Then, the event consisting in Sam's *inventing* the corkscrew at *t* would be determined; but it still would be logically incompossible for it to have been predicted under that description. The lesson to be learned here is not that inventions are undetermined actions, but that the alleged entailment between determinism and predictability is not an entailment at all. At any rate, the fact that an event is determined under a given description does not entail that it is *logically compossible* for it to be predicted under that description.[9]

The case of decisions can be handled similarly. It seems to me quite possible that a person's passing into a state of knowing, or intending, to do *A* be deducible from laws and antecedent conditions. But although this event would be determined (under the given description), it would not be logically compossible for Sam to have predicted it (under that description) and continued to know it until *t*.

IV

I turn now to physical possibility. Is it physically possible to make scientific predictions of human actions? Here the emphasis should be placed on the qualifier "scientific." Although it may well be physically possible to make "lucky guess" predictions, or perhaps even predictions based on "intuition," it is not obvious that predictions can be made by deducing an action from laws and antecedent conditions. And this is the only kind of prediction that bears on the issue of determinism.

Anti-predictionists might claim that it is physically impossible for human beings to make scientific predictions of actions, because human beings cannot learn enough antecedent conditions to deduce what will be done. But it is inessential to the predictionist's position to restrict the range of predictors to human beings. In order to avoid theological or supernatural issues, we may require that any predictor be a finite entity operating within the causal order of the universe. But apart from this, no arbitrary limits should be placed on admissible predictors.

Popper[10] has tried to show that there are certain limitations of the predictions that can be made by "classical mechanical calculating" machines. But to restrict the range of predictors to calculating machines is an important restriction; even if Popper is right about the prediction limitations of machines of the sort he discusses, there may be other beings that can make predictions his machines cannot. Another limitation on Popper's discussion is that much of it is aimed at establishing the physical impossibility of a *single* being, like Laplace's demon, making scientific predictions of *all* events or of a very large number of events. But the fact that all events cannot be predicted by a single being is compatible with the proposition that every event can be predicted by some being or other.

Anti-predictionists might proffer the following arguments for saying that it is physically impossible for *any* finite being, not just human beings, to make scientific predictions of human behavior. Scientific predictions, they might claim, require knowledge of infinitely many facts, but it is physically impossible for a finite being to know infinitely many facts. The infinity requirement seems necessary because in order to *deduce* that even a certain finite system will yield a given result, one must know that no interfering factors will intrude from outside the system. And knowing this may involve knowing *all* states of the world at least at one time.

This argument is of questionable force. It is far from clear that the deduction of actions from antecedent conditions and laws requires knowledge of infinitely many facts. Nor is it clear that no finite being could know infinitely many facts. Even if the argument is correct, however, it would seem to prove *too much*. For if the knowledge of infinitely many facts is required in order to make scientific predictions of actions, the same would be true for scientific predictions of physical events. Thus, the above argument would fail to establish any special immunity of human action to prediction. Finally, even if it is physically impossible for any finite being to make scientific predictions of actions, this would not prove that actions are undetermined. Here too, as above, we have a sense of "possibility" in which determinism does *not* entail the possibility of prediction. The proposition that an event is (formally) deducible from laws and antecedent conditions does not entail that it is physically possible for any being to come to know these laws and antecedent conditions and to deduce the event from them. Hence, even if the anti-predictionist could establish that it is physically impossible to predict actions scientifically, he would not thereby establish that actions are undetermined.

We have not conclusively shown either that it is physically possible for some beings to predict actions scientifically or that it is not. But unless we assume that this is physically possible, we cannot turn to the other interesting issues that surround the problem of the prediction of human behavior. Unless we assume this, the question of the causal compossibility of predicting actions cannot even arise. In order to explore these important issues, therefore, I shall henceforth assume that scientific predictions of actions (like scientific predictions generally) are physically possible.

V

Perhaps the anti-predictionist would think it obvious that it is causally incompossible to predict actions scientifically. He might argue as follows: "Let us grant, as is likely, that there have never been any genuine scientific predictions of voluntary human actions. If, as my opponent claims, determinism is true, then it is causally incompossible for any predictions to have been made of these actions. For every actual action *A*, there is an actual event $\bar{P}_A$, the *absence* of a prediction of *A*. Because each of these events $\bar{P}_A$ is actual, and because determinism is true, each of these events $\bar{P}_A$ must be causally necessitated by some set of actual events prior to it. But if each of these events $\bar{P}_A$ is causally necessitated by actual prior events, then each event P_A —the prediction of *A*—is causally incompossible relative to some actual events. In other words, for each actual action *A*, it is causally incompossible for *A* to have been predicted."

This argument, like a previous one, proves too much. The anti-predictionist is right in saying that non-actual predictions of actions are causally incompossible with the actual prior events in the world. But this is true simply because, assuming determinism, every non-actual event whatever is causally incompossible with some set of actual prior events. Thus, using the notion of causal-compossibility-relative-to-all-actual-events, we can establish the impossibility of predicting physical phenomena, as well as human behavior. We can point to an action that was never predicted and say that, in this sense, it "could not" have been predicted, because its non-prediction was causally necessitated by other actual events. But by the same token, we can point to a physical event that was never predicted and say that it "could not" have been predicted, because its non-prediction was also causally necessitated by other actual events. Using this notion of "possibility of prediction," the anti-predictionist again fails to establish any special immunity of action to prediction.

Apart from this point, however, the notion of causal-compossibility-relative-to-all-actual-events does not seem to be a pertinent kind of possibility for our discussion. We have seen that determinism does not entail the possibility of predicting actions in *every* sense of "possible." And here, I believe, we have still another sense of "possible" in which determinism does not entail that it is possible for every action to be predicted. Determinism does not say that, relative to all actual prior events, it is causally compossible for a prediction of an action to be made *even if* those actual prior events causally necessitate that no prediction occur. Thus, the fact that it is impossible, in this sense, for actions to be predicted does not conflict with the thesis that actions are determined. Nor is it surprising that the sense of "possible" here under discussion is not important. Using the notion of causally-compossible-relative-to-all-actual-prior-events, it turns out, assuming determinism, that only actual events are possible. But it is a strange and unduly restrictive notion of "possible" according to which only actual events are possible!

We need, then, a broader notion of possibility, one that allows for non-actual possibles while also taking into account the notion of causal necessity. We can discover a more relevant notion by examining what is often meant in ordinary contexts when we say, counterfactually, "*e* could have occurred." Suppose we say, counterfactually, "The picnic could have been a success." This sort of statement would normally be made with a suppressed "if" clause. We might mean, for example, "The picnic could have been a success if it had not rained." Now if the only thing that prevented the picnic from being a success was the rain, we are also likely to say, "The picnic *would* have been a success if it had not rained." In the first case we mean that the substitution of non-rain for rain in the course of events would have *allowed* the picnic to be a success; in the second case we mean that this substitution would have *ensured* the success of the picnic. In both cases we are saying that a certain event could have or would have occurred *if* the prior course of the world had differed from its actual course in specified ways.

Although in ordinary contexts we might not pursue the matter further, in order to be systematic we must inquire further: "*Could* it *not* have rained?" "Could non-rain have occurred instead of rain?" The actual rain was causally necessitated by actual events prior to the rain. If we are to suppose that it did not rain, we must also make changes (in our imagination) of still-earlier events. Carrying this argument to its logical conclusion, it is obvious that whenever a determinist says that a non-actual event *e* "could have" occurred, he must imagine *an entirely new world*. For the picnic to have been a success, it is required that it not have rained. For it not to have rained, the cloud formation would have had to be different. For the cloud formation to have been different, it is required that the wind velocity (or some other factor) have been different, etc.

Not only must we change conditions prior to *e*, if we are to suppose *e* occurs, but we probably[11] must change events after *e* as well. Had it not rained, a certain other picnic group near us would not have ended their picnic just then. And had they not ended their picnic just then, they would not have left for home just then. And had they not left for home just then, they would not have had an automobile accident when they did,[12] etc.

The determinist who says, counterfactually, "*e* could have occurred," must construct a whole world to justify his claim. Nevertheless, this gives him a sense of "possible" that allows non-actual possibles. For a determinist, "*e* could have occurred" may be translated as "a causally compossible world can be imagined in which *e* occurs." Normally the determinist will be able to construct worlds resembling the real one to a large extent. But these worlds will never be exactly like our world except for one event only. Any such imagined world will differ from the real world by at least one event for every moment of time. This will be true, at any rate, if the laws governing these imagined worlds are identical with those of the real world. And I shall assume throughout that these laws (whatever they are, exactly) are held constant.

VI

We can now give what I regard as a reasonable formulation of the question: "Is it possible, in principle, to make scientific predictions of voluntary actions?" The formulation is: "Can one construct causally compossible worlds in which scientific predictions are made of voluntary actions?" In saying that this is a "reasonable" formulation of the question, I do not mean that a negative answer to this question would entail that voluntary actions are not determined. I have already pointed out that determinism does not entail that it is physically possible to make scientific predictions of events, including actions. Hence, neither does determinism entail that there are causally compossible worlds in which scientific predictions of actions occur. However, because we are assuming that scientific predictions are physically possible, it would be an important negative result to discover that one cannot construct causally compossible worlds in which scientific predictions are made of voluntary actions. This might not prove that actions are undetermined, but it would suggest a disparity between actions and physical phenomena. For, assuming that scientific predictions are physically possible, it does seem that there are causally compossible worlds in which scientific predictions are made of physical events.

Similar comments are in order on the question "Can one construct causally compossible worlds in which scientific predictions are made of voluntary actions and in which the agent learns beforehand of the prediction?" Determinism does not entail that there must be such causally compossible worlds. But if no such worlds are constructible—worlds in which "books of life" are found, or things comparable to books of life—one might well claim a disparity between voluntary actions and physical phenomena.

Fortunately, I believe that there *are* causally compossible worlds in which scientific predictions are made of voluntary actions and in which, moreover, the agent learns of (some of these) predictions before he performs the predicted actions. I believe that there are causally compossible worlds in which books of life are written before a man's birth. Inscribed in these books are predictions of the agent's actions, predictions based on laws and antecedent conditions. These predictions are correct even though the agent sometimes reads them before he performs the predicted actions. I shall support my claim that there are such causally compossible worlds by giving a sketch of such a world. Before giving my sketch, however, I wish to examine the structure of prediction-making where the prediction itself has a causal effect on the predicted event. This will be essential in understanding how a "book of life" could be written, even though the writer knows that the agent will read it.

Consider the problem of an election predictor. He may know what the precise results of the upcoming election are going to be, if he makes no public prediction of the election. If he publishes a prediction, however, some of the voters, having found out what the results will be, may change their votes and thereby falsify his prediction. How, then, can a pollster make a genuinely scientific and accurate prediction of an election? Can he take into account the effect of the prediction itself? Herbert Simon has shown that, under specifiable conditions, a predictor can do this.[13] Essentially, what the predictor must know is the propensity of the voters in the community to *change* their voting intention in accordance with their expectations of the outcome. If persons are more likely to vote for a candidate when they expect him to win than when they expect him to lose, we have a "bandwagon" effect; if the opposite holds, we have an "underdog" effect.

Let us suppose that a given pollster has ascertained that, two days before the election, 60 percent of the electorate plans to vote for candidate A and 40 percent for B. He also knows that, unless he publishes a prediction, the percentages will be the same on election day. Further suppose he knows that there is a certain "bandwagon" effect obtaining in the voting community.[14] When the original intention of the electorate is to vote 60 percent for A, this bandwagon effect can be expressed by the equation, $V = 60 + .2(P - 50)$, where P is the percentage vote for A publicly predicted by a pollster, and V is the actual resultant vote for A. Clearly, if the pollster publicly predicts that A will receive 60 percent of the vote, his prediction will be falsified. Putting $P = 60$, the equation tells us that $V = 62$. In other words, the effect of the prediction, combined with the original voting intention of the electorate, would result in a 62 percent vote for A. However, the pollster can easily calculate a value for P that will make $P = V$. He need only solve the two equations, $P = V$ and $V = 60 + .2(P - 50)$. Such a solution yields $P = 62.5$. Thus, the pollster can publish a prediction saying that 62.5 percent of the electorate will vote for A, knowing that his own prediction will bring an additional 2.5 percent of the electorate into the *A* column, and thereby make his prediction come true.

Notice that all the antecedent conditions relevant to the outcome cannot be known until it is known what prediction (if any) the pollster will make. His prediction (or lack of prediction) is itself an important antecedent condition. However, one of the crucial determinants of the outcome—viz., the original voting intention of the electorate—is given independently of the pollster's prediction. Thus, while holding that factor constant, the pollster calculates what the outcome of the election *would* be, *if* he were to make certain predictions. By solving the equations given above, he discovers a prediction that, if published, would be confirmed. He thereupon forms an intention to publish that prediction and proceeds to fulfill that intention. Until he forms this intention, he does not know what prediction he will make, and therefore does not know all the requisite antecedent conditions from which to deduce the election outcome. But at the same time he makes the prediction (and perhaps even earlier), he does know all the relevant antecedent conditions and has deduced from these conditions what the results will be. Thus, his prediction of the outcome is a truly scientific prediction.

If someone wishes to predict a single person's behavior and yet let him learn of the prediction, the predictor must employ the same sort of strategy as the pollster. He must take into account what the agent's reaction will be to the prediction. There are several kinds of circumstances in which, having made the appropriate calculations, he will be able to make a correct prediction. (A) The agent learns of the prediction but does not want to falsify it. (B) Upon hearing the prediction, the agent decides to falsify it. But later, when the time of the action approaches, he acquires preponderant reasons for doing what was predicted after all. (C) Having decided to refute the prediction, the agent performs the action conforming with it because he doesn't realize that he is conforming with it. (D) At the time of the action, the agent lacks either the ability or the opportunity to do anything but conform with the prediction, though he may have believed that he would be able to falsify it. In any of these four kinds of cases, a predictor would be able to calculate that his prediction, together with numerous other antecedent conditions, would causally necessitate that the agent perform the predicted action. In a case of kind (B), for example, the predictor may be able to foresee that the agent will first read his prediction and decide to falsify it. But other factors will crop up—ones that the agent did not originally count on—that will make him change his mind and perform the predicted action after all. And the predictor also foresees this.

In the first three kinds of cases, (A), (B), and (C), the agent performs the predicted action *voluntarily* (though in (C) he does not realize that what he is doing falls under the description "what was predicted"). In other words, in each of these three kinds of cases, the agent *could have* acted otherwise, in at least one sense of "could have," which some philosophers think is relevant to free will. Thus, the possibility of a scientific prediction does not require that the agent be *unable* to act in any way different from the prediction. All that is required is that the agent will not *in fact* act in any way different from the prediction. A predictor might know that an agent will in fact act in a certain way, not because he knows the agent will be incapable of doing otherwise, but because he knows that the agent will *choose* or *decide* to act as predicted. This point will be clarified at the end of the chapter in a brief discussion of the indicated sense of "could have."

I shall now give a sketch of a causally compossible world in which a large number of correct predictions are made of an agent's behavior. Because I imagine this world to be governed by the same laws as those of the real world, and as I do not know all the laws of the real world, I cannot *prove* that my imagined world really is causally compossible. But as far as I can tell from commonsense knowledge of psychological and physical regularities, it certainly seems to be causally compossible. In this world, predictions of a man's life are made in great detail and inscribed in a "book of life," (parts of) which the agent subsequently reads. Obviously, I cannot describe the whole of this world, but I shall describe some of its most important and problematic features, namely the interaction between the agent and the book. Unfortunately, I shall have to omit a description of another important part of the world, the part in which the predictor (or predictors) gathers his data and makes his calculations. I am unable to describe this part of the world, first, because I do not know all the laws that the predictor would have at his disposal, and second, because I am not able to say just what the structure of this being would be. However, the main features of his modus operandi should be clear from our discussion of the pollster, whose technique is at the heart of such predicting.

VII

Now to the description of the world. While browsing around the library one day, I notice an old dusty tome, quite large, entitled "Alvin I. Goldman." I take it from the shelf and start reading. In great detail, it describes my life as a little boy. It always gibes with my memory and sometimes even revives my memory of forgotten events. I realize that this purports to be a book of my life and I resolve to test it. Turning to the section with today's date on it, I find the following entry for 2:36 p.m.: "He discovers me on the shelf. He takes me down and starts reading me. . . ." I look at the clock and see that it is 3:03. It is quite plausible, I say to myself, that I found the book about half an hour ago. I turn now to the entry for 3:03. It says: "He is reading me. He is reading me. He is reading me." I continue looking at the book in this place, meanwhile thinking how remarkable the book is. The entry reads: "He continues to look at me, meanwhile thinking how remarkable I am."

I decide to defeat the book by looking at a future entry. I turn to an entry eighteen minutes hence. It says: "He is reading this sentence." Aha, I say to myself, all I need do is refrain from reading that sentence eighteen minutes from now. I check the clock. To ensure that I won't read that sentence, I close the book. My mind wanders; the book has revived a buried memory and I reminisce about it. I decide to reread the book there and relive the experience. That's safe, I tell myself, because it is an earlier part of the book. I read that passage and become lost in reverie and rekindled emotion. Time passes. Suddenly I start. Oh yes, I intended to refute the book. But what was the time of the listed action, I ask myself. It was 3:19, wasn't it? But it's 3:21 now, which means I have already refuted the book. Let me check and make sure. I inspect the book at the entry for 3:17. Hmm, that seems to be the wrong place, for there it says I'm in a reverie. I skip a couple of pages and suddenly my eyes alight on the sentence: "He is reading this sentence." But it's an entry for 3:21, I notice! So I made a mistake. The action I had intended to refute was to occur at 3:21, not 3:19. I look at the clock, and it is still 3:21. I have not refuted the book after all.

I now turn to the entry for 3:28. It reads: "He is leaving the library, on his way to the president's office." Good heavens, I say to myself, I had completely forgotten about my appointment with the president of the university at 3:30. I suppose I could falsify the book by not going, but it is much more important for me not to be late for that appointment. I'll refute the book some other time! As I do have a few minutes, however, I turn back to the entry for 3:22. Sure enough, it says that my reading the 3:28 entry has reminded me about the appointment. Before putting the book back on the shelf and leaving, I turn to an entry for tomorrow at 3:30 p.m. "He's still riding the bus bound for Chicago," it reads. Well, I say to myself, *that* prediction will be easy to refute. I have absolutely no intention of going to Chicago tomorrow.

Despite my decision to refute the book, events later induce me to change my mind and to conform to it. For although I want to refute the book on this matter, stronger reasons arise for not refuting it. When I get home that evening, I find a note from my wife saying that her father (in Chicago) is ill and that she had to take the car and drive to Chicago. I call her there and she explains what has happened. I tell her about the book. Next morning she calls again with news that her father's condition is deteriorating and that I must come to Chicago immediately. As I hang up I realize that the book may turn out to be right after all, but the situation nevertheless demands that I go to Chicago. I might still refute it by going by plane or train. However, I call the airlines and am told that the fog is delaying all flights. The railroad says that there are no trains for Chicago until later in the day. So, acquiescing, I take a bus to Chicago and find myself on it at 3:30.

VIII

Let me interrupt my narrative here. I have given several cases in which the book is not refuted, and the reader should be convinced that I could easily continue this way. But it is important now to reply to several objections that the anti-predictionist is anxious to make against my procedure.

(1) "Your story clearly presupposes determinism. But whether or not determinism is true is the central matter of dispute. Hence, you are begging the question." Admittedly, my story does presuppose determinism. Unless determinism were true, the imagined predictor could not have figured out what actions the agent would perform and then written them in the book. However, I do not think that this begs the question. For I am not here trying to prove that determinism is true. I am merely trying to show that the thesis of determinism is quite compatible with the world as we know it and with human nature as we know it. The world depicted in my story seems to be very much like the real world except that it contains different antecedent conditions. The fact that this imagined world is determined and contains predictions of actions, and yet resembles the real world very closely, suggests to me that the real world may also be determined. At any rate, this supposition seems quite tenable, and its tenability is what I seek to establish in this chapter.

(2) "The story you told was fixed. Events might have been different from the way you described them. For example, the fog might not have curtailed all air traffic." No, events could not be different in the world I am imagining. That is, in my world all the events I described were causally necessitated by prior antecedent conditions. I did not describe all the antecedent conditions, so perhaps the reader cannot see that each event I did describe was causally necessitated by them. But, because it is a deterministic world, that is so. No one can imagine my world and also substitute the negation of one

of the events I described. I'm not "fixing" the story by saying that the fog curtailed air traffic; that is just the way my imagined world goes.

(3) *"But I can imagine a world in which some putative predictions of actions are refuted."* I have no doubt that you can; that is very easy. You could even imagine a world *somewhat* like the one I have just described, but in which putative predictions are falsified. But this proves nothing at all. I would never deny that one can construct some causally compossible worlds in which putative scientific predictions of actions are not successful. I have only claimed that one can (also) construct *some* causally compossible worlds in which genuine scientific predictions of actions are made (and are successful). The situation with predictions of action is no different from the one with predictions of physical events. We can construct causally compossible worlds in which predictions of physical phenomena are correct. But we can also construct worlds in which putative scientific predictions of physical phenomena are incorrect. If our ability to construct worlds in which predictions are unsuccessful proves the inherent unpredictableness of the kind of phenomena unsuccessfully predicted, then we can prove the unpredictableness of physical phenomena as easily as the unpredictableness of human action.

(4) "The world you have described, though possible, is a highly improbable world. Worlds in which putative predictions of actions are falsified are much more probable." The notion of one possible world being "more probable" than another seems to me unintelligible. Surely the statistical sense of probability cannot be intended. There is no way of "sampling" from possible worlds to discover what features most of them have. Perhaps the anti-predictionist means that we can imagine more worlds in which putative predictions of actions are falsified. But this too is questionable. I can imagine indefinitely many worlds in which successful predictions of actions are made.

Perhaps the anti-predictionist means that it is improbable that any such sequence of events as I described would occur in the *real* world. He may well be right on this point. However, to talk about what is probable (in the evidential sense) in the real world is just to talk about what has happened, is happening, and will happen *as a matter of fact*. But the dispute between predictionists and anti-predictionists is, presumably, not about what *will* happen, but about what *could* happen *in principle*. This "in principle" goes beyond the particular facts of the actual world.

(5) "The difference between physical phenomena and action is that predictions of actions can defeat themselves, but predictions of physical events cannot." This is not so. One can construct worlds in which the causal effect of a putative prediction of a physical event falsifies that prediction. Jones

calculates the position of a speck of dust three inches from his nose and the direction and velocity of wind currents in the room. He then announces his prediction that five seconds thence the speck will be in a certain position. He had neglected to account for the wind expelled from his mouth when he made the prediction, however, and this factor changes the expected position of the speck of dust. Perhaps one can imagine a wider variety of cases in which predictions affect human action more than physical phenomena. But this is only a difference of degree, not of kind.

(6) "Predictions of physical events can refute themselves because the predictor may fail to account for the effect of his own prediction. But were he to take this effect into account, he would make a correct prediction. On the other hand, there are conditions connected with the prediction of action in which, no matter what prediction the predictor makes, his prediction will be falsified. Here there is no question of inaccurate calculation or insufficient information. Whatever he predicts will be incorrect. Yet this situation arises only in connection with human action, not physical events."

This is an important objection and warrants detailed discussion.

IX

Suppose that I wish to predict what action you will perform thirty seconds from now, but that I shall not try to change or affect your behavior except by making my prediction. (Thus, I shall not, for example, predict that you will perform no action at all and then make that prediction come true by killing you.) Further suppose that the following conditions obtain. At this moment you want to falsify any prediction that I shall make of your action. Moreover, you will still have this desire thirty seconds from now, and it will be stronger than any conflicting desire you will have at that time. Right now you intend to do action *A*, but you are prepared to perform $\bar{A}$ (not-*A*) if I predict that you will perform *A*. Thirty seconds hence you will have the ability and opportunity to do *A* and the ability and opportunity to do $\bar{A}$. Finally, conditions are such that, if I make a prediction in English in your presence, you will understand it, will remember it for thirty seconds, and will be able to tell whether any of your actions will conform to it or not. Given all these conditions, whatever I predict—at least, if I make the prediction by saying it aloud, in your presence, in English, etc.—will be falsified. If I predict you will do *A*, then you will do $\bar{A}$, while if I predict that you will do $\bar{A}$, you will proceed to do *A*. In other words, in these conditions no prediction of mine is causally compossible with the occurrence of the event I predict. Let $C_1, \ldots C_n$ be the (actual) conditions just delineated, let P_A be my predicting you will do A (announced in the indicated way), and let $P_{\bar{A}}$ be my predicting you will do $\bar{A}$ (announced in the same way). Then *both* sets $\{C_1, \ldots, C_n, P_A, A\}$ and $\{C_1, \ldots, C_n, P_{\bar{A}}, \bar{A}\}$ are causally *in*compossible sets of events.

Notice that this example does not prove that it is causally incompossible "simpliciter" for me to make a scientific prediction of your action. All that it proves is that I cannot make such a prediction *in a certain manner*, viz., by announcing it to you in English. The events P_A and $P_{\bar{A}}$ include this particular manner, and that they do so is important. If I predict your action in some other manner, by thinking it to myself or by saying it aloud in Hindustani, for example, the effect on your action would not be the same as if I say it aloud in English. Assume that, if you do not hear me make any prediction or if you hear me say something you fail to understand, you will proceed to perform action A. Then it is causally compossible for me to predict your action correctly by announcing the prediction in Hindustani. In other words, letting P_A be my predicting that you will do A by announcing this in Hindustani, then the set of events $\{C_1, \ldots, C_n, P_A{}', A\}$ is a causally compossible set.

In determining whether or not a certain set of events, including (1) a prediction, (2) the event predicted, and (3) certain other assumed conditions, is a causally compossible set, it is essential to specify the manner of the prediction. This is true *in general*, not just in the case of predictions of action. A prediction that is "embodied" or expressed in one way will not have the same causal effects as the same prediction expressed in another way. We can see this in the case of the speck of dust. Jones predicted the position of the dust by announcing it orally, and this resulted in the falsification of the prediction. But had he made the same prediction in another fashion—say, by moving his toes in a certain conventional pattern—his prediction would not have been falsified, for the position of the dust would not have been affected.

What is the significance of the fact that it is causally incompossible, in some circumstances, for a (correct) prediction of an action to be made in a specified manner? First, this unpredictability does not prove that these actions are undetermined. Indeed, the very construction of the case in which no prediction is possible *presupposed* the existence of laws of nature that, together with a given prediction, would result in a certain action. In short, the case under discussion should, if anything, support rather than defeat the thesis that actions are determined. The only reason one might have for thinking the contrary is the assumption—which should by now appear very dubious—that determinism entails predictability. What our present case shows, I think, is that under some circumstances, even a determined event may not be susceptible to being correctly predicted in a specified manner. This fact can be further supported by adducing a similar case connected with purely physical events. And this brings me to my second point: The case produced above does not reflect a peculiarity of human action, because parallel examples can be found among physical phenomena.

Imagine a certain physical apparatus placed in front of a piano keyboard. A bar extends from the apparatus and is positioned above a certain key. (Only white keys will be considered.) If the apparatus is not disturbed, the bar will strike that key at a certain time. Now let us suppose that the apparatus is sensitive to sound and, in particular, can discriminate between sounds of varying pitches. If the apparatus picks up a certain sound, the position of the bar will move to the right and proceed to strike the key immediately to the right of the original one (if there is one). Specifically, if the sound has the same pitch as that of the key over which the bar is poised, the bar will move. If the monitored sound has any other pitch, the bar will remain in its position and proceed to strike that key.

Now suppose that someone (or something) wishes to make predictions of the behavior of the apparatus. He wishes to predict what key the bar will strike. But the following restriction is made on the *manner* in which the prediction is to be made. The prediction must be expressed according to a specific set of conventions or symbols. To predict that the bar will strike middle C, for example, the predictor must emit a sound with the pitch of middle C. To predict that the bar will strike D, he must emit a sound with the pitch of that key, etc. All sound emissions are to be made in the neighborhood of the apparatus. Given this restriction on the manner of prediction, it will be causally incompossible for the predictor to make a correct prediction. For suppose that the bar is poised above middle C. If he predicts that it will strike middle C—that is, if he emits a sound of that pitch—the bar will move and proceed to strike D. But if he predicts any other behavior of the bar, for example, that it will strike D, the bar will remain in its original position and strike middle C.

Admittedly, the manner of prediction I have allowed to the predictor of this physical phenomenon is much more narrowly restricted than the manner of prediction allowed to the predictor of human action. But we could imagine physical apparatuses with a greater degree of complexity, able to "refute" predictions made in any of a wider variety of manners. In any case, the principle of the situation is the same for both physical phenomena and human actions, though the manners of prediction that affect one phenomenon may be different from the manners of prediction that affect the other. The latter difference simply reflects that fact that physical objects and human beings do not respond in precisely the same ways to the same causes. But this is equally true of different kinds of physical objects and different pairs of human beings.

The reader should not suppose that the present discussion in any way vitiates my description of the book of life in section VII. Our present discussion shows that under *some* conditions, it is *not* causally compossible to predict a man's action in a way that allows him to learn of the prediction. But there are *other* conditions, such as the ones described in section VII, in which such predictions *are* causally compossible. The existence of the latter conditions suffices to establish the possibility (in principle) of scientific predictions of voluntary actions that the agent hears or reads. Admittedly, it is not always possible to make predictions in this manner. But even when it is impossible to let one's prediction become known to the agent, it does not follow that it is impossible to make the prediction "privately." Thus, suppose you are trying to write a book of my life before I am born. Your calculations might show that if you inscribe certain predictions in the book, they will be confirmed. For these calculations might reveal that I shall not read the book, or that I shall perform the actions despite the fact that I shall read the book. If so, you may proceed to write the book, having (scientific) knowledge that it will be correct. On the other hand, your calculations might reveal that, no matter what prediction you inscribe in the book, I shall refute it. In this case, you will be unable to write a book of my life. But you may nevertheless have scientific knowledge of what I shall do. Your calculations may reveal that I shall do a certain sequence of actions, as long as I do not come across any (putative) book of my life. If you decide not to write such a book yourself, and if you know that no one else will, you may conclude (deductively) that I shall perform the indicated sequence of acts.

X

In the previous section we saw that, under certain conditions, it may not be causally compossible to predict a certain action in a specified manner. Recently, however, Michael Scriven has claimed that human behavior exhibits an even more important unpredictability.[15] Scriven writes: "So far we merely demonstrate that human choice behavior can be made at least as unpredictable as any physical system. In an important class of examples . . . , a stronger conclusion is demonstrable."[16] Scriven's example consists in imagining an agent, *X*, who is contra-predictively motivated relative to a certain predictor, *Y*. That is, *X* wants to defeat any prediction *Y* makes about his actions. Scriven further supposes that *X* knows everything that *Y* knows about him. From this information, *X* figures out what *Y* will predict, or will have predicted, about *X*'s action. In other words, *X* "replicates" *Y*'s prediction; he comes to know what *Y*'s prediction is even though *Y* does not announce his prediction. After figuring out *Y*'s "secret" prediction, *X* proceeds to act otherwise. Scriven concludes: "... the present case is more interesting. The idea that human behavior is "in principle" predictable is not seriously affected by the recognition that one may not be able to announce the predictions to the subjects with impunity (nor, more generally, can one allow them to be discovered). For one can make the predictions and keep them from the subjects. But in the present case, *one cannot make true predictions at all.* Secret predictions are still predictions; unmakable ones are not."[17]

We must first note that Scriven has given a misleading account of his example in saying that "in the present case, *one cannot make true predictions at all*" (emphasis original). True, a particular person, *Y*, is unable to make correct predictions of *X*'s behavior. But *X*'s behavior is not completely immune to prediction. Scriven's case leaves open the possibility that there is, or was, some other being, *Z*, (who may have lived long before *X*) who predicted *X*'s behavior without *X* knowing of this. In order for *X*'s behavior to be *completely* immune to prediction, *X* would have to know with respect to *every* potential predictor (i.e., everyone who lived during or prior to *X*'s lifetime) what predictions, if any, be made about *X*'s behavior. Anything short of this state of knowledge by *X* would leave open the possibility that some being or other correctly predicted what *X* would do, indeed predicted it scientifically.

Second, it is questionable whether Scriven's example shows that human behavior is more unpredictable than physical systems, as he suggests in the passage on page 413. Admittedly, Scriven's case goes beyond my previous example in one respect. In my example, only certain manners of prediction lead to the performance of a different action. In Scriven's example, *any* manner of prediction leads to a different action. This is because even the *minimal* manner of "making" a prediction (i.e., having a future-looking *belief*) is self-defeating. Nevertheless, it may still be possible to duplicate Scriven's human behavior case with physical systems. Suppose, for example, that we found neurophysiological states that correlated with beliefs. That is, suppose we found one-to-one correlations between a person's believing certain propositions and his being in certain neurophysiological states. We might then "hook up" a physical system to a potential predictor in such a way that the state of the system is causally affected by the beliefs (or their neurophysiological correlates) of the predictor. The physical system would be arranged so that whenever the potential predictor had a belief about a future state of the physical system, this belief would cause the system to go into another state instead.

The third and most important point I wish to make is based on a criticism of Scriven by David K. Lewis and Jane Shelby Richardson.[18] In the competition between Scriven's agent and predictor, each is trying to get sufficient information about the other and to calculate from this data just what the other will do (or believe). Let us combine both of these factors—data and calculation—and call them "knowledge," because the function of the calculations is to add to the calculator's knowledge. Lewis and Richardson argue forcefully that it is impossible for *both* the predictor and the agent to have sufficient, or complete, knowledge of his opponent. We can construct two sorts of cases. We can endow the predictor with complete information about the agent, but this forces us to deny to the agent complete knowledge about the predictor. Or we can endow the agent with complete knowledge about the predictor, but in so doing we must deny complete knowledge about the agent to the predictor. Scriven does the latter, although he is not quite aware of this. In saying that the agent is able to "replicate" the predictor's prediction and then decide to act otherwise, he is in effect saying that there is some aspect of the agent's motivational structure that is *unknown* to the predictor. But if there is some fact relevant to the prediction that the predictor does not know, it is not surprising that he makes an incorrect prediction. He simply is not in a position to make a scientific prediction. This hardly shows that the agent's behavior is inherently immune to scientific prediction. It merely shows that this particular predictor, *Y*, does not know enough about *X* to make a scientific prediction.

For this reason, Scriven's case is less interesting than the one I presented in the foregoing section. In my example, the predictor is unable to make a correct prediction (in a specified manner) even though he has all relevant information. In Scriven's example, the predictor lacks some relevant information. Scriven's example, then, hardly warrants our concluding that human behavior is undetermined. The fact that someone with insufficient knowledge is unable to predict an event correctly does not at all suggest that the event is undetermined. Of course, Scriven does not claim that his example shows behavior to be undetermined. I say this merely to remind the anti-predictionist that he can take no comfort from Scriven's case.

XI

I have shown that there are causally compossible worlds in which voluntary actions are scientifically predicted. Let us now see whether there are causally compossible worlds in which a person scientifically predicts one of his *own* actions. I think that there are such worlds, and I shall illustrate by continuing the description of the world I was sketching earlier.

Having tested my book of life on a very large number of occasions during many months and failed to refute it, I become convinced that whatever it says is true. I have about as good inductive evidence for this proposition as I do for many other propositions I could be said to know. Finally, I get up enough courage to look at the very end of the book and, as expected, it tells when and how I shall die. Dated five years hence, it describes my committing suicide by jumping off the eighty-sixth-floor observation deck of the Empire State Building. From a description of the thoughts that will flash through my mind before jumping, it is clear that the intervening five years will have been terrible. As the result of those experiences, I shall have emotions and desires (and beliefs) that will induce me to jump. Because I trust the book completely, I now conclude that I *shall* commit suicide five years hence. Moreover, I can be said to *know* that I shall commit suicide.

As described so far, we cannot consider my prediction of my suicide a "scientific" prediction. To be a scientific prediction, the predicted event must be *deduced* from laws and antecedent conditions, whereas, as I have described the case, no deduction was involved. However, we might supplement the situation in order to include a deduction. The book may be imagined to list the relevant physical and psychological laws (in a footnote, say) and the relevant conditions that determine my committing suicide (my intention to commit suicide, my proximity to the fence surrounding the observation deck, the absence of guards or other interfering factors, etc.). From these laws and conditions I actually deduce my future action.[19]

This example shows, contrary to the view of some authors, that we can have inductive knowledge of our own future actions, knowledge that is not based on having already made a decision or formed an intention to perform the future action. Stuart Hampshire, for example, has recently written, "... I cannot intelligibly justify a claim to certain knowledge of what I shall voluntarily do on a specific occasion by an inductive argument; if I do really know what I shall do, voluntarily, and entirely of my own free will, on a specific occasion, I must know this in virtue of a firm intention to act in a certain way."[20] The case outlined, I believe, shows that Hampshire is mistaken. In that case, there is a time at which I do have certain knowledge of what I shall do (at any rate, about as "certain" as one can be with inductive evidence), and yet I have formed no intention nor made any decision to perform that action. At the time I read the book's prediction, I do not intend to commit suicide. But although I do not intend to commit suicide, I fully believe and know that, five years later, I shall intend to commit suicide. I firmly believe that, at that later time, I shall feel certain emotions and have certain desires that will induce me to jump off the Empire State Building. At the time of my reading the book I do not feel those things, but I commiserate with my future self, much as I commiserate with and understand another person's desires, beliefs, feelings, intentions, etc. Still, my understanding of these states of mind and of the action in which they will issue is the understanding of a spectator; my knowledge of these states and of my future action is purely inductive. Moreover, this knowledge is of a particular *voluntary* act to be performed at a specified time. Though the suicide will be a "desperate" action, it will in no sense be "coerced" or done unknowingly; it will flow from a firm intention, an intention formed very deliberately. But that intention will not be formed until after I have had certain experiences, experiences that, at the time I am reading the book, I have not yet had.

We can imagine two alternative series of events to occur between my reading the book and my suicide. First, I might *forget* what I have learned from the book and later decide to commit suicide. Second, while I never forget the prediction, the knowledge of my future suicide may gradually change from more inductive knowledge to knowledge based on intention. In this second alternative, there is never any "moment" of decision. I never pass from a state of complete doubt about committing suicide into a sudden intention of committing suicide. Rather, there is a gradual change, over the five-year period, from mere inductive knowledge that I shall commit suicide to an intention to commit suicide. When I first read the book, I am fully prepared to assent to the proposition that I shall commit suicide. But I am saddened by the thought; my heart isn't in it. Later, as a result of various tragic experiences, my *will* acquiesces in the idea. I begin to welcome the thought of suicide, to entertain the thought of committing suicide with pleasure and relief. By the time the appointed time comes around, I am *bent* on suicide. This gradual change in attitude constitutes the difference between the kind of knowledge of my future suicide, the difference between mere inductive knowledge and knowledge based on intention. Hampshire claims that the first kind of knowledge of one's own action is impossible. The present case, I believe, shows this claim to be mistaken.

Many philosophers seem to be very uncomfortable with the idea of a book of life. They believe that the existence of such books—or of foreknowledge of actions in any form—would deprive us of all the essential characteristics of voluntary behavior: choice, decision, deliberation, etc.[21] I do not think this fear is warranted, however. I have just shown that even if a person reads what a book of life predicts, and believes this prediction, he can still perform the indicated action voluntarily. Moreover, the existence of predictions that the agent does *not* read leaves ample opportunity for deliberation and decision. An agent may know that a book of his life exists and yet proceed to make decisions and to deliberate as all of us do now. The agent's belief that there is such a book, and his belief that the book's existence implies that his actions are causally necessitated, is compatible with his deliberating whether to do one action or another. Although his future action is causally necessitated, one of the antecedent conditions that necessitate it is his deliberation. Indeed, the prediction in the book of life was made precisely because its writer knew that the agent would deliberate and then decide to do the predicted action. Thus, the book of life can hardly be said to preclude deliberation. Nor does the book of life imply that the agent's deliberation is "for naught" or "irrelevant." On the contrary, his deliberation is a crucial antecedent condition: Were he not to deliberate, he probably would not perform the action he eventually does perform. Deliberation and decision are perfectly compatible with the existence of books of life; and they are perfectly compatible with the thesis that they, and the actions in which they issue, are determined.

XII

If actions are determined, there is at least one sense in which an agent "cannot" act other than he does: His actual action is causally necessitated by actual prior events, and hence any other (incompatible) action is causally incompossible with actual prior events. It is precisely because his action is causally necessitated that it is amenable to scientific prediction. But it is generally accepted that if an agent does an action *A* voluntarily, he also "can" do otherwise. So far we have only provided a sense of "can" in which he *cannot* do otherwise. We are therefore obliged to identify some *other* sense of "can" or "could" in which an agent "can" do actions he does not in fact do. Such a sense of "can" has long been in the literature, having been defended by Hobbes, Edwards, Hume, Mill, Moore, Nowell-Smith, Stevenson, and others.

Suppose that John does *A* voluntarily. To say that he "could have" done otherwise indicates that he had another alternative open to him, the alternative of doing $\bar{A}$. But what is meant by saying that $\bar{A}$ was "open" to John? What is meant, I believe, is that, *if*, contrary to fact, John had wanted or chosen or decided to do $\bar{A}$, then he would have succeeded in doing $\bar{A}$. The "alternative" open to John is not the alternative of doing $\bar{A}$ given his (actual) desire or decision to do *A*, but the possibility of doing $\bar{A}$ relative to a (non-actual) desire to do $\bar{A}$. Here, as elsewhere, the determinist's sense of "could have" involves the supposition of counterfactual conditions, indeed, if taken to its logical conclusion, a whole counterfactual world. In the "could have" pertaining to action, the main counterfactual feature pertains to the agent's desire, intention, or decision. But the analysis of "could have" is not wholly counterfactual. To say that John "could have" done $\bar{A}$ is also to make a categorical assertion about the real world, viz., that John *was able* to do $\bar{A}$, that he had the ability and the opportunity to do $\bar{A}$. John's actual ability and actual opportunity were such that, if his desire or intention to act had been different, his action would have been different. Thus, the analysis of "John could have done $\bar{A}$" would be formulated as "John had the ability and opportunity to do $\bar{A}$, and (therefore) if he had decided to do $\bar{A}$, he would have done $\bar{A}$."

A scientific predictor who predicts that someone will perform a voluntary action A may thus be justified in accepting the following two propositions: (1) The agent *will* (certainly) do A, and (2) the agent *could* do $\bar{A}$. Proposition (2) is warranted because, as the predictor realizes, the agent will have the ability and opportunity to do $\bar{A}$. Proposition (1) is also warranted, however, because, as the predictor knows, the agent will *in fact* choose to do A, not $\bar{A}$. The predictor knows that he will choose to do A because he has deduced that he will so choose from antecedent conditions and psychological laws having choices (or desires) as their dependent variables.

The question remains whether the specified sense of "could" is *the* sense relevant to the traditional problems of freedom and responsibility. My own opinion is that this is the relevant sense, but I have no new arguments to give on this score. Here I wish not to join the fray, but to leave it.

NOTES

1. Karl Popper, for example, defines "determined" as "predictable in accordance with the methods of science," in "Indeterminism in Quantum Mechanics and in Classical Physics," *The British Journal for the Philosophy of Science*, vol. I (1950–1951); see p. 120.
2. There are, of course, numerous problems associated with the concept of a law of nature. But a detailed discussion of these problems would go beyond the scope of this chapter.
3. For a defense of this view, see William P. Alston, "Wants, Actions, and Causal Explanations" in H. N. Castañeda, ed., *Minds, Intentionality, and Perception* (Detroit: Wayne State University Press, 1967); and R. Brandt & J. Kim, "Wants as Explanations of Actions," *The Journal of Philosophy*, vol. 60 (1963), 425–435.
4. The term "event" is here used to designate event *kinds*, not necessarily ones that have been actualized. The term "law," on the other hand, will be used only to designate actual laws (i.e., laws that obtain in the real world, and not merely possible

laws). For the most part, I shall consider events with built-in time references. Sam's jumping rope at 10:35 will be treated as a distinct event from Sam's jumping rope at 10:45. This is very natural in the present context, as a given set of events may be causally incompossible with Sam's jumping rope at 10:35 but causally compossible with Sam's jumping rope at 10:45.

5. *Freedom: A New Analysis* (London: Longmans, Green, 1954), 169.
6. "Can the Will Be Caused?" *The Philosophical Review*, vol. 71 (1962), 49–55.
7. One might challenge Ginet's argument by criticizing this definition of "deciding." This criticism has implicitly been made, along with other criticisms of Ginet's position, by various writers. For example, see John Canfield, "Knowing about Future Decisions," *Analysis*, vol. 22 (1962); and J. W. Roxbee Cox, "Can I Know Beforehand What I Am Going to Decide?" *The Philosophical Review*, vol. 72 (1963). Here I shall waive these criticisms, however, and accept Ginet's claim that it is impossible to predict one's own decisions. I shall then ask whether this proves that decisions are undetermined and whether they are intrinsically different from physical phenomena.
8. That this is an error has also been claimed by Arnold S. Kaufman in "Practical Decision," *Mind*, vol. 75 (1966); see p. 29.
9. It is also an error—committed at least as frequently—to think that determinism entails the possibility of retrodicting or explaining every event under any description. Suppose that Sam thinks of the corkscrew at *t* and that no one ever thinks of the corkscrew after *t*. Suppose, moreover, that both of these events are deducible from laws and antecedent conditions. Now let us introduce the expression "postventing *x*" to mean "thinking of *x* for the *last* time" (just as "inventing *x*" means "thinking of *x* for the *first* time"). Clearly, we may say of Sam that he "postvented" the corkscrew and that this action of his is determined. However, it is logically incompossible for anyone to *retrodict* Sam's postventing the corkscrew. To do so, the retrodicter would himself have to think of the corkscrew, and, ex hypothesi, Sam thought of the corkscrew for the *last* time at *t*.
10. Op. cit.
11. I say "probably" because the definition of determinism does not entail that every event is a determinant of some subsequent event. Thus, if not-*e* actually occurred but had no effect on any subsequent event, then we might substitute *e* for not-*e* without changing any subsequent events. However, though determinism does not require it, it is reasonable to assume that every event will have some differential effect on *some* later event or events.
12. This is all plausible, at any rate, if we deny fatalism. Fatalism, which is by no means implied by determinism, is the view that certain events will happen at certain times *no matter what* antecedent conditions obtain.
13. "Bandwagon and Underdog Effects of Election Predictions," reprinted in *Models of Man* (New York: Wiley, 1957). The requisite condition is that the function relating the actual outcome of the voting to the predicted outcome, given the electorate's original voting intention, be *continuous*.
14. That this bandwagon effect holds in the community could be discovered either by studying previous elections or by deducing it from "higher-level" generalizations found to be true of the community.
15. "An Essential Unpredictability in Human Behavior" in *Scientific Psychology: Principles and Approaches*, eds. B. B. Wolman & E. Nagel (New York: Basic Books, 1965).
16. Ibid., 413.
17. Ibid., 414.

18. "Scriven on Human Unpredictability," *Philosophical Studies*, vol. 17 (1966).
19. That these conditions will actually obtain is, of course, open to doubt. Moreover, I have not learned of *them* by scientific prediction. I have simply "taken the book's word" that these conditions will obtain; I have not deduced them from other, still-earlier, conditions. However, there are no restrictions on the manner in which a predictor comes to know antecedent conditions. One way predictors might learn about antecedent conditions is by using various measuring devices and instruments, the reliability of which is supported by inductive evidence. My book of life may be regarded as such a device, and my inductive evidence supporting its reliability may be as strong as that supporting the reliability of various other devices that scientists commonly use for obtaining knowledge of antecedent conditions.
20. *Freedom of the Individual* (London: Chatto & Windus, 1965); see p. 54.
21. One such philosopher is Richard Taylor. See his "Deliberation and Foreknowledge," *American Philosophical Quarterly*, vol. I. (1964), 73–80. Many others could also be named.

NAME INDEX

Note: Italicized letter *n* designates footnotes.

NAME INDEX

Note: Italicized letter *n* designates footnotes.

SUBJECT INDEX

Note: Italicized letter *n* designates footnotes. Headings which appear periodically throughout the page range are designated with *passim*.